THE WORLD OF CAROLUS CLUSIUS: NATURAL HISTORY IN THE MAKING, 1550–1610

Perspectives in Economic and Social History

Series Editors: *Andreas Gestrich*
Steven King
Robert E. Wright

Titles in this Series

1 Migrants and Urban Change: Newcomers to Antwerp, 1760–1860
Anne Winter

2 Female Entrepreneurs in Nineteenth-Century Russia
Galina Ulianova

3 Barriers to Competition: The Evolution of the Debate
Ana Rosado Cubero

4 Rural Unwed Mothers: An American Experience, 1870–1950
Mazie Hough

5 English Catholics and the Education of the Poor, 1847–1902
Eric G. Tenbus

Forthcoming Titles

The Determinants of Entrepreneurship: Leadership, Culture, Institutions
José L. García-Ruiz and Pierangelo Toninelli (eds)

London Clerical Workers, 1880–1914: Development of the Labour Market
Michael Heller

Financing India's Imperial Railways, 1875–1914
Stuart Sweeney

Mining and the State in Brazilian Development
Gail D. Triner

Global Trade and Commercial Networks: Eighteenth-Century Diamond Merchants
Tijl Vanneste

THE WORLD OF CAROLUS CLUSIUS: NATURAL HISTORY IN THE MAKING, 1550–1610

BY

Florike Egmond

Routledge
Taylor & Francis Group
LONDON AND NEW YORK

First published 2010 by Pickering & Chatto (Publishers) Limited

Published 2016 by Routledge
2 Park Square, Milton Park, Abingdon, Oxfordshire OX14 4RN
711 Third Avenue, New York, NY 10017, USA

First issued in paperback 2015

Routledge is an imprint of the Taylor & Francis Group, an informa business

BRITISH LIBRARY CATALOGUING IN PUBLICATION DATA

Egmond, Florike.
The world of Carolus Clusius: natural history in the making, 1550–1610. – (Perspectives in economic and social history)
1. Botany – Europe – History – 16th century. 2. Natural history – Europe – History – 16th century. 3. Clusius, Carolus, 1526–1609.
I. Title II. Series
580.9'4'09031-dc22

ISBN-13: 978-1-138-66119-6 (pbk)
ISBN-13: 978-1-8489-3008-7 (hbk)

Typeset by Pickering & Chatto (Publishers) Limited

CONTENTS

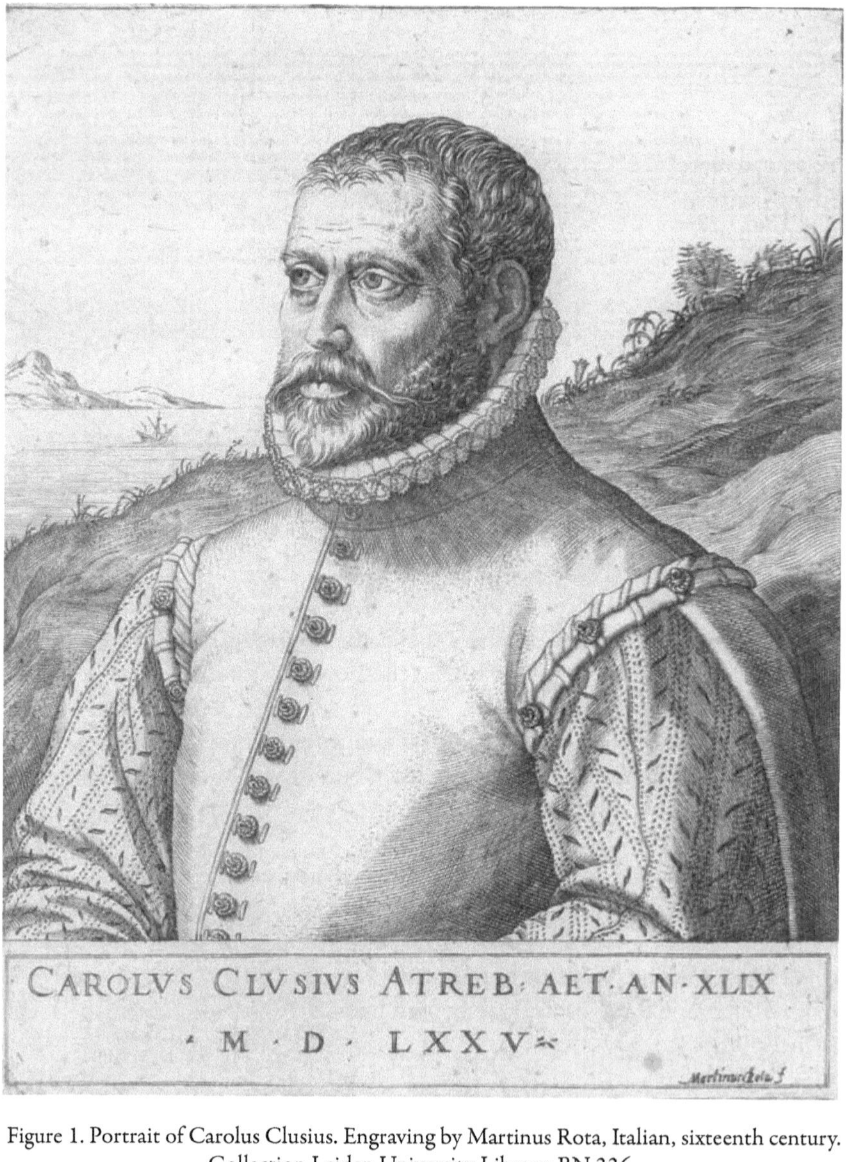

Figure 1. Portrait of Carolus Clusius. Engraving by Martinus Rota, Italian, sixteenth century.
Collection Leiden University Library, BN 336.

PREFACE

My research in the field of early modern natural history – after many years spent in the domain of historical crime and punishment – started in the early 1990s and was triggered by 'meeting' Adriaen Coenen, a sixteenth-century Dutchman whose marvellous illustrated manuscripts on marine life continue to fascinate me. The humble Coenen was a contemporary of one of the most famous naturalists of the period, Carolus Clusius. They had some acquaintances in common and both spent the last years of their lives in Leiden, though not in the same period. It is thanks to Coenen that I became interested in Clusius and his world, and decided to focus – if I ever would have the opportunity – not so much on Clusius himself but on the men and women, famous or not, who formed part of his world and of the history of natural history.

That opportunity came in 2005 thanks to a grant from the Netherlands Organization for Scientific Research for the Clusius Project (2005–9), which was appropriately based at the Scaliger Institute of Leiden University; besides myself it comprised Professor Paul Hoftijzer and two PhD students, Esther van Gelder and Sylvia van Zanen. Shortly before the official start of the Clusius Project I moved to Rome, and was fortunate enough to find an apartment with a small garden. Most of the research and writing for this book was thus accompanied by putting into practice (on a small scale) what I read about in the sixteenth-century sources: trying out plants in a new setting; experimenting with bulbs and tubers transferred from Holland to a Mediterranean setting; importing exotica, such as a dragon tree and many succulents from the Canary Islands; and occasionally transplanting indigenous wild plants found on village walls or along the roadside. Excursions to the plateaus of Norcia and Castelluccio followed, to discover where sixteenth-century naturalists went botanizing.

While both the transfer of plants and my regular trips back and forth between Italy and the Netherlands drove home the importance of 'place' and setting in the widest sense of the word, the practice of learning about plants showed me more clearly than anything else could have done the difference between myself as an amateur gardener and Clusius and his friends: the real plant experts and collectors. But the pleasure was probably much the same for all of us, and I have done my best to ignore the advice to keep work and pleasure separate.

For Clusius and friends the study of nature was in most respects a collective enterprise, involving many of them in exchanges with fellow enthusiasts all over Europe. The web of relations connected with the Clusius Project grew in a similar way and reached – at least in a geographical sense – similar dimensions. I found myself involved in weekly and sometimes daily exchanges of information about natural history in Clusius's age with colleagues and friends all over Europe. Those exchanges were enormously stimulating and the help I have received was invariably generous. It is impossible to thank all here with whom I have been involved in such exchanges during the last five or so years, but I would like to name in particular Karen Reeds, Deborah Harkness, Kim Sloan, Dirk van Miert, Roelof van Gelder, Francisco Bethencourt, Diogo Ramada Curto, Rea Alexandratos, Ian Rolfe, Michiel Verweij, Hugo Soly, Giuseppe Olmi, Lucia Tongiorgi Tomasi, Giuseppe Sandrini, Andrea Ubrizsy Savoia, Irene Baldriga, Harold Cook, Sven Dupré, Sachiko Kusukawa, Emma Spary, José Pardo Tomás, Sabine Anagnostou, Christoph Friedrich and Arthur MacGregor. The constant possibility of 'talking' on email to colleagues and friends in many different places and from many different disciplinary backgrounds about small and big questions has helped to give density to my research. Of course none of them is responsible for the errors which inevitably will have found their way into this study. Sincere thanks also go to the following institutes for their invitations to present parts of my research and test out ideas – sensible or far-fetched – in workshops, conferences or other sessions: the European University Institute, Florence; the Descartes Institute, Utrecht; the Huygens Institute, The Hague; the Free University, Brussels; the Institut für Geschichte der Pharmazie, Marburg; the Fundacion Canaria Orotava de Historia de la Ciencia, Tenerife; the University of Toronto, Canada; the Wellcome Institute and the Warburg Institute, London; St John's College, Oxford; and the Society of the History of Science and Technology, Barcelona.

Special thanks go to the Netherlands Organization for Scientific Research (NWO), whose project grant made my dream come true to be able to spend four years working on this subject, and to my colleagues in Leiden: Paul Hoftijzer, friend and excellent director of our project; Kasper van Ommen, key figure of the inspiring Scaliger Institute; Esther van Gelder and Sylvia van Zanen, the PhD students with whom it has been a pleasure to work; Gerda van Uffelen, Carla Teune, Jan de Koning and Pieter Baas at the Hortus Botanicus and the Nationaal Herbarium in Leiden for their expert advice on all plant matters – modern and historical – and much else; Harm Beukers and Rob Visser for their initiative in setting up the project and their continued support. I would also like to thank the staff of Leiden University Library for their assistance and their important role in making digital images of the complete Clusius Correspondence in Leiden freely available on internet. The Scaliger Institute and Leiden University Library were also crucially important to the international conference in 2004 that opened the Clusius Project and the exhibition in Leiden to commemorate the 400th anniversary of Clusius's death in April 2009. The resulting publications (*Carolus Clusius: Towards a Cultural History of a Renaissance Naturalist*, 2007; and *The Exotic World of Carolus Clusius 1526–1609: Catalogue of an Exhibition on the Quatercentenary of Clusius' Death, 4 April 1609*, 2009) are very much the result of joint efforts.

New friendships have originated in the course of the project, often via long-distance mail, and old ones were strengthened. I would specially like to mention Arthur MacGregor, José Pardo Tomás, Giuseppe Olmi, Esther van Gelder and Paul Hoftijzer – all of whom have been kind enough to read the whole manuscript. Their advice and support and that of Lucia Tongiorgi Tomasi, Andrea Ubrizsy Savoia and Sabine Anagnostou were invaluable. Nell Riviere-Platt was kind enough to read and comment on the French chapters.

This book is dedicated to Mayke de Jong, my best friend, without whom neither history writing nor gardening would be the same, and to Peter Mason, companion in life, research and other adventures. He has read every letter of the many versions of this book and we have so often discussed ideas, compared notes and invented new plans that it has been a joint project in many respects. I hope the adventure continues.

TECHNICAL NOTE

Geography

During Clusius's lifetime the Netherlands split up as a consequence of the revolt of the northern provinces against Spanish rule. The Northern Netherlands became the Dutch Republic, while the southern provinces remained part of the Habsburg empire. I use the Netherlands or the Low Countries for the Southern and Northern Netherlands together; the Southern Netherlands (and the adjective Netherlandish) for the southern provinces, and either the Northern Netherlands or the Dutch Republic (and the adjective Dutch) for the Northern provinces. Holland was then and is now used both for the most powerful province of the Dutch Republic and for the whole of the Dutch Republic. I use it only for the province of Holland.

Plant Names

In this study, which focuses more on the fascination with plants than on the plants themselves, plant identification has not been my purpose. Where possible, I have tried to use common plant names (hyacinth, lilac, horse chestnut, sunflower), which are often close to the contemporary sixteenth-century ones, rather than their modern scientific (Latin) names. For less commonly known plants that are mentioned in the original texts by a name that allows identification, I have used the modern scientific name or both the old name and the modern scientific one. For these identifications I have used botanical handbooks; publications on Clusius, Dodoens, Lobel and several other contemporary naturalists; and especially the recent major publication which identifies the plants depicted in the sixteenth-century *Libri Picturati*: J. de Koning et al. (eds), *Drawn after Nature: The Complete Botanical Watercolours of the 16th-Century Libri Picturati* (Zeist: KNNV-Publishing, 2008). In some particularly tricky cases, I have had invaluable help from Gerda van Uffelen and Carla Teune from the Leiden Hortus Botanicus. Where identification was not possible or less relevant to my argument, I have generally kept the original name found in the sources, whatever the language used (lilium susanum, tusai, zomboul arabique, Christusogen, Schmalzblümchen, etc.). It should be emphasized that terms like narcissus or hyacinth were used at the time for a much larger category of plants than at present.

The Clusius Correspondence: Sources and References

The roughly 1,300 extant letters sent to Clusius from all over Europe by more than 300 correspondents, and the roughly 200 extant ones sent by Clusius himself, form the principal source of this study. The vast majority of these manuscript letters are in the University Library of Leiden. Most of them bear the same shelfmark VUL 101. All original letters to and from Clusius in the possession of Leiden University Library have been digitized. The digital images of the original letters in Leiden can be freely consulted and downloaded via the website of Leiden University Library (http://www.bibliotheek.leidenuniv.nl/ with a link to the Digital Special Collections), or directly via: https://socrates.leidenuniv.nl/exlibris/dtl/u3_1/dtle/www_r_eng/js/correspondents.html. Details concerning the language in which a particular letter is written (which is mentioned here only if relevant to the argument), and concerning the place of dispatch and the destination of each letter can be found there. This special situation enables everyone who is interested in the original sources to read them at her or his leisure – taking into account, however, that the letters are written in six different languages: Latin, French, Italian, Spanish, German and Dutch (in order of importance), and that the handwriting of the period is not always easy to decipher.

Thanks to the fact that the vast majority of the letters used here belong to one collection, share the same shelfmark and were addressed to Clusius, it was possible to use an abbreviated reference system to the letters throughout this book. In both text (after direct quotations) and notes, all manuscript letters used here are referred to by the surname of the sender (or surname plus initials or first name in cases of possible confusion) and the date of dispatch: e.g. Pona, 15 August 1596, or James Garet Jr, 13 September 1599. Unless otherwise indicated such letters belong to the Leiden University Library collection, have the shelfmark VUL 101 and were sent to Clusius. In all other cases – whether concerning letters to or from Clusius or other manuscript sources – full details are given in the relevant notes.

It is not always known whether the dating of letters is in old or new style; if indications are available these are included in the reference as OS or NS. Roman dates (such as 3 Id. Martius 1592) have been converted into modern ones.

For the proper names of Clusius's correspondents I have generally used the vernacular rather than the Latinized version – Levenier and not Venerius, Roels and not Roelsius, Belli and not Bellus – unless this could lead to misunderstandings or when a correspondent is better known under the Latinized version.

Unless otherwise indicated I have consulted the original letters. All translations are mine with the constant advice and help of Peter Mason; certainly none of the errors are his.

ABBREVIATIONS

Curae	C. Clusius, *Curae posteriores, seu plurimarum non antè cognitarum, aut descriptarum stirpium, per egrinorumque aliquot animalium novae descriptiones* (Leiden and Antwerp: Officina Plantiniana, 1611).
Exoticorum	C. Clusius, *Exoticorum libri decem: quibus animalium, plantarum, aromatum, aliorumque peregrinorum fructuum historiae describuntur* (Leiden: Officina Plantiniana, 1605).
Rariorum	C. Clusius, *Rariorum plantarum historia* (Antwerp: Officina Plantiniana, 1601).

LIST OF FIGURES

INTRODUCTION

'The true monarch of flowers' is how in 1602 Emanuel, Prince of Portugal, addressed the old and famous Carolus Clusius (1526–1609) – naturalist and at that moment prefect of the botanical garden at the university of Leiden in Holland. The prince asked Clusius for a grape hyacinth on behalf of his wife Emilia of Orange-Nassau, who had picked up 'the gardening curiosity in which she takes a singular pleasure' from Marie de Brimeu, Princess of Chimay and a lifelong friend of Clusius.[1] And it was Marie de Brimeu – known in her time for the gardens which she designed and planted with rare flowers in the Netherlands – who hailed Clusius as 'the father of all the beautiful gardens in this country'.[2]

These courtly praises which acknowledge Clusius's merits and fame in flowery terminology direct us to a number of topics about which we will hear much more in the course of this book, such as long friendships celebrated and indeed formed by the gift exchange of plants; women and their involvement in gardens; the fashion of rare bulbs; the passion for plants and gardening, curiosity and collecting; and the fascination with rarity in Europe during the period 1550–1610. Those topics in their turn steer us towards themes that are at the heart of this study. We will be looking at the importance of passion, fashion and display in the formation of botany as a scientific discipline; at the role of men and women whose knowledge was mainly based on practice in the creation of a specialized field of expertise; and at the practices in which they engaged – in other words, at natural history in the making in the practical sense of the term.

Clusius himself was the first explicitly to recognize that his great expertise – universally acknowledged as scientific then and now – was rooted in pleasure. From Leiden, at the age of eighty and looking back on his long life and the origins of his interest in plants, he wrote to his young friend Matteo Caccini in Florence that he had never obtained any university degree,

> but I have merely followed studies for my own delectation, and since from boyhood I have always delighted in plants, I have now wanted to publish the history of all the rare plants which I have observed in my journeyings that have not been described by others.[3]

The many hundreds of letters to Clusius which constitute the most important source for the present study and at the time formed the web of exchange in which and on the basis of which botanical knowledge was being created, likewise abound in expressions of pleasure, love and fascination, but also of intense rivalry and jealousy centred on the discovery and possession of rare plants. It is these letters that allow us to reconstruct Clusius's world – what we might now be tempted to call a virtual network of relations criss-crossing the whole of sixteenth-century Europe, interlinking and coinciding, clashing and forming part of other individuals' networks or worlds. The need for that reconstruction follows from my two very simple points of departure: that the study of 'science before science' cannot proceed by focusing only on the recognized categories belonging to science (publications, universities, learned men, scientists, ideas) but needs to cast its net far wider to include categories, sources, practices and persons that are usually regarded as outside or beyond its domain given the fact that this domain was still being demarcated at the time; and, secondly, that knowledge of the relevant context (in the wide sense of place, social and cultural and sometimes political setting) is essential to our understanding of early modern natural history.

In the present exploration of early modern natural history I hope to join authors such as Lucia Tongiorgi Tomasi, Harold Cook, Giuseppe Olmi, Karen Reeds, Deborah Harkness, Stephen Pumfrey, Emma Spary, Paula Findlen, Pamela Smith, Brian Ogilvie and Mario Biagioli. By their inspiring work – which is constantly present in this study, even if it is not continually mentioned in the notes – all of them have thrown a great amount of new light on early modern research practices, practitioners and their relevance to the formation of early modern science. If by investigating the world of Carolus Clusius we find that the boundaries of the new history of science itself are being stretched and that it is turning into a cultural history of knowledge and its formation, so much the better.[4]

People and Practices in the Botanical Renaissance

Knowledge of nature had always formed part of everyday life. But in the sixteenth century – precisely during the lifetime of Clusius – a number of new phenomena emerged which together are known as the Botanical Renaissance and produced the greatest changes in European knowledge of living nature before the age of Linnaeus, or perhaps even before Darwin.[5] The face of the gardens of Europe changed with the introduction of a vast number of exotic plants from the Middle East, Africa, the Far East and the Americas. The indigenous flora of Europe itself was for the first time investigated by means of field trips and documented in text and image. The passion for gardening and the cultivation of rare plants spread like wildfire throughout Europe; the first university botanical gardens and academic chairs in medicinal botany were created; and the first richly illustrated surveys of the world of plants and the world of animals were published. Clusius was a key figure in each and every one of these developments, and has been responsible more than anyone else for making exotic nature known to Europeans.

The major changes of the Botanical Renaissance were strongly stimulated by the discovery of exotic nature via the European voyages of exploration, and influenced by new and critical approaches (originating in Renaissance humanism) to classical descriptions of living nature. Many of the above-named innovations were unthinkable, moreover, without a complex phenomenon which we will try to explore in this book. In the sixteenth century the world of nature came for the first time to lie at the centre of interest of the European social and intellectual elites. It became a fashionable subject, a respectable pastime and even a passionate interest, manifesting itself in the cultures of collecting and gardening, and the fascination with the exotic and rarity. Attention has been paid to all of these phenomena, but they have usually been treated separately. We will see in the course of this study that they were indelibly linked together, and formed one of the driving forces of the Botanical Renaissance.

The focus on practices and practitioners follows directly from the attempt to study 'science before science'. It is the process of emerging expert knowledge in a phase before it had actually solidified into 'science' that interests us here, and more specifically the question of which practices generated which kinds of expertise in the field of natural history and which persons were involved. It seems unlikely that great expertise concerning living nature could have come to attain that quality exclusively on the basis of the learning of a relatively small group of men who published (mostly in Latin), had studied and had generally been trained as humanists. In this respect – but it is an important one – my point of departure is thus almost diametrically opposed to Ogilvie's, although many of our findings are not, and our paths often run parallel. Ogilvie looks at practices as well as texts, but in a social sense explicitly focuses his study on 'the Latin-writing, humanistically educated elite of the sixteenth century' and defines Renaissance natural history as the discipline developed in the community of this elite from which many people and their knowledge were excluded.[6] Precisely given the fact that the sixteenth century saw the origins of natural history as a discipline, it is unwarranted to assume that the roots of any field of scientific expertise (and that concerning living nature perhaps least of all) can be found exclusively in the knowledge circulating among the intellectual elite of the period. This exclusion seems to me a historiographical rather than a historical fact. In this respect I find myself much closer to the approach followed by Findlen in her investigation of Italian naturalists and collectors:

> The intellectual problem of understanding what Renaissance naturalists meant when they 'did science' also has social consequences. The indistinguishability of natural history from other aspects of learned and courtly culture certainly calls into question the notion of a scientific community, as historians have commonly understood it.[7]

My emphasis will therefore be on those persons who had a great amount of expertise concerning living nature to offer to Clusius, but most of whom would, according to Ogilvie's criterion, be outside the margins of the community of naturalists or at best on its edge. Few learned humanists will therefore be discussed here – even though Clusius maintained friendly relations with some of the major humanists of his age and was trained in this tradition himself. Universities will be looked at as only one (albeit an important) factor in the shaping of knowledge about nature and investigative methodology.[8] And printed works on animals and plants, even those by Clusius, will only rarely be discussed. If they occur, it is usually as instruments – often subsidiary ones, compared to practice – in the formation of expertise rather than as major stepping stones in the history of knowledge. Our social focus is wide and covers a heterogeneous mixture of apothecaries, aristocratic women, merchants, voyagers, physicians and local herborists as well as courtiers, high officials in the administration of the Habsburg empire, physicians and even a few princes, military men and revolutionaries. And we will try to show that a domain of knowledge in which even today laymen and women may develop very high-quality expertise, was shaped in a major way by practice-based knowledge and by the experience, methods, observations and types of insight developed by people who had no university background or humanist training.[9]

Again it is Clusius himself who pointed the way, in a remark made just a few months before his death in which he reminisced about his days as a young naturalist. Referring to the nobleman Jean Boisot in Brussels, he put their respective interest (*curiosité*) in and expertise concerning plants on a par, while explicitly rooting the knowledge of both Boisot and himself in practices – investigation by travel and growing plants in the garden:

> That gentleman was my very great friend, a learned man, and I think that he and I have been the first *curiosi* to understand the variety of plants, but he did not undertake any journeys but cultivated them in his garden.[10]

Clusius has been called the first scientific botanist for his extensive knowledge of living nature, his precision, his detailed descriptions – in which a critical comparison of information and personal observation of plants played a key role – his interest in ecology and his fascination with exotic, non-European nature. As we will discover, those characteristics by no means applied only to Clusius: many of his friends and correspondents had exactly the same interests. Their expertise was relevant to Clusius precisely because it was based on and tested in practice, while first-hand observation, detailed description and a critical evaluation of observed evidence played a crucial part in its formation. We thus cannot prise the characteristics of Clusius's work loose from the much wider set of interests and practices to which they belonged at the time. Nor can we – when looking at practices relevant to the study of nature – impose an *a priori* distinction between those that have later come to be regarded as 'scientific' and others. We will look therefore at horticultural techniques, as well as fashions in the collecting of rare plants and in garden display, plant-hunting expeditions, the fascination with colour, botanical experiment, acclimatization, methods of evaluating evidence, modes of exchanging knowledge, styles of reporting and the idiom of friendly collaboration in the pursuit of knowledge.

Seeing that parts of natural history (botany first and foremost) were being turned into natural sciences in the course of Clusius's lifetime, we should ask ourselves to what extent we are dealing here with the professionalization of a certain domain of knowledge. Clusius, who studied medicine but never obtained his degree or practised as a physician, was regarded in his age as one of the very few and first professional naturalists: he earned his living as an expert in these matters; he helped to set the standards of professionalism with respect to the knowledge of plants; he was renowned for his expertise in this field; and he divulged his knowledge. Throughout our investigation we will explore, in so far as possible, which terms his correspondents used to describe their field or expertise concerning living nature, and look at their professions to see for how many members of Clusius's world that type of expertise formed a source of income, whether in the form of patronage or other earnings. And we will explore to what extent Clusius's friends too were involved in setting the standards of botanical expertise, by special expertise, research methods or divulging their knowledge. At the time, however, experts by no means only gained a reputation through publications, but also – and often more quickly and efficiently – by word of mouth, personal exchanges and correspondence.

Clusius and Friends: Correspondence and the Myth of the 'Isolated Genius'

Although Clusius is the central figure in this study, not he but his world is its subject. One of the purposes of studying that world is to (further) undermine a myth that has characterized the history of science in various more or less sophisticated forms since at least the nineteenth century: that of the isolated genius who almost single-handedly changed the direction of scientific progress. Without detracting from the fame or relevance of renowned and innovative men such as Darwin, Linnaeus, Galileo or Da Vinci, the attention is shifting to 'famous men in their context' in order to reach a different understanding of how (scientific) innovation is connected with a socio-cultural background.[11] The myth of the isolated genius has been under attack for many years now from various directions, but by no means all of the implications have yet been explored. To investigate the men and women who belonged to the context of famous scientists is by no means always possible given the nature of historical sources, and it is generally even more difficult to trace their expertise and involvement in research.[12] In this case it is Clusius himself who allows us to do so. He kept their letters; he put himself on a par with the men and women with whom he investigated nature; and he included hundreds if not thousands of references to their information and knowledge in his printed works.

In particular Clusius's generosity as a man and a scholar opens up the possibility of tracing in some detail what his less famous friends contributed to natural history. He lived and worked in many different European countries, and in the course of his long life built up a correspondence network which covered most of Europe while spanning a considerable part of the social spectrum. Unusually, he preserved a very large number of letters sent to him by correspondents during almost half a century (*c.* 1560 until his death in 1609) – even though the extant letters can be but a fraction of those sent to him: there are some 1,200 extant letters written to Clusius by some 330 correspondents in six languages (Latin, French, Italian, German, Spanish and Dutch in order of importance).[13] Clusius's correspondents lived all over Europe, from England to Hungary and Austria, from Greece and Italy to Poland, and from Spain and Portugal to the Northern Netherlands, France, Germany and Norway. Some were his social equals, but there were also many persons of a higher or lower social position, ranging from ruling princes and the very top of the European aristocracy to diplomats and famous humanists, and from fellow physicians or naturalists to printer-publishers, artists and apothecaries. The better known Clusius became as a leading botanical expert, the more his epistolary contacts proliferated. Since many of Clusius's correspondents had their own epistolary networks of exchange and had access, moreover, to the partly overlapping networks of yet other friends with whom they maintained relations of exchange, it is no exaggeration to claim that he could be in touch, directly or indirectly, with all the then relevant experts on living nature in Europe. That is why the present study is called Clusius's world rather than Clusius's network, and why it does not aim at a network reconstruction in the strict sense of the term.[14]

The many, meticulous and usually generous references in Clusius's printed works to his friends and their gifts of plants, animals, information and knowledge are almost as important to the tracing of their expertise and contributions. If the correspondence did not already show us as much, these references by themselves tell us how the expertise of Clusius as an erudite, Latin-writing and publishing naturalist was inextricably linked with that of experts from extremely diverse backgrounds. As will become clear, the 'practical experts' among his friends by no means merely acted as a kind of intellectual servant – carrying information to a scientist who then elevated that information by means of natural philosophy, classification, Latin, jargon, *et cetera* to the higher plane of 'science'.[15] Although correspondence networks were perhaps equally important to both Clusius and Darwin, Clusius was no Darwin-style scientist who fitted information carried to him by hundreds if not thousands of correspondents together into a grand theory. Cooperation and exchange in Clusius's world affected the very character of knowledge about living nature.[16] By exploring the expertise of Clusius's correspondents we will be able to trace which practice-based insights were suggested to Clusius by his partners in exchange and helped to shape his own way of dealing with nature, thereby contributing to the formation of the new scientific discipline of botany.

Pride of Place

The emphasis on the social and cultural construction of knowledge in the new cultural history of science entails paying explicit attention to contextualization and circulation as well as reception or permeation of knowledge. Not only people but also locations – or 'place' in the sense of geography, location and setting – have become immediately relevant, therefore. And knowledge is no longer disembodied, as shown in particular by Pamela Smith, but intrinsically linked with the body and the senses.[17] The attention to place, the importance of the body, circulation and reception, and the increasing attention to visual aspects (the latter an important topic which will, however, not be discussed in this book)[18] in the new cultural history of science reflect similar, often earlier, interests in the wider domain of cultural history, which was influenced in the course of the 1980s and 1990s by literary studies and reception theory, visual studies, the history of the body and historical anthropology. With a background in cultural history and historical anthropology rather than in the history of science, for me the focus on place as a crucial element of context is not only linked, therefore, to the new demands of the history of science but also to my earlier research in cultural and microhistory in which one of the recurring questions has been to find out which contexts can be regarded as relevant.[19] But well before that, Richard Cobb's *A Sense of Place* (1975) and several of his other books were the first historical studies to show me that it is indeed possible to get a historical sense of place and come to know even the not so famous in their setting.[20]

Place, interpreted as shifting geographies of expertise, to adapt slightly Lissa Roberts's phrase 'shifting geographies of skill', structures this book.[21] Moving from one European area and social setting to another, but in each part of Europe always keeping the focus on Clusius's many friends, we will explore the particular practices and forms of expertise concerning living nature and the ways in which they were embedded in those settings. The Southern Netherlands – as part of the Habsburg empire closely connected with both Spain and Austria at the time – and Italy receive special emphasis as the core areas of gardening and botanical culture in Europe. From there we will make brief excursions to Austria, Spain and Crete. Subsequently, we move northward along the Atlantic coast to France, England and the Dutch Republic. In each geographical section we will be dealing with a slightly different socio-cultural constellation: court and landed aristocracy in the Southern Netherlands and related Habsburg court circles in Austria and Spain; aristocratic women in the Habsburg countries; princes, aristocrats, members of religious orders and urban professionals in Italy; clergymen, aristocrats, physicians and apothecaries in France; physicians, merchants and apothecaries in the Dutch ports; intellectuals, merchants and other members of the urban elite in the Dutch university town Leiden; aristocrats, merchants, apothecaries and physicians in England.

Two important territories with which Clusius was closely connected play a very minor role in this study: the German-speaking world (i.e. Austria and the German states) and Spain. Vienna and its court-connected nobility figure mainly in the chapters on the Southern Netherlands and on Habsburg women. Spain too, which Clusius visited on an extended journey cum field trip in 1564–5 and with whose expert physicians in Valencia and Seville he established contacts in the course of the 1590s, mainly figures in the chapter on the Southern Netherlands, while the letters of these physicians are incidentally used in other chapters. The reasons are simple. While the letters by Clusius's Iberian correspondents are most interesting in terms of their contents concerning natural history, as shown for instance by Barona and Ramón-Laca, they do not provide enough material to form an image of how a cluster of people was 'making natural history' in Spain during the second half of the sixteenth century.[22] Politics and warfare interrupted and impeded exchanges by post between those places where Clusius lived and the Spanish naturalists. As a consequence the correspondence offers us only two, barely connected glimpses which are some forty years apart (the 1560s and around 1600) of a small number of naturalists in Spain who were in contact with Clusius. This problem of interrupted or obstructed contact did not only apply to Clusius, of course. Indeed, the political decisions taken at the time in Spain are to a large extent responsible for the fact that it is still under-represented on the scientific map of Europe for that period.[23] With respect to the German-speaking world the situation is almost the opposite. More literature is available on it than on any other part of Europe in connection with Clusius, while several new publications focus specifically on the connections between Clusius, Austria-Hungary and the German states.[24] In the light of that relative abundance, it seemed more important to concentrate here on other parts of Europe.

In the final chapters before the conclusion we step 'out of place', not by leaving a specific setting – here London and Amsterdam – but by looking at those locations as connecting nodes for relations at a local, regional, national, international and even intercontinental level. Rather than providing marked boundaries or enclosing spaces, place and setting should be seen to open up possibilities. Similarly, widening the history of science to a cultural history of knowledge formation, and stepping outside its usual boundaries into the domains of passion, curiosity, collecting and fashion, can lead to unexpected findings, which call for a rethinking of the concept of early modern science itself. But the most relevant act may be the transgression itself of those boundaries, and the rediscovery of the relevance of pleasure and passion to discovery – whether botanical or historical.

I THE SOUTHERN NETHERLANDS

1 THE GARDEN OF EUROPE: BOTANY AS A COURTLY FASHION IN THE SOUTHERN NETHERLANDS

Cultivating Plants

According to the botanist Matthias de Lobel (1538–1616), his native Southern Netherlands occupied the first rank in botanical and horticultural matters in all of Europe. In this single country more plants, shrubs and trees could be found than in any other – not excluding 'ancient Greece, spacious Spain, the whole of Germany, England and France, and even Italy which is so well cultivated'. He particularly emphasized that plants and flowers from all over the world were brought to this 'most famous market [*emporium*] of Europe', and were cultivated there, in spite of the adverse climate, but thanks to the hard work, diligence and perseverance of the inhabitants.[1]

Lobel was neither the first nor the last to remark upon the special reputation of the Southern Netherlands in this respect. There are references to the enormous quantities of fruit trees, flowers, shrubs, trees and medicinal herbs that could be seen there in the early sixteenth century; every foreigner visiting this part of the world admired them. By the 1560s inhabitants of the Southern Netherlands were said to have developed a real taste for flowers and to spend great sums of money on plants just for pleasure.[2] And there is other evidence as well, such as the number of celebrated gardens of this period; the importance of Flemish printers-publishers – most famously Plantin – who produced works on plants; the prominence of the Flemish 'fathers of botany', Lobel (Lobelius),

Dodoens (Dodonaeus) and Clusius; and the importation and expert cultivation of rare plants from Turkey, the Middle East, Spain, Asia and the New World.[3]

With respect to the introduction in Europe of plants from the Middle East during the late 1550s and early 1560s, the role of the diplomat Ogier Ghislain de Busbecq (1522–92) and the physician Willem Quackelbeen (1527–61) was especially important. Originally from the Southern Netherlands, they were based at Emperor Ferdinand I's court in Vienna, and Busbecq was sent as ambassador to Sultan Suleiman the Magnificent. The trips to Turkey and Asia Minor of Busbecq and Quackelbeen resulted in the introduction – generally first in Vienna and subsequently in the Southern Netherlands and Italy – of a considerable number of plants. Among the ones now best known are the tulip, horse chestnut, lilac, mock orange, gladiolus, iris tuberosa and oriental plane tree.[4] These and later introductions from the Middle East in the course of the sixteenth century caused a rapid increase in the number of plant species that could be found in European gardens north of the Alps. Until *c.* 1550 that number is said not to have exceeded some 600, of which the majority were indigenous, a few were Mediterranean and a very few American or Asian.[5]

In the Southern Netherlands the phenomenon of the country house surrounded by extensive grounds was already widespread around the middle of the sixteenth century, while the earliest ones have been traced to the 1520s. Most of these *maisons de plaisance* belonged to aristocrats or members of the upper strata of the bourgeoisie, such as merchants or bankers. Inventories and visual representations indicate that the surrounding grounds consisted of much more than a simple kitchen or pleasure garden, and could comprise tree-lined lanes, a bridge, moat, lake or stream, meadows, a grotto, spring, island, sports area (such as an archery), labyrinth, fountains, herb garden and ornamental garden with shrubs in pots. Flowers and ornamental plants were never the only component. Such grounds generally combined zones of pleasure and zones of utility, such as a poultry run, fish pond, fruit trees, kitchen garden and an area for medicinal herbs.[6]

The sixteenth-century Southern Netherlands was also known for the development of special techniques of plant care, such as the cold frame, designed for protecting less hardy plants during the northern winters. Gardening utensils, plants and the expertise itself were exported to other parts of Europe – which, of course, did not exclude contemporaneous opposite movements. Throughout the second half of the sixteenth century and the early years of the seventeenth century, expert gardeners and botanical advisors from the Southern Netherlands were employed by princes in various parts over Europe. In Italy, for instance, we hear of Giuseppe Casabona (*c.* 1535–95), who became court botanist in Florence and prefect of the botanical garden of Pisa. His original name was Joost Goedenhuyse and he came from Flanders or Maastricht.[7] In sixteenth-century Europe the boundaries between the horticultural consultants, some of whom had a medical university training, and the expert gardeners and architects-garden designers who generally had a practical training, were fluid in terms of their practical involvement with plants and gardens, even though their social status differed on account of their training.[8] The expert gardeners had generally received a practical training, but the highest ranking ones acted both as head-gardeners who supervised a considerable staff and as skilled landscape and garden designers.

Members of the Holbecque family, who worked for Emperor Charles V and his son Philip II, provide a well-documented example. When retiring to Yuste (Spain) in June 1556, Charles V discharged all of his servants except for a small staff which included his gardener from Brussels, François Holbecque, who was also an expert distiller and perfumer. François later returned to the Netherlands, where he acted as gardener of Cardinal Antoine Perrenot de Granvelle, then Bishop of Arras, before leaving for Spain once more in 1565, after his brother Jean Holbecque had been appointed by Philip II. The latter had been so impressed by the garden culture of the Southern Netherlands which he had seen as a young man during his two journeys through Europe in 1548–51 and 1554–9, that he hired a whole team from the Southern Netherlands to help create the court gardens of Aranjuez: it consisted of eight gardeners, besides masons, agricultural labourers and a dike engineer. Together with the team, a large transport consisting of seeds and many different types of gardening tools was sent to Aranjuez. There, Jean Holbecque soon became more than head gardener: as one of the architects of the Aranjuez gardens he had regular consultations with Philip II about their design, and he ordered large quantities of plants, shrubs and trees to be brought to Spain from the Southern Netherlands. François Holbecque eventually succeeded his brother as head gardener of Aranjuez in 1579, and from 1587 to 1594 yet another François Holbecque occupied the same position.[9]

Such expert gardeners did not only operate in the Habsburg countries. The head gardener Joachim Gille, to name but one example, was for some twenty years (from *c.* 1569 to at least the late 1580s) responsible for the private botanical garden of Wilhelm IV, Landgrave of Hesse-Kassel. Gille and Wilhelm IV regularly corresponded about garden design, planting and cultivation methods, and they discussed experiments with plants. Gille was sent several times on trips to the area of Venice in order to buy exotic plants and trees for Wilhelm's garden, and to supervise their transport across the Alps. His son studied medicine and eventually became court physician annex botanist to Moritz (Wilhelm's son) – he called himself 'botanologus' and 'botanographus' – as well as supervisor of the garden.[10] And we will come across more examples of father–son upward social mobility in which the father had a practical training while the son went to university, but both acted as expert gardeners cum naturalists.

Interest and expertise in the Southern Netherlands extended from the plants and gardens themselves to their images. Some of the most consummate painters and illustrators of *naturalia* during the second half of the sixteenth century – all of them personally known to Clusius – were from the Southern Netherlands: Jacques vanden Corenhuyse; Peter van der Borcht; Anselmus de Boodt; and Joris (Georg) and Jacob Hoefnagel. It was also in the Southern Netherlands that the art of tapestry weaving, in particular the huge *verdures* with plants and (exotic) animals, reached its height around the middle of the sixteenth century, with the work of the artists-designers Michiel Coxie and Pieter Coecke van Aelst.

This is the setting of Clusius and his circle of friends and contacts in the Southern Netherlands. That setting itself is by no means self-evident, however. How and when did this part of Europe become a prime zone for horticulture and botany? And can we connect the emergence of the Flemish trio of botanical experts – Clusius, Lobel, Dodoens – during the second half of the sixteenth century with the prominence of the Southern Netherlands in horticulture and the artistic representation of *naturalia*, beyond pointing, rather vaguely, to contemporaneity?

Gardens and Collections

Lobel singled out a number of individuals in the Southern Netherlands of the 1560s and 1570s on account of their impressive gardens and special contributions to the knowledge of plants. Clusius himself was seen by Lobel as the greatest botanical expert of all – with hindsight of course, since Clusius was still an up and coming botanical expert in need of patrons during the late 1550s and early 1560s. Four men mentioned by Lobel as great plant experts already belonged to the past by the time his work appeared in 1581: the Bishop of Tournai Gérard (Gilbert) d'Oignies, and the noblemen Reynoutre, Brancion and Van der Delft. Thereafter came a list of seventeen 'modern' experts, of whom the following seven also were friends and correspondents of Clusius: Charles de Croÿ; Charles de Houchin; Jean Boisot; Philippe Deurnaghele; Philips de Marnix, Lord of Saint Aldegonde; Mathias Laurin; and Jean Mouton.[11] Nearly all belonged to the overlapping circles of the landed aristocracy and the higher strata of government officials and other experts in the service of the Habsburg empire. All were rich. But politics, armed revolt and religious warfare affected all of them and theirs is a story of loss and destruction as much as of the creation of great gardens and collections.

That is especially true of several garden owners and collectors based in the Bruges-Antwerp-Malines area. In the course of the 1560s Charles de Saint Omer (1533–69), Lord of Reynoutre or Dranoutre and Moerkercke, became Clusius's very first patron. Apart from the fact that he was a wealthy young nobleman from the highest circles of the aristocracy, little is known about Saint Omer's life prior to the mid-1560s. His health was not strong and he gave up his military career, probably even before he was thirty years of age. Upon retiring to his estate and castle at Moerkerke close to Bruges he devoted himself to the arts and sciences, in particular to the study of nature. Saint Omer's connection with Clusius – and perhaps even his patronage – dated from before Clusius's journey to the Iberian peninsula of 1564–5, and Clusius's own botanical interest plus the fact that he sent seeds to Saint Omer from Spain certainly stimulated the latter's fascination with plants, as we know from a mutual friend in Bruges.[12]

> I have delivered a bunch of seeds together with a letter [i.e. from Clusius] to the illustrious lord of Dranoutere in Moerkercke and somehow the opportunity provided by this letter enormously seduced him to the love of botany, and now for a fortnight the Lord and myself are doing nothing but that he keeps me occupied in identifying the simples (although it is the heart of winter), arranging into a certain order that wonderful book of which I believe you have seen a sample according to the Dioscoridean method. (G. Laurin, 25 November 1564)

After his return from Spain Clusius stayed for months on end at the castle of his patron at Moerkerke, where he saw various exotic substances, such as true cinnamon. In 1566 he gave seeds of the Brazilian pepper to Saint Omer, who managed to cultivate a pepper plant in his garden which even flowered that autumn. Another one of Clusius's gifts was a shoot of the aloe Americana, which had come from a big plant with some thirty small side-shoots in the garden of Clusius's host in Valencia. Clusius brought two of these back to the Netherlands.[13]

The setting at Moerkerke must have been more than congenial to Clusius, since Charles de Saint Omer was the creator and owner of one of the earliest and most remarkable collections of the *Kunst- und Wunderkammer*-type in Europe north of the Alps.[14] Living nature, dried *naturalia* and depicted nature played a crucial role in his collection. A lengthy inventory made shortly after Saint Omer's death in 1569 gives us some idea of his impressive possessions. Besides the estate and castle of Moerkerke, Saint Omer owned extensive gardens and parks, farms, mills, various feudal landholdings and rights, as well as a town house in Bruges. Around the castle of Moerkerke there were gardens, an orchard and a menagerie with not only horses, a mule, sheep, goats, swans and pheasants, but also the more special 'African chickens' (possibly guinea fowl), an eagle, a stork, a bear, a parrot and some other animals. Indoors a rich wardrobe, jewellery, tapestries (including *verdures*), silverware and paintings testified to his wealth. Possessions specifically pointing to Saint Omer's interests as a collector were his cabinet containing curiosities, and his collection of weapons and books. In a second cabinet, explicitly listed as belonging to his wife Anne d'Oingnies, seeds were kept as well as various albums with 'painted plants', fish and other animals.[15]

Albums of watercolours had been commissioned by Saint Omer, 'a man', according to Clusius, 'not only most expert in matters of plants and who had the plants themselves, birds, and quadrupeds depicted with outstanding skill in lively colours, but also very interested in the nature of all sorts of wonderful things'.[16] They formed the core of the famous zoological and botanical watercolours known as the *Libri Picturati* A.16–30, one of the most remarkable European collections of watercolours depicting *naturalia* of the sixteenth century. Much is still unknown about the history of the *Libri Picturati* watercolours, especially for the period after Saint Omer's death in 1569 until the early seventeenth century. There continued to be links with Clusius, however: some 122 illustrations in Clusius's printed works are directly based on watercolours in the *Libri Picturati* collection. Sheets with watercolours were added to it in the course of the century, some of which bear the names of Clusius's friends, and by the end of the century the whole collection came into the possession of an aristocrat of even higher rank than Saint Omer, yet another friend of Clusius: the Count of Arenberg.[17]

It seems likely that the botanical watercolours commissioned by Saint Omer served various purposes, but they did not form a visual handbook for the identification of medicinal plants, and very few of the annotations evince an interest in the medicinal use of the plants.[18] However, the brief annotation on the watercolours regularly mentions the natural habitat of the plants and provides ecological information concerning the parts of Europe where they could be found, as in the example of a small wild hyacinth: 'They grow spontaneously on the edges of fields preferably in a light and sandy soil, and are frequent near Malines'.[19] Indeed, we know that Saint Omer was interested not only in rare exotics but also in the local flora of his own area: comfortably seated in a carriage, Saint Omer and Clusius explored the marshy area of Maldegem not far from Bruges and discovered a pseudoasphodel.[20] Many of the annotations to the botanical watercolours of the *Libri Picturati* look like personal observations by a commentator who knew about both wild plants and horticultural practice, a person who was first and foremost interested in how and where plants grew and how they could be identified by their flowers, roots and leaves. Some of the watercolours probably served as documentation of plants grown in Saint Omer's own garden, but by no means all watercolours depicted *naturalia* that actually belonged to Saint Omer, so they clearly did not simply constitute a catalogue. Yet, they formed an integral part of Saint Omer's collection: as a painted collection on paper they complemented his live and dead animals and plants on the one hand, and his printed works about natural history on the other.

The *Libri Picturati* watercolours demonstrate an intimate link between the history of collecting of the *Kunst- und Wunderkammer*-type, botanical interest and scientific illustration. The key to an understanding of the interest in nature as manifested by Clusius's patron Saint Omer – comprising exotica as well as indigenous flora and fauna, living plants and animals as well as their pictures, cultivation as well as ecology and classification – lies in the phenomenon of collecting that was a vital element in his region in a cultural movement of the 1560s and 1570s known as Bruges Humanism. Two of Clusius's and Saint Omer's close friends, the brothers Guy and Marc Laurin, played a central role in it. They belonged to an illustrious local family: their father had been treasurer and mayor of Bruges, while another relative had acted as Erasmus's patron. The two brothers combined an interest in nature with a passion for antiquity. They acted as patrons to the artist, numismatic expert and printer Hubert Goltzius (1526–83), and Marc Laurin's collection of antique coins was one of the best in Europe. Hubert Goltzius provides the link between the Bruges humanists and a circle of famous artists related by both family ties and artistic interests. His wife's sister married Pieter Coecke van Aelst. Among the most famous of his creations are tapestries, *verdures*, which show birds, plants and beasts often based upon the woodcuts in contemporary scientific works on nature, and also demonstrate the widespread interest in exotic *naturalia*. Goltzius and Pieter Coecke were therefore brothers-in-law, while Goltzius's niece married Pieter Breughel. In 1558 the brothers Laurin invited Goltzius to come and live in Bruges, and they eventually appointed him as head of their private printing press. Their country house Laurocorinth became a centre of studies for the circle of Bruges humanists.[21] There were, therefore, striking similarities between the circle around the Laurins and Goltzius, with its focus on antiquities and classical learning, and that around Saint Omer and Clusius, which concentrated more on *naturalia*: both united artists, collectors and scholars in the region of Bruges; in both knowledge and representations of nature were regarded as important. The chronological overlap of the circles and the friendship between Clusius, the Laurins and Saint Omer indicate that contacts and interaction between members of these circles would have been frequent and close.

In his *Descrittione di tutti i paesi bassi* of 1567 Lodovico Guicciardini said of Saint Omer's garden that it was 'admirable with an infinity of excellent plants'. He put it in the same category as another famous garden of the period by praising both as incredibly rich in species.[22] That other garden belonged to the Antwerp apothecary Peeter van Coudenberghe (1517–99) and had, according to Guicciardini, more than 400 exotic plants besides many kinds of ordinary ones. A few years later Coudenberghe himself spoke of some 600 exotic plants.[23] In fact, Coudenberghe's garden was one of the early private botanical gardens in Europe: he started it in 1548 in Borgerhout, just outside the city walls of Antwerp. Visits by Guicciardini were not the only link between Saint Omer and Coudenberghe: while Saint Omer had in 1565 been the recipient of one of Clusius's shoots of the American aloe from Valencia, a second shoot was given by Clusius to Coudenberghe – and this one did indeed survive the harsh northern winters to serve as the model for the illustration in Clusius's *Rariorum*.[24]

Coudenberghe belonged to the patriciate of Brussels and established himself as an apothecary in Antwerp, but his elite background and impeccable Latin point to the circles of scholarly apothecaries and rich garden owners rather than to those of middle-class shopkeepers or merchant-entrepreneurs. He was, in fact, well known as an erudite apothecary who corrected a manual of herbal remedies by Valerius Cordus for the Plantin press, and maintained personal contacts and correspondence with a network of persons of international reputation such as Ortelius, Camerarius, Gessner, Goropius Becanus, Posthius, Fugger, Plantin, Lobel and Dodoens. The fame of his garden was indeed such, that some of the most famous naturalists of that period came to see it: Gessner from Switzerland and Turner from England. Gessner, Dodoens, Lobel and Clusius too refer in their printed works to botanical information and plants received from Coudenberghe. Coudenberghe's garden was a vast botanical collection rather than a practical garden of medicinal simples, and deserves to be studied in direct comparison with the more famous Italian botanical gardens. Among the rare plants in his collection, besides the aloe, were the dragon tree from the Canary Islands or Madeira, as well as plants from the New World (such as Brazilian pepper, tobacco, ipomoea, tomato and guaiacum), the East-Indies (such as cotton and zizyphus or jujube) and the Mediterranean (pomegranate, aubergine, a type of gladiolus, cypress and artichoke). Coudenberghe had grown a number of plants from seeds received from Italy. He learned how to acclimatize plants from warmer parts of the world, and developed special spaces, such as the so-called 'temperate or Flemish' protective structure and the basement in which plants were put in their pots to survive the winter.[25]

After Saint Omer's death, Clusius found a new friend and patron in the nobleman Jean de Brancion (*c.* 1520–75) at Malines who belonged to the Habsburg court circle.[26] During the years 1568–73 his house at Malines formed an almost permanent base for Clusius, and Brancion introduced Clusius to many persons with whom he was to exchange botanical information for decades. In Malines, Brancion created a rich garden with both indigenous and exotic plants, perhaps as early as the 1550s. Some of the very first American plants in these parts of the world could be seen in his garden. The first description of the American sunflower in the botanical literature is by Dodoens, who also lived in Malines and saw it flower in Brancion's garden in or before 1568. It reached 'only' 10–11 feet, while sunflowers in the royal garden in Spain could grow as tall as 24 feet. In 1565 Brancion sent seeds of 'panicum caeruleum indicum' (perhaps a type of millet) to Lobel in Montpellier, which he had apparently received straight from America. And Clusius saw two small plants in Brancion's garden which had apparently grown from seeds of a tree that grew in Peru.[27]

The Spanish diplomat, humanist and naturalist Benito Arias Montano (1527–98), who lived for eight years in Antwerp during the late 1560s and early 1570s, praised Brancion highly and was very interested in receiving some of his beans, bulbs and irises. Elegantly thanking Brancion for plants and seeds, Arias Montano wrote that these gifts were treasures which reminded him of the giver, especially when they flourished and flowered (Arias Montano, 22 April 1569, 7 August 1569). Brancion was in contact – either in person or via correspondence – with Lobel, Dodoens and, of course, Clusius, and he used his large international network for the exchange of seeds and bulbs.[28] His contacts with Italian naturalists were especially fruitful. The naturalist Ulisse Aldrovandi in Bologna, for instance, thanked both Clusius and Brancion in February 1570 for the seeds they had sent, and asked to share in the rare *naturalia* which they received. As Aldrovandi said, Brancion 'grew an infinite number of plants in his most fertile green court' (Aldrovandi 6 February 1570). From Padua, Giacomo Antonio Cortuso (later prefect of the *hortus* there) sent Brancion a kind of muscari known as 'dibcadi', and Brancion received a heart fern from the Roman papal physician and ichthyologist Ippolito Salviani.[29] But the Italian to whom Brancion probably owed most was Alfonso Pancio, from about 1550 to 1574 professor of medicine at Ferrara, private physician to Duke Alfonso d'Este and supervisor of several private gardens of the Duke. Thanks to Pancio's gifts, various kinds of anemones, a special cyclamen and the ornithogalum arabicum could be seen for the first time ever in the Southern Netherlands in Brancion's garden. Gifts travelled from Brancion to Italy as well, and resulted in the naming of at least two kinds of tulip after him. The planting plan of 1625 for the flower garden near Cisterna (now Ninfa) which belonged to the Duke of Sermoneta, Francesco Gaetani (1594–1682), lists both the plain 'Brancion' tulip and the 'Testament Brancion' tulip, which was apparently so called because Brancion named it in his will.[30]

Brancion's contacts also extended to Habsburg Spain and Vienna: he managed to grow the asphodel from seeds received from Spain, and via his cousin, a diplomat who visited Italy and spent much time in Austria, received rare plants that had reached Vienna from Constantinople. The first introduction of the lilium susanium in the Southern Netherlands also took place via Brancion: from Constantinople the plant had reached Venice, where Brancion's cousin received a bulb as a gift and sent it to Malines. Some of the most important diplomats who worked for the Habsburg emperor in Constantinople were among Brancion's connections: in the autumn of 1572, for instance, 'a basket full of bulbs and seeds of Turkish delights' was sent to Brancion by Charles Rym, imperial diplomat in Constantinople between 1569 and 1574 (J. De Croix to Brancion, 5 October 1572). After Brancion's death in 1575 his nephew Johan van der Delft – another person on Lobel's list of plant experts – managed to get this lilium susanium to flower in Malines in 1576. The artist Raphael Coxie made a coloured drawing of the flowering plant, which he sent to Clusius in Vienna. And the picture in Clusius's printed work is based on that portrait, thus epitomizing the highly collaborative and international character of natural history.[31] Not only the plants themselves travelled and passed through many hands, but so did texts and illustrations, while expertise accumulated.

Brancion had a precious collection of books, fossils and curiosities as well as a garden.[32] The theme of collecting thus links Brancion, Saint Omer, the Laurins and Coudenberghe, but each had his own style of collecting and his own preferences. While Saint Omer's garden and watercolours depicting plants and animals formed an integral part of his *Kunst- und Wunderkammer*, for the Laurins plants seem to have been more of a side-interest compared with their collection of antique coins. Brancion combined his living plants with dead *naturalia* and books, whereas Coudenberghe appears to have concentrated completely on his botanical garden, which he catalogued and treated as a living collection. None of these collections met with a happy end. Saint Omer died childless at the age of 36. Of his great collection only the volumes of watercolours remain, and he himself was almost completely forgotten until a renewed interest in *naturalia* and collections stimulated research during the 1980s and 1990s. Coudenberghe's great garden was destroyed during his lifetime, in the siege of Antwerp, most likely in 1585 by the troops of General Alessandro Farnese, himself a plant and garden lover.[33] Marc Laurin had to flee Bruges in 1581 on account of local troubles connected with the Dutch Revolt. On his way to Calais his luggage with almost the whole of his collection of antique coins was stolen. He died shortly afterwards in Calais.[34] After Brancion's death his nephew took good care of the garden and collection, but when he too died in 1579 the garden was destroyed. The most precious items were stolen, many others sold.[35]

The presence of a considerable number of wealthy men with gardens and collections in the Southern Netherlands in the 1550s, 1560s and 1570s, plus the fact that so many of those mentioned as experts by Lobel were in personal contact with Dodoens and Lobel as well as Clusius, help us to understand how closely interlinked the work and interests of these three botanists were with the general cult of plants and gardens in the Southern Netherlands. The 'Flemish fathers of botany' were socially, culturally and intellectually embedded in this setting, and depended on its patronage, wealth, contacts and customs. In the most practical sense, the cult of rare plants created employment as well as a readership interested in information about *naturalia*. Rich garden owners needed both expert gardeners and expert botanical consultants. The latter had to have an education that allowed them to read the classics in Latin and contemporary works in both Latin and other languages, such as Mattioli or Fuchs on plants, Belon on fish or Gessner on animals. Such consultants also had to have contacts which provided access to new and rare *naturalia*. Lobel, Clusius and Dodoens fulfilled all these requirements. In so far as natural history was concerned, all three worked for most of their lives as private advisors on botanical and gardening matters to princes and noblemen, who also formed a readership for their publications. The pervasive passion for gardens and plants among the elite of the Southern Netherlands thus goes a long way towards helping us understand how three such eminent experts could emerge within one generation in this part of the world. The fact that none of them stayed there had nothing to do with a decline of that botanical interest, but with the warfare, religious persecution and plundering that accompanied the first phases of the Dutch Revolt and the splitting of the Netherlands into two separate states.

The Burgundian Tradition

To understand when and where this special botanical culture emerged in the Southern Netherlands, we have to go back to the fifteenth and even the late fourteenth centuries. Busbecq and Quackelbeen had brought rare plants to Austria and the Southern Netherlands from their mission to Constantinople during the 1550s. But they were not the first diplomats to return with plants, either from the Mediterranean or from the Levant. During 1545–6, Gerard van Veltwijck (*c.* 1505–55) had also been on missions from Brussels to Constantinople. He was known as an extremely learned man, fluent in Greek, Latin and Hebrew, and a botanical expert. Whether or not he taught the taste for plants and flowers to Mary of Hungary, regent of the Southern Netherlands from 1531, there is little doubt that he explored the Alps and other mountainous regions *en route* for new plants and brought back plants from his various missions.[36]

The passion for plants among the elite in the Habsburg countries was thus already pronounced before the middle of the sixteenth century, and it was the imperial family that set the example. After retiring to Yuste, Charles V liked personally to take care of his garden with the assistance of his Flemish gardener François Holbecque.[37] Charles V's sister Isabella, wife of the Danish King Christian II, in her turn invited Flemish gardeners to Denmark, where they transformed the environs of Copenhagen.[38] And Mary of Hungary – to whom Dodoens dedicated the first edition (1554) of his *Cruijde Boeck* – must have been a culturally defining figure for both Clusius's own generation and that of his parents and patrons, given the fact that she ruled the Southern Netherlands as regent in 1531–55. Although her personal preferences seem to have been for hunting and music rather than gardening, she is still known for her residence in the Coudenberg palace in Brussels with its adjacent park, her Renaissance castle at Binche with a hunting lodge and enormous royal domain at Mariemont (all in the neighbourhood of Tournai), and her castle at Turnhout which was surrounded by a park with marble fountains and exotic plants. Her collection too was of special interest. She owned the most important collection of tapestries in Europe besides that of the French King François I, and many contemporaries praised her collection of rarities from the New World. It included exotic *naturalia* as well as *ethnographica*, such as the famous 'headdress' of Moctezuma. Mary had, in fact, inherited most of these New World objects, which had been greatly admired by Dürer, from her predecessor Margaret of Austria. She in turn had received them from her nephew Emperor Charles V, to whom they had been given in or shortly after 1519 by Hernan Cortés himself.[39]

With regard to the culture of collecting the role of Margaret of Austria seems to have been even more crucial than that of Mary of Hungary. In a remarkable reconstruction of the use of the rooms, interior decoration and collections at Margaret's Court of Savoije in Malines, Eichberger has shown that a small room on the ground floor was used during the mid-1520s to house some of Margaret's rare *naturalia*: the corals. This 'Cabinet next to the garden' is mentioned directly after the Studiolo, an intimate room devoted to study. Accounts for the years 1515–28 furthermore show that there was a herb garden, with vines and herbs as well as ornamental plants, such as oregano, rosemary, roses and stock. This garden was used for leisure, and in summertime for eating outdoors.[40] Compared to Brancion's garden in the same town and other extremely rich gardens in the Southern Netherlands of the 1560s and 1570s, this may sound like a rather modest affair for the person who ruled the country, but that only serves to underline the incredible expansion of gardening and the introduction of vast numbers of new plants during the intervening forty or so years. Moreover, this modest garden located in the courtyard of the Court of Savoije was a *novum* in the history of gardens in Northern Europe, where gardens until then had, it seems, never been directly adjacent to living quarters. Margaret of Austria's cabinet room with the corals in Malines was therefore a very important and innovative space: linking inside and outside, living nature and collected *naturalia*, nature and art, and the cultures of gardening and collecting.[41]

One more step back in time is needed – into the age of Charles V's Burgundian forebears. The Coudenberg palace with its great park the Warande at Brussels was the central residence of the Burgundian rulers and had been expanded during the rule (1419–67) of Philip the Good, whose court is regarded as the most splendid of Europe in his day. A menagerie was added to the palace and grounds during this period: it housed a wolf, lynx, fox, ibex, tabby cat, wild boar, stags, monkeys, hares and deer. In 1461 Philip also ordered a (live) lion from Venice. By the 1520s, when Dürer made a drawing of the park, it had a bear, lion and lioness, as well as hares, pheasants, deer, peacocks, ducks and swans.[42] Such detailed information is lacking for the plants, which were not only less spectacular, but also less closely connected with the passion for hunting that seems to have characterized most of the Burgundian-Habsburg princes and princesses. But we can find other kinds of visual evidence for the Burgundian interest in plants thanks to a significant phenomenon connected with the Burgundian elite: the emergence around 1390 of the so-called 'international Gothic style with its new-found and fearless joy in Nature'. Blunt and Stearn describe how 'naturalism [was] breaking out almost simultaneously in the art of Italy, Germany, Flanders and France'; by the fifteenth century illuminated manuscripts were characterized by representations of flowers, insects and birds 'drawn with much realism and thrown into relief with strong shadows', while the earliest such images can be found in the work of the master of Mary of Burgundy (*c.* 1480) and in a manuscript from Bruges (1482).[43] In 1950 Pächt too pointed to the almost contemporaneous emergence of naturalistic nature representation in Northern Italy and Burgundy in a pathbreaking essay on the history of nature studies and the connections between the natural sciences and art:

> The Northern Schools approached the problem from an altogether different angle: zoological and botanical specimens were not studied by them and portrayed as isolated objects as was done by the Italian specialists, but animal or plant was seen as part of and inseparable from its natural setting, its living space, its home in nature.[44]

For Pächt this formed a step in the emergence of landscape painting. But in the present context we might also read it as a suggestive statement foreshadowing the ecological interest and the attention to the natural habitat of plants manifested by Clusius and his fellow botanical enthusiasts of the second half of the sixteenth century. A long and unbroken Burgundian-Habsburg elite tradition of interest in nature and its representations at least dates from the second half of the fifteenth century, and so does the emergence of naturalistic artistic representation of nature. Both were closely related, and this evidence strongly suggests that the emergence of the 'scientific' type of botany as represented by Lobel, Clusius and Dodoens was predicated upon the fashionable, aristocratic and courtly passion for nature transmitted from the Burgundians to the Habsburgs, and the equally court-inspired traditions and aesthetics of naturalistic representation in art.

2 EXPERT GARDENERS: GROWING RARE PLANTS DURING THE LATE SIXTEENTH CENTURY

Returning to Clusius's own age, we find two more significant clusters of wealthy garden lovers in the Southern Netherlands – all of them close friends of Clusius, while several also figure on Lobel's list of experts. Most of these men belonged to a slightly younger generation than Saint Omer, Brancion or Coudenberghe, but the various clusters overlapped in terms of generation as well as time. Interestingly, their letters tell us not only about their fascination with plants and gardens, but occasionally also reveal how they regarded themselves and each other in so far as expertise was concerned.

The apothecary Jean Mouton (d. 1589) was the link between these overlapping clusters. He knew both Brancion in Malines and Marc Laurin in Bruges, and introduced Clusius to new plant lovers. Mouton owned a rich garden in Tournai which seems to have been on a par with those of Coudenberghe and Brancion. He was in close contact with Lobel and Clusius, who both mention him frequently in their printed works as a source of information and plants. Mouton, for instance, sent Clusius pictures of a pseudonarcissus that flowered in his garden in 1576, of a 'hyacinthus stellaris' and an eryngium. And he showed Clusius an anemone that closely resembled one which Clusius had observed many years earlier in Portugal and which Mouton had grown from seed received from there. His interest was not limited to plants in his garden. Mouton also explored the area surrounding Tournai and observed the wild cyclamen growing there in the woods.[1] Every time he read Clusius's work about the flora of Hungary, as he wrote, 'I become as longing and full of immoderate desire for plants as a pregnant woman is for certain types of meat et cetera' (Mouton, 5 May 1585). His strong, emotional attachment to his garden lasted until his dying day. Even in Mouton's final months, when he was extremely weak and paralysed from the neck down, one of his main delights and preoccupations in life was his garden: 'still he was taking care of his garden and had plants moved because of the cold and other things done as if he was not ill at all' (De la Fosse, 31 August

1589). The double, perhaps equally important aspects of use and pleasure were expressed well by his nephew, the apothecary Thomas de la Fosse: 'it is all our pleasure, and also there is nothing that fits better with our work as apothecaries' (De la Fosse, 12 July 1596).

From the way in which Mouton introduced new acquaintances with gardens to Clusius, and his emphasis on their love for plants and dedication to gardening, it is clear that these aspects were crucial for acceptance as a member of the community of exchange in which all of these men operated: 'the pleasure and delight they take in plants make them deserve to be given plants' (Mouton, 5 May 1585). Jean Bosquiel, Seigneur Desplanckes, used much the same terms of pleasure and delight in rarity in the letter in which he asked to become a member of Clusius's circle of friends. He was careful to emphasize that he regarded the involvement with plants and gardens as a serious occupation (he used the French term 'profession'):

> I believe you will have heard from Sr Jean Mouton about the pleasure that I take in simples, having no other or more recommended activity to which I devote myself since several years, and via friends I have obtained some rarities. Since you have rightly acquired the title of father in this profession, and several of your friends in this region enjoy various rarities thanks to your favour, I would like to ask you very affectionately to do me the favour of enrolling me among their number and to make me share in these singularities as well, assuring you that you will oblige a gentleman to serve you in everything within his power. (Bosquiel, 5 November 1584)

Friends of both Mouton and Clusius who deserve special attention on account of the high quality of their expertise concerning plants, which was also recognized by Lobel, are Charles de Houchin, Seigneur de Longatre (d. 1607); and Jean Boisot (d. in or shortly before 1602), a nobleman and owner of a private botanical garden in Brussels. The fiercely Protestant Houchin lived near the village of Annezin not far from Bethune in the Pas de Calais, where he had a castle and a church built. Given the proximity of his region to Arras, Clusius's birthplace, it is possible that the two men had known each other since their youth, but the extant letters from Houchin to Clusius all date from the last twenty years of their lives. They testify to Houchin's erudition in botanical matters and his long-standing interest in rare plants and his garden – from which Houchin was occasionally distracted by politics, religion, his new wife or in later years by gout. Mouton's praise of Houchin gives us some idea of the latter's wide-ranging interest in plants:

> I love him all the more since he is a universal lover of plants, while the majority of the others in this region are particular: they love only those with a beautiful flower, according to their judgement, and do not take the rest into account. (Mouton, 21 December 1584)

Houchin attempted to create a garden with as many of the rare plants that he could obtain which had been described by Clusius: a botanical garden, in other words, in the form of a living collection:

> I have grown so fond of my garden that for the moment I do not have any other pastimes and to the extent that I have banished all kitchen herbs in order to provide more space for my plants, and my garden is not that small but is 100 paces long and 60 paces wide ... But we still lack a considerable number of the singularities described in your books. (Houchin, 23 February 1585)

As befitted a decent collection, Houchin had a catalogue of his garden, which he sent in 1585 to Clusius in Vienna, offering him any plant in his garden which the latter might not yet have. And he did indeed send Clusius several special bulbs or information concerning their flowering period and colour: a colchicum, narcissus, a 'bulbus eriophorus' and two bulbs which Houchin had received from Spain: a hyacinthus hispanicus and a white hyacinthus stellaris.[2] Houchin was pleased when Clusius asked him about the appearance, colour, seeds and flowering time of a number of plants (Houchin, 30 April 1585).[3] He read all of Clusius's works, knew the work of Mattioli and did not hesitate to question some of Clusius's identifications of trees on the basis of his own experience and observation.

According to Houchin, Clusius had made some mistakes in his Spanish flora of 1576 (Houchin, 7 February 1585).[4] In particular he doubted the correctness of the descriptions of various types of conifers. Comparing what he saw in living nature with what he saw and read in Clusius's work, he found that the two did not match. Houchin had tentatively identified two types of 'wild pines' in his garden (one with big and one with small cones) as Pinaster 1 and 2 in Clusius's Spanish flora. According to Clusius, Pinaster 2 was rarely taller than two adult men and had a twisted trunk. But in his own garden Houchin saw that the 'wild pine' with the small cones had a straight trunk and was tall, sometimes even taller than the others. He also had many exemplars of the 'pirea Mattioli' in his garden, which he would have identified – on the basis of Clusius's printed text – with 'your pinaster tertius, if its portrait had not so clearly opposed this, but there is no resemblance whatsoever between that portrait and my saplings' (Houchin, 7 February 1585). Whether Clusius was indeed wrong is hardly relevant here, but it is important to note that first-hand observation of nature was clearly regarded as a sufficient basis to criticize the work of a great botanical expert. Both Houchin and Clusius shared that point of view and their friendly exchanges continued for another 25 years until Houchin's death in 1607.[5]

The nobleman Jean Boisot was much more centrally located, in Brussels, socially as well as internationally even better connected, more erudite and even more expert in matters of plants.[6] He occasionally quoted in Greek in his letters, read Latin, and was in touch with the famous scholar Justus Lipsius – another friend and correspondent of Clusius.[7] Boisot knew Plantin and occasionally discussed publishing issues and the potential market for Clusius's printed works with the latter, stating *en passant*: 'The [i.e. your] book will be much in demand given the fact that the number of us gardeners increases by the day' (J. Boisot, 4 August 1588). Boisot's long-standing friendship and plant exchanges with Clusius are reflected in the fact that he is the second most quoted source in Clusius's *Rariorum*.[8] All of that already puts him into the category of key figures in the Southern Netherlands in so far as rare plants are concerned. But Clusius went much further in his appraisal of Boisot, as we have seen in the introduction, and put him on a par with himself in so far as their type of interest in plants and their role as pioneers of botanical investigation were concerned.[9] Boisot's principal botanical interest seems to have concerned bulbous and tuberous plants such as lilies, tulips, crocus, colchicum, narcissi, fritillaria, iris and cyclamen, besides the equally fashionable anemones and ranunculus. In the course of the late 1570s and the following decades he sent numerous special bulbs or their seeds to Clusius – among them a rare type of fritillary which may originally have come from Aquitaine or the Pyrenees, various types of crocus and colchicum, double narcissus, hyacinthus stellatus and several types of iris bulbosa.[10]

Like Brancion, Boisot had many international contacts. He received rare plants via Holland and Zealand, merchants in Antwerp, friends and relatives in the Southern Netherlands, and occasionally from Italy. In August 1590, for instance, Boisot sent a small but strong-smelling branch of 'absinthium ponticum' to Clusius. It had been given to him by an Italian who himself had received it in Florence 'as a great gift' from Giuseppe Casabona, the court botanist at Florence. According to Boisot this was by no means the common Roman absinth: 'I have compared the one with the other and have found the Roman one much stronger and whiter in colour, the other with a more agreeable smell' (J. Boisot, 29 August 1590). According to Clusius, Boisot was the first among the Belgians to receive a certain kind of Italian cyclamen from Alfonso Pancio in Ferrara, who was also one of Brancion's main sources of Italian plants. Boisot had various types of myrtle and a ribes-like bush in his garden as well, and a laurus silvestris (Viburnum tinus L.) grown from seeds which he had received from Italy, an exotic type of bean grown from seed received from Spain, and lychnis Byzantina. Lipsius praised Boisot's garden as even richer than Clusius's own: 'You are our commander, to whom Clusius himself yields the place of honour. Because I have seen his [i.e. Clusius's] garden in Frankfurt, and I know that yours is more richly provided.'[11] Jean de Maes, a military man and relative of Clusius who had then only recently begun to 'floriser', was equally impressed by Boisot's plant collection. Like Bosquiel, he spoke (in French) of their interest in plants as a 'profession'.

> In this profession, friends are not really friends; on the contrary, nowadays people only seek profit there. Even Monsieur Boisot himself, who has such a great abundance of plants that it seems as if the Orient has showered his garden with everything, is so stingy that he would sooner give away an *écu* than a hyacinth bulb. (De Maes, 8 April 1599)

Boisot's fifteen extant letters to Clusius, which span the period 1582–97, are long, detailed, lively and full of captivating detail that throws light on his care of the plants in the garden, his successes and losses, fascination with bulbs and their propagation, experiments with plant care during the winter and in different soil types, involvement in identification and plant comparison, growing bulbs from seeds, grafting and in particular on his constant interest in colour stability, all of which led him to long-term observation and experiment. Boisot noticed, for instance, that his blue crocus vernus, which he had kept for years in pots just like his other most beautiful bulbs, did not flower in 1584 and that the bulbs diminished in size and produced leaves but no flowers. He decided to put them out in the open in his garden, and added soil from the fields to see if that gave better results. Elsewhere in his garden he found that a gladiolus did not flower well because it was too close to a peony plant and situated almost underneath a big rose bush. After some problems with tulip bulbs, he decided that it was better not to move them around too much and that they were not affected by the cold but by excessive humidity. Straw covers did not help: 'experience showed us that this weakens them a lot, obstructing the air and light from which they take their force' (J. Boisot, 17 May 1590). Looking among his aristolochia plants, he marked the one with the largest and blackest flowers, to see whether its fruit would also be the roundest, in order to send its seeds to Clusius. And in May 1586, he reported to Clusius that he had followed the latter's advice and grafted his rare horse chestnut on a common chestnut.[12] Practical experience, based on observation and experiment, thus went together with erudition, an interest in plant identification and a great curiosity concerning new plants.

In his detailed descriptions concerning his experience with the propagation of bulbs, Boisot uses some interesting terms, which suggest that he – and probably many of his contemporaries – thought, implicitly or not, about the propagation of bulbs in terms of human family descent. He had received, for instance, the 'child' (i.e. an offset) of a double crocus from a friend, 'but both the mother and the child are single this year'; in the same year his tulips 'took after their parents in colour', but other bulbs did not flower at all even though the plants were big enough and came back for four to five years. These were, as he called them, 'sterile … I do not know what went wrong' (J. Boisot, 17 May 1590). Boisot shared a predilection for rare white variants of flowers with several of Clusius's correspondents; they seem to have regarded white as the most rare and noble colour. If a plant which had first had a white flower reverted to another colour, or if its 'children' had flowers that were no longer as white as its 'parent', Boisot uses the term degenerate (*forligner*). Obviously, that notion had no direct connection with the same term used in modern plant genetics, but it does fit in with descent, family and the concept of honour, which in the circles of sixteenth-century noblemen were indissolubly linked:

> Concerning the gladiolus with a whitish flower ... Perhaps some of the plants that I have sent you have grown from the seeds that fell from the white ones and have changed kind, which as you know several other plants do as well, such as the completely white oriental hyacinth which also degenerates, as I have seen in four or five plants that grew out of its seeds, all of which produced blue flowers this year. (J. Boisot, 19 July 1591)

For Boisot these seeds had thus 'changed kind' when they produced blue instead of white flowers. For sixteenth-century plant experts, different colour variants of the same species – as they would be called today – *could* belong to different species, even if they still belonged to the same family in the old sense of the term. That makes the experimentation with colour variation by sixteenth-century plant lovers even more exciting. They, after all, may have had the feeling that they were literally creating new, and therefore rare types of plants – almost like advanced plant alchemy.[13]

Seen in the context of this view of propagation and plant species and of the passion for rarity, it becomes clear why many correspondents of Clusius paid so much attention to colour in their letters. Boisot wrote down his observations on colour in detail and kept Clusius informed of changes over time. In May 1582, for instance, he reported:

> The most beautiful sort of tulip that has emerged this year is a multicoloured one with white, red and yellow ... I have another, completely red, yet another, below white and at the tip red, which has come from the seeds of the white Eduarde, from which has also come one that is almost violet. My sister de Tisnac has had one completely white with the tips red, another one with more red at the top, and yet another late tulip which is mixed white, yellow and orange, one of the most beautiful that I have ever seen. That one came from one of the rare late tulips which was first grown by my niece ... From its seeds many different coloured flowers have come ... summing up, this sort is beautiful for the variety of colours that one batch of seeds produces ... I had another late tulip a year ago of which just half of the flower was yellow and the other half red mingled with yellow, with the bottom part a very dark green. I have another yellow late one, with the bottom part light green. *Voilà*, so much with regard to tulips this year. (J. Boisot, 7 May 1582)

Already in the 1570s Boisot was growing tulips from seeds in his garden. Only this method makes it possible to obtain new, unpredictable varieties, but it often takes five or more years before the new bulbs flower. It requires considerable expertise – for instance concerning the choice of which tulips to cross-fertilize, the best temperatures, humidity and types of soil – while the duration of the whole process demands much patience and induces tulip growers to keep track of their yearly results and of colour changes from one generation of bulbs to the next, over long periods of time. Passages like the one quoted above can thus be read both as purely descriptive statements and as reports on long-term experience with colour and propagation. Tulips were not the only flower to be experimented with. Boisot also remarked on the different colour patterns of irises, and had various types of anemone, of which he sent several to Clusius, carefully wrapped in paper marked to indicate the colour: 'by sowing them, one daily obtains new colours' (J. Boisot, 10 July 1597).[14]

The quest for rare colour varieties was not confined to the garden. The wild anemone with a double red flower, of which Boisot sent Clusius a sample, had long been known in the Southern Netherlands, as he wrote, but it was considered such a great rarity that those who saw it growing in the wild not far from Louvain 'refuse to pass this knowledge on, in spite of the fact that it is wild' (J. Boisot, 19 July 1591). Boisot had nonetheless managed to obtain one from friends in that area. During the 1580s Boisot had received a white double anemone from one of the highest noblemen in the Southern Netherlands, the Prince of Chimay, a great lover of plants himself, who had found it growing on his father's estates not far from Louvain (J. Boisot, 19 July 1591). Boisot was curious enough to investigate further, and went to interview a plant lover who had formerly lived in Louvain in order to learn more about 'our wild anemone that they call *hanevoet*' (J. Boisot, 3 September 1591). The person in question told him that he had explored the woods near Louvain several years earlier to find this anemone, and that he had indeed found two such plants near the monastery of Florival – a location very close to the one where the Prince of Chimay's double white anemone had come from. Boisot wondered why such a beautiful plant had not attracted attention before, and came up with two reasons:

> One, that those who had seen it first regarded it as extremely precious and did not want to share it with anyone else; the other that some had planted it in their gardens and seeing that it became single (as it does there), they dismissed it; for it has to be planted in a pot, where it always produces a profusion of double flowers. If you were to plant it in the garden and surround it with tiles or crocks the result would be the same. I think that in woods the roots of the tree have the same function. (J. Boisot, 3 September 1591)

The quest for rarity here intersects with specialized knowledge in which attention to habitat in the wild and horticultural expertise fused.

We do not know what Jean Boisot's profession was or indeed if he had one. His family connections and social network all point in the direction of the highest levels of Habsburg government officials and of excellent contacts with

Spain.[15] Boisot's direct and indirect connections with the gardens of the Spanish king were especially frequent. Almost every year between 1584 and 1592 Boisot received one or more packages with seeds, bulbs and possibly whole plants or cuttings, either directly from the head gardener at Aranjuez, François Holbecque, or from the wives of high functionaries from the Southern Netherlands working in Spain. In 1584, for instance, Boisot received a crocus vernus from Holbecque which flowered with a kind of bluish-white colour 'rather displeasing but I could not but regard them as highly fantastic' (J. Boisot, 20 November 1584). In fact, Boisot showed himself not impressed by this particular Holbecque's capacities as a gardener. Holbecque did not pay much attention to the lists with requests that Boisot sent him but, as the latter complained, sent packages with such a large quantity of seeds and other botanical items – some superfluous and all mixed up together – that it was impossible to know what was what (J. Boisot, 1 February 1589, 3 August 1590).[16] A package with a large quantity of seeds which arrived in Brussels in 1595 had survived the long journey by sea, but the seeds were 'ordinary ones, since he persists in his usual negligence, and will from now on have less opportunity to send bulbs or seeds because (as he writes) he has lost almost everything due to a great overflowing and inundation of the rivers' (J. Boisot, 25 June 1592).

Via several female relatives Boisot was connected with the Tassis (or Thaxis) family which played a crucially important role in matters of transport and postage in the Habsburg empire. Besides being ideal connections for the always complicated and often expensive transport of botanical material, several members of the Tassis family were prominent garden lovers and collectors of rare *naturalia* themselves. They were also correspondents of Clusius in their own right, such as the brothers Jean Baptiste (1553–86) and Charles de Tassis (d. 1610) in Antwerp, both of whom inherited the function of postmaster general of Antwerp from their father.[17] In his role as garden lover Charles de Tassis asked Clusius's advice about how to grow certain rare plants in his own garden, and mentioned some of his special rarities – such as the 'fantastic' iris susiana, the 'strange and rather ugly' Lady's Slipper orchid, and his double delphiniums. He also showed his deep appreciation of the many botanical gifts he received from Clusius, 'of which I am aware every time I enter my small garden where I have had the last few days the most beautiful tulips, which had come from you, that you could find in this town' (C. de Tassis, 20 May 1590).[18] On the other hand, he acted as intermediary in postal and transport matters with various plant lovers in England, such as James Garet Jr, with whom he also exchanged botanical information, and from whom he obtained exotic plants for his own garden, such as gooseberries and the potato. Charles also offered the help of his brother-in-law François Damant who would soon return from Spain in order to obtain rarities from Clusius from there, mentioning that he was studying the best ways of wrapping bulbs and cuttings in order to protect them during their journey (C. de Tassis, 25 January 1589, 5 April 1589).

The passion for plants in the Tassis family was transmitted from one generation to the next, as was their bond with Clusius. In 1602 Clusius's relative Jean de Maes testified to the credentials of Lamoral de Tassis as a real connoisseur of rare *naturalia*:

> who has many rarities, among which a jonquil with a double flower that he has in his garden which several [persons] have seen, I say trustworthy persons ... I have also seen in the hands of the said Sieur de Tassis the sowbread white as snow [*pan porcin* = wild cyclamen] that had been sent to him from Italy. Also the double hepatica with a celestial blue flower, a beautiful flower. He sends you a big bulb of the martagon pomponii in order to begin making your acquaintance, with his recommendations. He will shortly send you the portraits after life of the rare plants that he has. And they are true to nature. This is the first time in my life that I have talked to him in order to make sure whether he did have the double jonquil, as I had heard. After all, various people claim to have it, but he really does. I find him a very honest gentleman. (De Maes, 7 October 1602)

Lamoral de Tassis was well aware of the value of these rarities. When he sent a list of his special plants to Clusius he did not dare sign it with his name, fearing that it might fall into the hands of the wrong persons and that his garden might be robbed.[19] That fear was by no means exaggerated, given the numerous thefts of bulbs and rare plants that Clusius and many other garden owners reported, and the number of packages with bulbs that were plundered *en route* or never arrived. But Lamoral was happy to exchange information with Clusius about rare plants from Italy and Constantinople, the latter of which reached him via Hungary. He asked Clusius for 'a register of plants growing in Italy' and promised him in return a list of his own greatest rarities which he would gladly share (L. de Tassis, September 1602).

Indirectly, the passion for gardening and the emotional attachment to plants and gardens demonstrated by many of Clusius's correspondents must have been stimulated by the wars and general insecurity. The need for the garden as a safe haven, whether realistic or mainly symbolic, has been emphasized in other ages and countries as well – and there is no contradiction between this function and that of the garden as significant status symbol and form of conspicuous display. Another aspect is perhaps as important and more specific to the Netherlands during this age. There must have been a craving for a passionate interest that could bridge social gaps and, even more importantly, the political and religious divisions that were splitting the country, regions and even families down the middle. A shared interest in nature may indeed have done just that. Olmi has pointed out its supra-religious character, and in the following chapters we will see that there were similar effects with respect to gender barriers.[20] Members of Clusius's circle of garden owners and plant lovers could be found on both sides of the politico-religious divide in the Netherlands, and many were indeed all too familiar with the issue of divided loyalties. While Charles de Houchin's father-in-law had been advisor and *maître d'hôtel* of Emperor Charles V, Houchin himself was one of a group of Protestant noblemen who in 1565 joined the so-called Compromis and subsequently the rebellion against the Habsburg government.[21] All members of the Tassis family remained on the Habsburg and Catholic side – obtaining ever higher positions as postmasters in the Habsburg empire – whereas several of the Boisots rebelled. What Jean Boisot's political choices were we do not know. But his brother Pierre's sons Charles (*c.* 1530–75) and Louis de Boisot (1530–76) chose the side of the Prince of Orange and became famous commanders of the Sea Beggars, whereas their sister Louise married into the Tassis family. Both Charles and Louis were killed fighting.

We have seen that the gardens and collections of Saint Omer, Brancion and Coudenberghe were all destroyed before 1585, generally as a consequence of the warfare and rebellion related to the Dutch Revolt and the splitting up of the Netherlands. The next generation appears to have fared hardly better, and in this respect there was no difference between those who supported the rebellion and those who remained loyal to Habsburg rule. Little is known about the fate of the gardens of the Houchin, Boisot and Tassis families, but it seems likely that even if they were not destroyed by warfare they did not long survive their creators. That lack of information itself is related to another process of destruction. The devastation of the impressive gardens and collections of the sixteenth century was matched almost four centuries later by a another wave of demolition: during the First World War many Belgian archives, libraries, historical buildings and other historical records were destroyed, annihilating historical traces that had survived until then.

Perhaps the most immediate effects of warfare on botanical interest, however, are shown by Jean Baptiste de Tassis (1553–86) in Antwerp. Only six months after the fall of Antwerp in August 1585 he wrote to Clusius asking for seeds of plants for his kitchen garden – one of the very few of Clusius's more than 300 correspondents to show an interest in edible plants:

> You remember my garden which is in great need of your assistance and generosity in order to regain its original lustre ... I ask you for some seeds of plants for the kitchen, such as good and special lettuce, cauliflowers and savoy cabbage, good and rare onions, some red, white and black and winter radishes, cucumbers, pumpkin, squash, melons, artichokes and cardoons, sweet fennel, oregano and thyme, fine marjoram. (J. B. de Tassis, 28 January 1586)

Jean Baptiste de Tassis died of the plague a few months after writing this letter, possibly a victim of the collateral effects of warfare.

Patrons and Experts

The locations of the gardens discussed here as well as brief remarks by several of Clusius's correspondents suggest that the epicentre of gardening culture shifted in the last two decades of the sixteenth century – no doubt on account of warfare – from Antwerp and the coastal area, eastward and somewhat more to the south, to the environs of Lille, Tournai, Brussels and Liège. An elderly female friend from Clusius's youth, who lived in Antwerp, wrote, for instance, in 1605 that 'the heritages of the gardens do not recover as they do in other cities and we do not have the pleasure of the garden' (Maria van der Laen, 20 November 1605). But horticultural expertise did not diminish, nor was the tradition of gardening or collecting rare plants interrupted. In 1604, for instance, Petrus Bloeme, an émigré friend of Clusius in Frankfurt, wrote that he was importing most of his garden stocks straight from Brussels since they were not available in the Frankfurt area (P. Bloeme, 25 March 1604). And one of Clusius's relatives in Brussels said in 1606: 'we have so many herborists and florists here that I am sorry to see that the boys of tapestry weavers and second hand clothes vendors interfere in this honourable activity' (De Maes, 18 April 1606).

The creation of what must have been one of the most impressive collections and gardens of the Southern Netherlands belongs, in fact, to the very end of the sixteenth and the early part of the seventeenth century, although their creator already showed an explicit and expert interest in plants as a very young man. When only seventeen years old Charles de Ligne, Count and later Prince of Arenberg (1550–1616), sent a list of seeds to the Italian naturalist Aldrovandi. Arenberg must have known all or nearly all of the great gardens of the 1570s, 1580s and 1590s; he was a friend and correspondent of Clusius and Boisot, and in 1595–6 became the owner of the *Libri Picturati* watercolours, to which he added new sheets with portraits of plants which he commissioned. Arenberg belonged to the very highest circles of the aristocracy in the Southern Netherlands. Emperor Charles V himself had been his godfather. After having been involved for most of his life in military campaigns, politics and diplomacy, he retired to Enghien to the south-west of Brussels, where he bought the castle and park in 1605 and devoted himself entirely to collecting plants and other *naturalia*. Together with his wife Anne de Croÿ, the Duchess of Aerschot (1564–1635), he turned the grounds at Enghien from a neglected domain into one of the most beautiful gardens of Europe.[22]

In Arenberg we meet the aristocrat who was expert as well as rich enough to befriend, hire, work together with or otherwise employ the foremost experts of his age in order to create a collection of extremely high quality that covered the whole range of paintings, tapestries, coins, books, antiquities and plants. Arenberg's acquisition of the *Libri Picturati* collection, the inventory of his library

– which contained works by Clusius, Dodoens, Lobel, Joachim Camerarius, Dalechamps, Mattioli and Colonna – and his correspondences with Clusius and the Florentine garden owner and plant dealer Matteo Caccini all demonstrate his expert interest in *naturalia*.[23] Arenberg also knew how to obtain rare plants and the best advice possible. Like Boisot and Brancion, he made use of wide-ranging family and professional contacts with Spain and Italy. In 1595, well before his involvement with Enghien, he was already cataloguing his own plants and trying to obtain the catalogue of the royal gardens in Aranjuez. Holbecque was dead by then, but Arenberg wrote: 'since I have one of my men there I will order him to do his utmost to obtain it, and if there are any rarities there that he will bring me some of them'.[24] Relations with experts abroad were carefully cultivated. Arenberg not only requested seeds from Caccini – of impatiens balsamina, mirabilis jalapa and tropaelium major, which were apparently not easy to find in the Netherlands – but he also sent plants to Italy, probably as a counter gift or payment in kind.[25] It is understandable, therefore, that Arenberg very much regretted the death of the Spanish physician Simón de Tovar (d. 1596) in Seville with whom he had made plans for regular transports of exotic plants.[26]

> I had begun a great correspondence with him and already outlined a route so that we could have had everything here that comes from the East and West Indies, Portugal and Spain; now we have to be patient as with many other things. (Arenberg, 7 December 1596)

Thirteen years later, he was again trying to arrange for the transport of rare plants from Southern Europe: he had heard about a man in Venice who ordered plants every year and sold them for money. The Enghien gardens were still new, however:

> and the soil is still very rough, and there is in the environs here no soil that is suitable for bulbs and plants. It will take time for it to adapt, while my plants languish and die, which I very much regret. (Arenberg, 25 February 1609)[27]

The graciousness of Arenberg's letters to Clusius and his sincere offers of friendship and assistance do not disguise the fact that he was a patron of extremely high status. To Arenberg, who commanded persons all over Europe in order to have his wishes fulfilled, Clusius, like Tovar, was an expert advisor concerning plants and gardens whose assistance could be called in and who was rewarded by counter gifts or perhaps also by payment. That difference in position was also explicitly recognized by Clusius himself when he wrote to Caccini in Florence that it might not turn out to be so easy to obtain a package with plants or seeds which the latter had sent to Arenberg in order to be sent on to Clusius: 'because it is not often possible to recover those things that end up in the hands of the great'; Clusius also suggested that Arenberg's gardener (*hortulanus*) was probably robbing both of them, stealing the rare plants that arrived in these packages, selling them and telling Arenberg that they had not survived the long journey.[28]

The complex character of the rapport between a nobleman-patron who was an expert himself and a respected but – up to a point – dependent expert consultant is perhaps most clearly visible in the connection between Arenberg's brother-in-law, Charles de Croÿ, Duke of Aerschot (1560–1612), and Jacques Plateau (d. 1608) in Tournai. Very little is known about Plateau's background, except that he must have been wealthy, may have been a merchant or apothecary-merchant and certainly acted as local dignitary in Tournai during the mid-1590s.[29] He was certainly a great collector of and expert on both plants and exotic *naturalia* in his own right. He corresponded on an equal footing with Clusius for at least twenty years (1585–1605) and his 21 extant letters are crammed with observations on rare plants, exotic animals, horticultural detail, references to seeds and plants arriving from Spain and Italy, tentative identifications of plants, the gardens of friends and much more. Judging from the alphabetical catalogue of the plants in his garden for the year 1584 – which lists some 325 different kinds of plants, among which eight kinds of anemone, twelve kinds of hyacinth (a term that covered a much wider range of bulbs than today), fourteen types of iris, eight kinds of ranunculus, many narcissi and Mediterranean or exotic plants such as the oleander, tobacco, Judas tree, the tusai or crown imperial – Plateau indeed owned a private botanical garden which was almost on a par with that of Coudenberghe. Although bulbous and tuberous plants occupied an important place in it, he does not seem to have focused as much on exotica as Coudenberghe, and had many European indigenous plants.[30] Already in his youth Plateau used to make plant drawings, and for more than twenty years he regularly sent Clusius drawings of unusual, new or rare plants, many of which he made himself: 'Earlier I have taken pleasure in drawing those plants that seemed to me not to have been brought to light; then I had more spare time than now' (Plateau, 2 April 1590).

Plateau's museum, or 'Cabinet' as he himself called it, comprised shells, dried plants, minerals and stones, stuffed animals, drawings of *naturalia*, drugs, pieces of marble paper, *ethnographica* and probably a herbarium. It must have had a considerable and indeed international reputation, given the fact that the naturalist and astronomer Johann Schreck (1576–1630; also known as Terrentius), a German member of the Italian Accademia dei Lincei who later became a Jesuit and joined the mission in China in 1611, referred in the 1620s to the fact that he had (many years earlier) seen three kinds of armadillo in the 'museum of Jacques Plateau of Tournai which was well stocked with animals and in particular with birds'.[31] Plateau was a friend and advisor in gardening matters of the Boisot family, a friend of Lobel, who respected him highly, and the only person to be quoted more often than Boisot in Clusius's *Rariorum* (and extremely often in his *Exoticorum*) as a source of information on plants.[32] Plateau knew almost all of the great garden lovers and collectors in the Southern Netherlands mentioned in this chapter and he had contacts with botanical experts and collectors in France and the Northern Netherlands. Plateau rated Boisot's garden as the foremost in Brussels, and very clearly regarded himself (and Clusius) as real experts, as his rather critical remarks about some other plant lovers in the Southern Netherlands indicate:

> I am pretty certain that one would give them more pleasure by sending some oriental hyacinth, muscari, or some coloured tulip, crown imperial, or similar plants, as long as they can see the flower the next year, than any kind of rare seeds ... because several are impatient. (Plateau, 12 April 1586)

Real experts apparently distinguished themselves (or said they did) by an interest in rare plants that required patience and care but might not produce spectacular flowers in the short term.

After what must have been almost a lifetime of collecting *naturalia*, growing rare plants and becoming an expert on such matters, Plateau was in 1604 hired by Arenberg's brother-in-law, Charles de Croÿ, Duke of Aerschot, who was after Arenberg one of the highest and most wealthy noblemen of the Southern Netherlands, and a great collector like him.[33] Croÿ owned various estates and town houses, a castle at Beaumont and an estate with a country house, garden, and a *maison de plaisance* just outside Brussels. He collected paintings and prints, jewellery and precious stones, coins, statues and antiquities, curiosities, books, maps and manuscripts, furniture and precious textiles. Like Arenberg, Croÿ had a special interest in *naturalia*. The Brussels estate included gardens consisting of various sections – for decorative plants, for aromatic herbs and kitchen plants, and for fruit trees – and also comprised a grotto decorated with shells and pearls, an aviary, various waterworks and a pavilion. His encyclopedic interest as a collector is evident from the coherence between the various segments of his collection – wall decorations of wild animals complemented, for instance, his collection of living *naturalia* and natural curiosities (birds, fish, shells, trees, herbs and flowers) – but also from his instruction to plant several varieties of fruit trees (apples, pears, plums and cherries) in a row. The estate just outside Brussels also had a distillery, and there were rooms for his personal physician and his botanist-pharmacist, a certain Monsieur Balduin.[34]

It was not to the Brussels estate, however, that Croÿ summoned Plateau, but to the castle at Beaumont.[35] Given the fact that Plateau was no humble employee but a man with means and expertise of his own, this apparently needed some explanation:

> Monsieur de l'Ecluse, I wanted to let you know that I have summoned from Tournai to my castle Beaumont Jacques Plateau with the whole of his family, having put him in charge of both my Cabinet and the foreign plants in which I take a great pleasure

> and delight, admiring the works of nature. Although I have many other important occupations I occasionally take some time to visit them, which is why I ask you whether you can buy some rarities for me, whether plants, animals or other rarities. I will have you reimbursed.
>
> I think that you know at least in part what Plateau had in his Cabinet and garden. I quite expect that you will be astounded [to hear] that he has abandoned Tournai, his freedom, Cabinet and plants and conveniences in order to enter my service. I answer you that he is more free than he has ever been before, because I let him do everything according to his imagination. And as regards his food, he has no worse than I. I believe that he will never have reason to complain. (Croÿ, 25 September 1604)

A year later, in 1605, Plateau certainly did not sound dissatisfied with his new position. On behalf of Croÿ he had just been able to buy, from a collection that had belonged to an Antwerp merchant, two beautiful birds of paradise (a male and a female), 'Indian' weapons and rugs, some fruits and three to four shells. At that time Croÿ's collection included quite a number of live animals as well: a cockatoo that raised its neck feathers when it got excited or irritated, birds that could speak perfectly while others danced, a porcupine and genets. The nicest animal, though, was a seal that followed Plateau like a dog through the house using its flippers; it complained when it was away from the water for too long (Plateau, 20 April 1605).

The step from an independent and well-to-do expert collector to a hired consultant could thus be a relatively small one. Clusius himself manoeuvred for most of his life on the edge between courtier, expert, nobleman and client, and the question of who paid deference to whom was not always simple. Like Arenberg, Croÿ was extremely respectful towards Clusius. Having heard that the latter was preparing a book on exotica, he sent him portraits of a 'bécasse Indienne' (perhaps a woodcock) in his cabinet and of a plant named 'spiny Cretan stoebe' by Plateau that was in full flower in his garden. Clearly, not even a man of such high social standing as Croÿ was insensitive to the honour of being named in Clusius's work. But lest it be thought that he had no matters of importance to attend to, he immediately made clear that any further information and portraits would come from his consultant Plateau: 'he will give you a fuller description and hopes to send you other portraits when he has some more spare time' (Croÿ, 25 September 1604).

Everything we hear about Arenberg and Croÿ characterizes them as prince-practitioners in the sense in which the term is often used – a patron of experts who is an expert himself – except for the fact that they did not rule a state. Both were patrons, owned castles and large estates, and were heads of very extended households with numerous servants which may have been as impressive as those of ruling princes in Italy or Germany. In that sense they can be regarded as having had their own courts. Yet, while their social position was higher and their command of resources much greater than those of Houchin or Boisot, the types of expertise of these four noblemen concerning living nature probably hardly differed – and all were part of the court setting of those who ruled the Southern Netherlands. In the Southern Netherlands of the 1570s–1600s we are therefore not dealing with a single prince-practitioner and a ruler's court with a scientist in residence, but with a whole range of noblemen-practitioners and noblemen-naturalists.[36] Several of them could change back and forth between the position of client, friend and patron. That was the case with Jean Boisot, whom we have come to know as a plant expert with an international network of contacts, an aristocrat garden owner and patron in his own right. Boisot and Arenberg exchanged or shared rare plants, such as the double fritillaria, the 'fritillaria Holbecque', and an orange-red peony which had originally come from Clusius, on a basis of relative equality. And in 1596 they together visited the garden at Brussels of the widow Tisnacq (Boisot's sister Catherine), where they admired a very big double peony of a white-pinkish colour.[37] Yet, Boisot occasionally also acted as caretaker and gardening consultant for Arenberg, as the following case shows.

Clusius refers to the exotic Aztec lily as 'narcissus jacobeus' or 'narcissus latifolia indicus rubro flore' (Figure 2).[38] In spite of its name, this was a spectacular type of amaryllis, now known as Aztec lily, Jacobean lily or St James's lily (Sprekelia formosissima L. Herbert (Amaryllis formosissima L.)), native to Mexico and Guatemala. In June 1596, Dr Tovar sent Clusius a long and beautifully detailed description of the plant, which was then flowering in his garden in Seville. He also mentions its American name 'azcal xochitl' or 'bulb with the red flower'.[39] Tovar had baptized it narcissus jacobeus, because the flower reminded him of the shape of the sword carried by the Spanish knights of Santiago de Espada (St James of the Sword), an order that had been originally created to protect pilgrims to Santiago de Compostela from attacks by the Moors.[40] In 1595 or early 1596 Tovar had sent three bulbs of this rare plant to Arenberg, who in turn promised one to Boisot. However, the merchant in Antwerp responsible for the transport stole one of them, which meant that Boisot would have to wait until one of Arenberg's two plants had grown new 'children'. The biggest of Arenberg's plants flowered in the season of 1596, but the flower lasted only briefly because of the rain and did not produce seed. Arenberg wrote to Clusius on 29 January 1597 that he had asked Boisot to make a description for Clusius (probably not knowing that the latter had already received one from Tovar), while Arenberg himself had a portrait made and sent to Clusius.[41] It thus was Arenberg and not Clusius who received the actual plants from Tovar; who promised one as a sign of his largesse to Boisot; and who commanded portraits and descriptions to be made of it. Arenberg would hardly go so far as to make a description himself, and thus give more of his precious time than that needed for the writing of a letter, but asked – or commanded – Boisot to do so.[42]

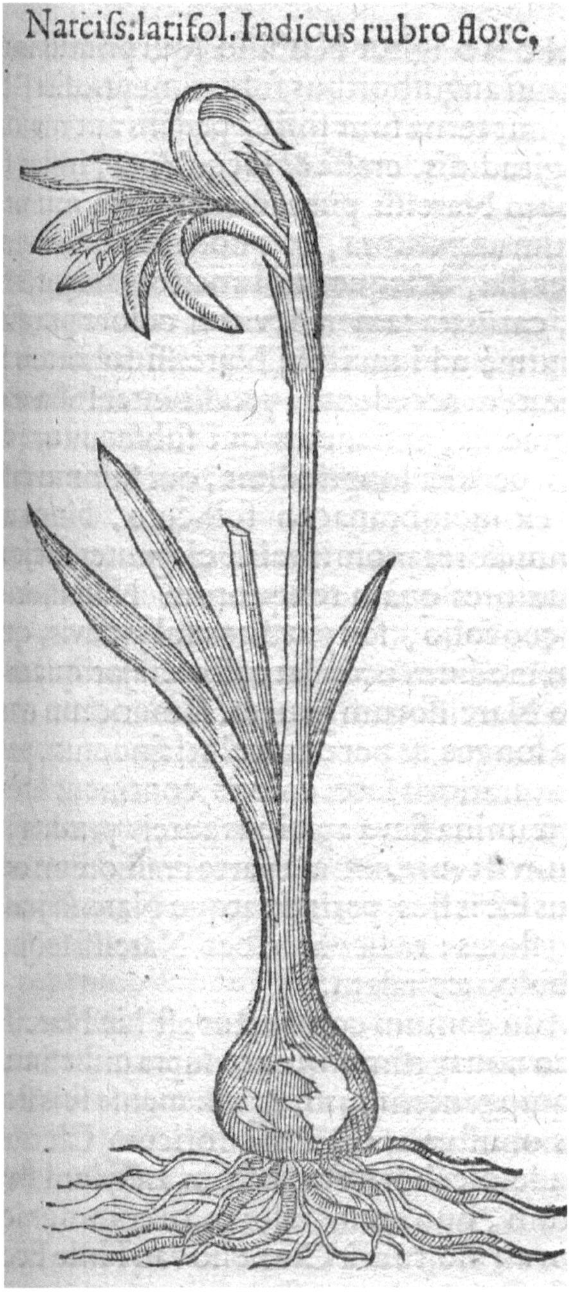

Figure 2. The Aztec lily or narcissus jacobeus. From Clusius, *Rariorum plantarum historia* (Antwerp, 1601), p. 157. Collection Leiden University Library, UBLWGW_661 A 3.

Only one person had a higher status in the country than Arenberg: Archduke Albrecht of Austria, at that moment governor of the Southern Netherlands on behalf of Philip II. It was to him that Arenberg planned to give the (second) Aztec lily – a truly princely gift – in the summer of 1597. Again Boisot was a close observer:

> So that the said Sr [i.e. Arenberg] only has two plants of which the biggest flowered last year and now is once again in flower, and has been carried in its pot to His Highness to be seen, and immediately brought back to my garden, since I am still taking care of it. Concerning the second plant, I think that the said Sr will give it to His Highness who is turning himself into a gardener, taking a marvellous pleasure in flowers ... He has seen my garden, and has asked me many questions concerning gardening, from which I could tell that he is very fond of it. I hope that we will be able to obtain rare plants through his favour and influence. (J. Boisot, 10 July 1597)[43]

Boisot had already explained to Clusius that the portrayal of the Aztec lily, though ordered by Arenberg, had taken place in his own house since he was taking care of the plant – another sign that Arenberg claimed the honour of ownership and gift giving, but that Boisot claimed the respect due to the expert. Boisot had made use of the opportunity to take the plant out of its pot with the entire clump of its fibrous roots. Boisot had obviously also been involved in the original planting of the bulb, since he writes that he had observed an excrescence at the base when first planting it; later it was no longer visible because hidden by the tangle of roots, 'just as the painter expresses it' (J. Boisot, 29 January 1597). And that is precisely what we see on the illustration of the Aztec lily in Clusius's *Rariorum*, which was based on the portrait of the plant sent to Clusius by Arenberg. That original portrait in watercolour still exists (Figure 3). It forms part of the *Libri Picturati* and was an addition by Arenberg to this collection, the core of which had originally been commissioned more than thirty years earlier by Clusius's first patron Charles de Saint Omer.[44]

The unusually well-documented vicissitudes of what was probably the very first Aztec lily to be seen in Northern Europe thus highlight the importance of the connections between the Southern Netherlands and Spain, the love of exotics, the intricate relationships between collectors and experts, the complicated routes through Europe followed by verbal and visual information, and the extent to which rare plants and gardening were shared passions in the highest court circles, even in the midst of warfare and revolution. Like alchemical experiments or astronomical observations of court scientists and prince-practitioners, rare *naturalia* formed part of what we might call spectacular science – the domain where a scientific interest in nature in the broadest sense of the term coincided with the need for display.[45] They also bring our story full circle, linking the gardens, collections, botanical fascination and elites of the Southern Netherlands of the mid-1560s with those from the end of the century.[46]

Figure 3. The Aztec lily. Drawing commissioned by the Count of Arenberg and added to the *Libri Picturati*. *Libri Picturati*, A21, f.85, Jagiellonska Library, Krakow, Poland.

Conclusion

We have gone back far into the fifteenth century to look at the archaeology of a courtly Burgundian-Habsburgian interest in nature, plants, gardening and collecting. Within the Southern Netherlands the geography of this interest follows the pattern set by this courtly tradition. Of the eight expert gardeners hired for Aranjuez by Philip II four came from the Tournai area, three from Brussels and one from Malines.[47] 'Botanophiles', as Lobel's biographer Louis calls them, gravitated around the famous gardens of Brancion at Malines, Mouton at Tournai and Boisot at Brussels, while for shorter periods centres of botanical excellence could be found near Bruges and close to Antwerp and Lier.[48] The core areas in this respect were thus connected with the residences of the Burgundian-Habsburg rulers, their governors or with noblemen who occupied high positions in the Habsburg administration and were part of court culture.

The fact that the Netherlands was part of a vast Habsburg empire and thus linked with Spain and the Austrian territories – which in their turn gave access to the New World and to the Middle East – turned the Southern Netherlands into a huge clearing house of both plants and botanical expertise, comprising everything from gardeners and gardening tools to the learned publications of Clusius or the watercolours of the *Libri Picturati*. This situation reached its heyday between the 1530s and the late 1580s. The intertwining fashions of gardening, botany and collecting created an increasing awareness that the activity of gardening and the botanical knowledge involved formed a special domain – not yet a discipline, but something variously called an art, a practice and a profession by those directly concerned. It also created an increasing demand for expertise and experts: consultants who became scholars, such as Lobel and Clusius, or who remained somewhere between a hired expert and a scholar, such as Plateau. At the same time, the noblemen involved in this practice themselves became experts, noblemen-practitioners or noblemen-naturalists. Their wives, as we shall see, did not lag behind.

II HABSBURG WOMEN

3 WOMEN IN THE GARDEN: NOBLE WOMEN AND THE PASSION FOR PLANTS

Mice and Ginger

During the second half of the 1590s the noblewoman Yzabeau (also Elisabeth) van Arkel (1536–1617) had a garden at her castle Merkenborch not far from Utrecht in the Northern Netherlands. The climate was inclement and she worried about problems with the insubordinate local population. But a more immediate danger were the mice, which ran over her table even while she was eating and threatened to destroy the rare plants that Clusius had sent her: a crown imperial, some fritillaries, martagon lilies, bulbs and a cyclamen. She told Clusius that her garden was her only solace, and this was more than a polite topos. Clusius's gifts helped to make her garden special, but she herself also went to a lot of effort and expense in order to obtain rare plants. Not that many people in that part of the world can have had green ginger growing in their gardens, let alone one that was already developing several buds. She had planted it specially in a pot, and mixed in some extra sand to improve the soil. Yzabeau van Arkel had received the ginger plant from a person who had obtained it in Middelburg, the port in Zealand where ships from the Indies arrived, and he had promised her more rarities. She was, as we will see, by no means the only woman who knew how to make plants come to her – even exotic ones from the other side of the world (Van Arkel, 19 October 1596, 29 April 1598).

While the presence of ginger in a woman's garden in Holland might come as a surprise, the involvement of women in gardening is not. We have already come across several examples of elite female interest in the collection of *naturalia* and gardening, and there is a tradition of references to late medieval and early modern noblewomen who personally took care of their herb gardens and supervised the preparation of medicines based on 'simples' for their households.[1] Were these women mainly interested in medicinal plants or were wealthy women collectors and garden owners in their own right, and did they actually create, plan or design gardens? Were those laid out along the lines of the 'hortus conclusus': private domains that were only open to a restricted circle? Did women cultivate plants themselves and instruct their gardeners, or were they only watching on the sidelines while their male relatives did so? Did they collect or buy plants, and perhaps even take part in botanizing trips? Did women describe or depict plants? And especially, what kinds of expertise did they have? Was their knowledge of plants and gardens taken seriously at the time, and did they play a part in the community of experts in natural history?

The literature about women and botany or gardens (as distinct from that concerning the growing of medicinal plants) largely concentrates on the eighteenth and nineteenth centuries.[2] There is an abundance of contemporary visual sources (paintings, frescoes, tapestries, embroidery and other floral decorative elements on furniture and household elements, female garments and jewellery) and textual sources of the fictional type, such as poetry, which connect women and flowers or gardens. But it is nearly always extremely difficult to deduce anything from these sources about the personal involvement of particular women with plants or gardens. Although attention has been drawn to some manuals written by early modern women about gardening and silk-worm production, non-fictional textual sources which throw light on women's involvement in gardening are very hard to find indeed.[3] Even the interest in plants and *naturalia* of Mary of Hungary or Margaret of Austria has had to be largely deduced from inventories, household bills and ground plans of houses, castles, parks and gardens. The large number of personal letters written to Clusius by female correspondents from several European countries is therefore a truly precious source.

Women Correspondents

Clusius never married and we know of no affairs, lovers or children, but it has long been known that there were several women among Clusius's close friends and correspondents. Of the 42 women known to us by name who were in direct contact with Clusius as either friends or relatives, 25 corresponded with him. Very few of their 110 letters have been published in full or even quoted, and as a corpus they have not been studied except by Clusius's biographer Hunger, who mainly used them for biographical purposes.[4] Such numbers in themselves are significant and should add a dimension to the study of correspondence in the circles of early modern scholars and scientists. Nearly all of Clusius's female correspondents (apart from his relatives) wrote to him about plants and gardening, and sent him packages with bulbs, seeds, cuttings or pictures of plants. And these were no marginal topics in their letters either. Often, at least half to almost the whole of a letter concerned plants, discussed the best ways of obtaining them and growing them in their gardens, and concerned exchanges of seeds, cuttings, bulbs and fruits. Nor did his female correspondents write only about common garden flowers or plants for kitchen gardens: useful medicinal and edible plants are completely absent in women's letters to Clusius, apart from an occasional reference to the layout of a new orchard.

The importance that Clusius attached to his exchanges with women is evident from their duration and frequency. Ten letters from Anna Aicholtz Starzerin to Clusius survive for the years 1588–92: she wrote to Clusius at least two to three times a year and, as the wife of one of the Viennese professors of medicine, had earlier been Clusius's landlady during a considerable part of his long stay in Vienna. The twenty-five letters (spanning the years 1588–1606) from the rich and aristocratic Viennese garden owner Anna Maria von Heusenstain Starhemberg to Clusius likewise testify to a very long-standing relationship, which must have started in the 1570s and probably ended only with her death. With the exception of the years 1593–1601 Anna Maria von Heusenstain wrote to Clusius up to seven or eight times a year. In some years her letters even travelled from Vienna to Frankfurt, and later to Leiden, on an almost monthly basis. In others, there seems to have been a more seasonal rhythm: a letter around New Year, one to report on spring events in the garden, a late summer letter and one or two autumn ones. Clusius replied somewhat less often, but still – it seems – two to four times a year. But Clusius's best-known and almost lifelong female friend was the notorious Princess Marie de Brimeu (*c.* 1550–1605). Twenty-seven of her letters to Clusius have been preserved, all but one from the period 1592–1605, and the frequency with which she wrote to Clusius appears to have been at least as high as in the case of Madame Heusenstain. Perhaps it is a coincidence, but few of his far greater number of male correspondents can boast such a large quantity of letters or a similar frequency of exchanges.[5]

One of the most touching proofs of long-standing personal friendship and mutual esteem can be found in the letters by Johanna van der Laen, dowager of Themseke. She belonged to one of the oldest families of Bruges, had been friends with Clusius when both of them were young and shared his pleasure in gardening. In fact, her house in Bruges was called La Pensee: The Thought, but also The Pansy. In 1603, when Clusius was seventy-six years old, she wrote about her 'indescribable joy seeing the memory that you have of me, your old and total friend, that neither distance, our long separation nor age makes you forget the true and sincere friendship that you have shown me since my youth' (J. van der Laen, 22 March 1603). Clusius had sent her a beautiful gilt cup to commemorate their long-standing friendship, as well as three excellent bulbs, and his portrait. She had retired to one of her houses not far from Bruges where she spent most of her time in the garden, and she was interested in 'something rare' for it – like nearly all women and men of Clusius's acquaintance.[6]

Whereas Clusius's male correspondents lived all over Europe, the large majority of his women friends and correspondents lived in the Habsburg countries: in particular in the Southern Netherlands and Austria. Obviously, Clusius also met women during his student days in Germany, Paris and Montpellier, his period in Hungary, his long stay in Frankfurt, his trips to England and his later years in Leiden: the many greetings that his male correspondents from these countries sent him on behalf of their wives and daughters tell us as much. Yet Clusius seems to have acquired almost no new female correspondents in those contexts. His main friendships with women either originated during his young years in the Southern Netherlands, in the elite circles around his aristocratic patron-friends Brancion and Boisot or in the fairly small circle of court-related Viennese women. The only women in Spain with whom he was in personal touch were aristocratic ladies from the Southern Netherlands whose husbands fulfilled high functions in the Spanish-Habsburg administration and court. Late in life, when he moved to Leiden, Clusius mainly turned to female friends with whom he had probably been already in contact in a much earlier period. Clusius's correspondence with women thus originated in personal encounters which took place in the social context of extended aristocratic households and court-related circles. However, the contents of their letters indicate that long-standing friendships developed only if these women shared Clusius's interest in plants and a curiosity for new *naturalia*. The fashion of gardening was already widespread in elite circles in the Southern Netherlands during Clusius's young years, and there is no reason to suppose that his female correspondents only developed such an interest through contact with Clusius.

None of the women who corresponded with Clusius did so in Latin. They used either French, German or Dutch. Clearly this reflects the fact that women at the time did not have access to universities or other formal types of higher education. Yet, it is extremely difficult to infer anything from this about their actual level of education, which seems to have varied enormously among Clusius's female correspondents and more generally among elite women in Europe. Some elite women received a home education which was on a par with that of their male siblings, and even included Latin and Greek. An example from Clusius's world is Mary Sidney Herbert, Countess of Pembroke (1561–1621), sister of Sir Philip Sidney (a correspondent of Clusius), who was especially known for her erudition. She was taught French, Italian, Latin and possibly Greek at home and came from a background strongly reminiscent of that of quite a few of Clusius's elite female friends: wealthy, enterprising, strongly anti-Spanish and either overtly Protestant or sympathetic to Protestantism. The Sidneys were involved in financing the exploration of the New World (Sir Walter Raleigh was a cousin of Robert Sidney's wife), and Mary Sidney personally invested in Edward Fenton's voyage of 1582–3 and was a stockholder in the Virginia Company. Mary could not go to university and was not allowed to travel on the Continent like her brothers, but she was one of the few women in England to publish her literary works, and after her marriage acted as a patroness in her own right. She created a circle of scholars, poets, musicians and naturalists-alchemists in her household known as the Wilton circle – a real 'college' with the name 'Te Musarum dominam'. She was also personally involved in chemistry and alchemical experiment, worked in her own laboratory with chemists, developed medicines, was interested in embryology and raised silk worms. Thomas Moffett, the entomologist-naturalist about whom more will be said in a later chapter, is regarded as her most important protégé.[7]

All but one of the thirty-five women who are known to us by name in Clusius's circle (and who were not his own relatives) belonged to high or even the highest strata of society in terms of both wealth and family.[8] Only in one case does this seem to have led to some tension. Whereas Princess Marie de Brimeu at certain times openly acted as his patroness – pulling strings at the court in The Hague to get Clusius appointed at the university of Leiden – she never patronized him in her letters, and always wrote to him in terms of great respect and friendship. That was much less the case with the English noblewoman Eleanor Zouche (d. 1611), since *c.* 1578 the wife of her cousin Edward, Lord Zouche (1556–1625).[9] Although the Zouches are reputed to have lived apart ever since 1582 and Eleanor's father complained bitterly about the treatment of his daughter, it seems that Zouche and his wife at least shared a passion for plants. Via a servant we hear that Lady Zouche 'begged him with insistence to ask Clusius whether it would be possible to send her a double red ranunculus either for money or the equivalent of money' (L. Le Myre, 20 February 1597); she included a special pair of gloves made of dog's skin as a present for Clusius. Lady Zouche seems to have been almost exclusively on the demanding side with respect to Clusius – concerning plants and seeds as well as expert advice. In the summer of 1597, she asked her servant to write again to Clusius, displeased at having been ignored by him:

> hearing that you have been sharing out your plants in many directions, she is rather annoyed that she did not know this, since she would have hoped to be among those to whom you were distributing plants. Nonetheless, after having changed gardens, and seeing that the soil is not very suitable for tulips ... she has requested me to write to you again in order to know whether either you or those friends whom you can command have anything else left. (L. Le Myre, 15 July 1597)

For her Clusius was more a resource to be commanded than an honoured friend or expert. It might not have been a coincidence that Clusius overlooked her in his own display of gift-giving.

Female Garden Owners in the Brancion and Boisot Circles

In the Southern Netherlands we encounter two interlinked groups of noblewomen with whom Clusius came in contact during the late 1560s and early 1570s. A few he may even have known since his youth, but he met several of them via his patron Jean de Brancion at Malines, or via Jean Boisot in Brussels. These women enable us to look from a different angle at the circles discussed in the previous chapters, and in particular at the issue of women's interest in plants and gardens. Brancion's letters give us a glimpse, moreover, of a young Clusius who was clearly, but without success, regarded as a possible husband for one of the many young noblewomen who visited Brancion's house and famous garden:

> My *cousin* and *cousine* de Renesse are here who are really our kind of people ... and on Wednesday the Chancellor of Gueldres arrives with his wife and three good-looking young women who are nubile and have a good conversation. And these weeks my *cousine* d'Opfen should arrive with her two daughters who are also marriageable, so that we are here engaged in high-level husbandry. I often wish you were here and also the good wine that one drinks in Paris. (Brancion, 3 August 1571)

More was going on than just partying in this almost Jane Austen-like setting. Brancion, as we have seen, was a passionate grower of special plants with an international reputation, but the ladies encountered in his house were growing bulbs in their gardens as well. At least ten women were named by Brancion in the letter quoted above as sending their thanks to Clusius for his gifts of bulbs, which in some cases comprised whole packages.[10] Interestingly, in exchanges between these gardening enthusiasts – male and female – bulbs sometimes literally served as a payment in kind, especially for outstanding debts in the gift exchange of plants:

> I am sorry that he does not know the *sommersotke* [lit. summer fool] because I had hoped that my garden and pots would have been repopulated with them, and that I would have the means to repay my debts. Before her departure Mademoiselle du Maisnil gave me a beautiful bulb of this type, with which I have paid the Mistress of the Post. (Brancion, 3 August 1571)

This 'Mistress of the Post' was none other than Louise Boisot (*c.* 1538–1610), niece of Jean Boisot, wife of Leonard I de Tassis, the Imperial and Spanish Head Postmaster of the Netherlands, and mother of the even more powerful Lamoral de Tassis whom we encountered in the previous chapter. She literally personified the imperial postal services in the Habsburg empire: her relatives and several of Clusius's other correspondents often simply referred to her as 'Madame des Postes'.[11]

Louise Boisot was very familiar with Brancion's garden: she commented, for instance, in a letter to Clusius on the sad loss of a white oleander, which had been sent to her uncle from Ferrara but had been spoiled during the voyage, 'which is a pity for something so rare' – although she added that she had actually seen several of these plants in Brancion's garden (L. Boisot, 20 May 1571). Like her uncle, Louise Boisot was a passionate gardener who corresponded with Clusius and used her international contacts, especially with Spanish court circles, in order to obtain rare plants for herself, her circle of friends and for Clusius. She was personally in contact with a considerable network of men and women (mainly based in the Southern Netherlands or Spain) who shared this passion, as is evident from the various correspondents of Clusius who name her as a donor of plants, and from her uncle's references. Hers was by no means a temporary interest either, as we can deduce from her observations concerning plants over the years. We have only six of her letters to Clusius, but these cover a time span of seventeen years (1571–88), while her uncle's fifteen letters (1582–97) add information and fill in some gaps. The latter clearly regarded Louise as a real expert from whom he and other garden lovers – male and female – regularly received rare plants. In 1591 he wrote, for instance, about an autumn flowering crocus:

> My niece of the Post has brought it to this city of Tournai in the year 1586, and she had received it from Jehan Mouton who said that he had received it from Spain. It is a plant that really satisfies me. (J. Boisot, 3 September 1591)

Boisot asked her for one of her three double narcissus as a present for Clusius, but all of these had already been promised to others and the plant had not produced any seeds. Some rare plants, such as a large type of iris, were, according to Jean Boisot, seen for the first time in this part of Europe in his niece's garden. He also mentions multicoloured late-flowering tulips – 'among the most beautiful that I have ever seen' – which grew in the garden of his sister Catherine (widow Tisnacq) and were the descendants of those in Louise's garden (J. Boisot, 4 August 1588, 19 July 1591, 7 May 1582). Again, Louise de Boisot had been the first garden owner in their part of the world to have them. Nearly every person who belonged to these circles of serious gardeners tried to grow tulips and other bulbs from seeds, and the women were as interested in colour variation as the men.[12] The results of the efforts involved in obtaining exotic bulbs were not always rewarding, however:

> My niece of the Post has had one of her bulbs in flower that have come from Spain, which she had been devotedly waiting for, but in the end she has found that it was only a type of narcissus medioluteus which produced two or three big flowers on a

> stem, one which has been very common here ... for quite a long time. (J. Boisot, 17 May 1590)

Louise Boisot herself regarded gardening and the care for plants not as a mere pastime, but as a special kind of expertise. It must have occupied a great deal of her time and probably involved great expense. In a letter from 1584, for instance, she explicitly recognized her godfather ('compère') Jacques Plateau – the garden owner and collector from Tournai who later became Charles de Croÿ's consultant – as 'her great guide concerning plants' (L. Boisot, 21 December 1584). Her remarks about a rare exotic plant in her garden which drew a lot of public attention inform us about the function of status symbols that exotic *naturalia* and special gardens had for their owners, but they show at the same time that her garden was certainly no *hortus conclusus* in the sense of a purely private, let alone female, domain. It was open to visitors, though certainly not a public space:

> I should not forget to let you know that le *zombol arabicque* [literally Arabian hyacinth] ... has grown a very beautiful flower ... and I assure you that it has been visited often on account of its rarity. I had presented one to my uncle Monsieur Boisot but he did not want to accept it, saying that it would not flower at the top any more, which we will see with God's aid ... It has had 14 flowers, but no seed at all. (L. Boisot, 22 December 1588)

Her uncle described this same flower in more detail to Clusius:

> As regards our gardens, the most beautiful plant that has flowered there again was one of the *zomboul arabi* that you had sent to my niece of the Post, and I have been pleased to see it and even more so to smell it, being as content with its fragrance as with that of a thousand others. It is a pity that it does not flower every year. It seems that the leaves of this one are more narrow than those of the type that we have had here before. I do not know whether it is a coincidence but the flower is quite similar, and so will be the condition, according to me, that it does not flower in this climate. Something that we should not find strange, given the fact that it is as fantastic in Spain, even in Madrid, as here, as I have been assured by Madame Hopperus, who is well provided with them. (J. Boisot, 4 August 1588)

Jean Boisot's many references to his niece, her garden, her knowledge of plants and the fact that *she* grew and distributed rare plants and tried to obtain them from abroad, all demonstrate an apparently completely self-evident acceptance of the role of women in such activities. She may not have been the owner in legal terms of her garden, but it was without any doubt *her* garden, and not her husband's. She was the person who did the planning, buying, ordering and the supervision of the care for plants – while she was guided in some matters by Clusius, Brancion, Plateau, her uncle and perhaps others. Precisely the fact that neither the female correspondents of Clusius nor any of the men seem to have found it strange or remarkable that women were thus actively involved, indicates that this practice must have been widespread and accepted in this age – at least in these circles. Jean Boisot's attitude is even more significant in the sense that he, like Clusius himself, explicitly recognized his indebtedness to various women as sources of information about plants, donors of exemplars and providers of exotic plants from abroad.

Women connected with the Boisot circle had access to plants from Constantinople, Italy and Spain. Madame Indevelde, for instance, a relative of the Boisots who also received plants from Clusius for her garden, mentioned that she had obtained a plant from Constantinople. We can guess how it reached her: her husband had studied in Padua with Ogier Ghislain de Busbecq, the Emperor's ambassador to Constantinople. In connection with another rare plant, tentatively called 'a kind of hemerocallis valentina' by Boisot, the latter wrote to Clusius that Jacques Plateau had one in his garden in Tournai which he had grown from the seeds that Mademoiselle Laurin had received straight from the Medici gardens in Florence (for both, J. Boisot, 25 June 1592). Plant lovers kept track of such filiations, especially where rarities were concerned. The interest in descent and heredity – as we have seen – at the same time functioned as recognition of the chain of gratitude and obligation linking donors and recipients of such botanical rarities, whether male or female.

Contacts with Spain and Spanish court circles were the most frequent in so far as the women associated with Boisot were concerned. In connection with the rare 'zomboul arabique' we have already met Christine Bertolf under the name of Madame Hopperus. She was based in Spain for many years and closely connected with the Spanish court via her husband, the jurist Joachim Hopperus (1523–76), councillor of the Grand Council of Malines and the Privy Council (early 1560s), and subsequently secretary in Spain to King Philip II. She was a keen gardener and had some plants in her Spanish garden which were regarded as rarities even over there. In fact, it was probably her initiative to send one of the very earliest known portraits of the American sunflower to Clusius's fellow botanist and friend Dodoens.[13] The latter, as he wrote in 1569, saw the live flower in Jean de Brancion's garden, but knew that it could grow as high as 24 feet in the royal garden in Spain. This information about the sunflower in Spain – and perhaps even the seeds from which the plants in Brancion's garden had grown – had come from Madame Hopperus. We know that sunflower seeds were handed around the following spring as a special gift in the area of Malines and Bruges: in April 1570, a close friend of the late Charles de Saint Omer had managed to obtain some eight seeds of the 'Chrisanthemum peruvianum or big flower of the Indies', four of which he sent from Bruges to Clusius in Malines (Van Heede, 10 April 1570).[14] Dodoens included the sunflower and published an illustration (the earliest one to circulate in Europe) in his *Florum et coronarium odoratumque nonnullarum herbarum historia* (Antwerp, 1569), which he dedicated to Madame Hopperus's husband.

This example shows that women did sometimes play a significant role in the spreading of knowledge concerning exotic *naturalia* in Europe. It also shows – for this particular plant – how that fact has come to be forgotten. Dodoens himself recognized the role of Madame Hopperus, and says that she sent the portrait of the sunflower to him after having received it from her husband, but he speaks of her only as the go-between.[15] The dedication of Dodoens's work to Hopperus explains why: Hopperus – as a man who occupied high functions in the Habsburg administration and, moreover, a relative of Dodoens – was far more important in terms of patronage than his wife. For his subservience to the governess of the Southern Netherlands Hopperus had even been nicknamed '*Oui Madame*' (Yes Madam).[16] But there is no evidence that Hopperus had any interest in gardening or plants at all. His name does not figure among the many plant lovers or donors of plants in the Southern Netherlands mentioned by Lobel, and he is mentioned only once by Clusius, not in connection with a plant but only on account of his function.[17]

His wife was a very different case. Jean Boisot asked her several times to send him plants from Spain. She had access to the royal gardens at Aranjuez – a relevant point given the remark about the sunflowers observed there – and in 1587–8 indicated to Boisot that she was not terribly impressed by the collection of plants that could be seen there, showing thereby that she had (or thought she had) considerable judgement in these matters, and trusted it enough to make such explicit statements to a known plant expert:

> [she] has written to me that she finds the garden of Aranjuez very beautiful and magnificent, but short on rare plants, not having seen a single tulip, crown imperial, or martagon from Constantinople, nor even an ordinary martagon or iris bulbosa, which must be the result of the negligence by Holbecque given the fact that the King very much loves beautiful flowers, as does the Infanta, his oldest daughter. Madame Hopperus said to me that many plants of this garden are lost through theft and corruption of the garden staff, who then convince the King that they have perished and died because of some mishap. (J. Boisot, 4 August 1588)

Clusius too mentions Christine Bertolf in connection with exotic *naturalia* in her garden: he saw an exotic bean in her garden in Cologne in 1581, and says that she had sent him another exotic bean from Spain in earlier years. Given its name, this 'frizole guateli', like the sunflower, had probably come from the New World.[18] All of this suggests that it was Madame and not Monsieur Hopperus who was the most relevant person in the transmission of information about the sunflower which contributed to its first publication in Europe.

The expertise of Barbara Brant, Madame Damant, must have been similar to Christine Bertolf's. As the wife of Nicolas Damant (d. 1616), Chancellor of Brabant, President of the Council of Flanders and Councillor of the Secret Council, she belonged to the same circles. Barbara Brant lived for several years in Madrid, where she died in 1591. Jean Boisot describes her as a lady who knew much about plants, had many special ones and took as much pleasure in rare plants as he did. After his bad experience with François Holbecque, it is significant that Boisot decided to trust to Madame Damant's botanical expertise for a selection of the plants to be sent to him in Brussels rather than rely any longer on Holbecque:

> I have written to my nephew Tisnacq who is still in Spain, saying that when he receives bulbs or seeds for me from the said Holbecque he has to hand them over to Madame Damant ... who is a gardener, so that she can reject the ordinary and superfluous ones and send the best ones in smaller portions, as she has promised to do. To this end I have sent her a list of the ones that I desire and those that I don't want, just as I have done several times to Holbecque, but he has not followed them ... The said lady has already sent me some seeds enclosed in a letter, but few, in a small quantity, and rather common ones. If there is anything to be had in Spain I hope that in the time to come it will be done through her. (J. Boisot, 1 February 1589)

Interest and Expertise

Intentionally, we have as yet only looked at those women who were *not* Clusius's two principal female correspondents. After all, the latter might be such exceptional cases that they would tempt us to stretch the evidence in favour of an important involvement of women in botanical matters beyond its credibility. Even on the basis of the evidence presented so far we can, however, already state that the fashion of gardening and the passion for rare plants affected both men and women from the Southern Netherlands who belonged to the elite and in particular to the circles of partly aristocratic and court-related high government officials around the Habsburg courts. Given the fact that men and women did not have the same means, liberties or functions, the similarities between their activities and types of involvement are striking: all created gardens, imported, exchanged and propagated rare plants, some described them and had them portrayed, and in doing so all developed expertise concerning identification and cultivation.

The involvement of these women with plants and gardens was direct and personal. They were in control of their gardens; they directed the planting, sowing, repotting and pruning, and kept a close eye on their rare favourites. Probably, they directed the layout of their gardens as well. Given their status, it is unlikely that they were actually doing any of the real work themselves: like the rich male garden owners, they must have had garden staff. But they personally selected and distributed seeds and plants to their friends and fellow gardeners. Those comprised both women and men, and the men seem to have recognized these women as more or less equal partners in such exchanges. On the same basis women participated in international exchanges in order to obtain rare plants from exotic places. Like the men, they grew precious exotica such as white oleanders, tulips and Spanish lilies, and they showed off their rare plants (plus their close connection with Clusius as expert) to male and female visitors alike, some of whom were no doubt rival garden owners as well as fellow enthusiasts. Like the men, they experimented with colour changes by propagating tulips via seeds as well as bulbs. While all of Clusius's female correspondents received rare plants from him, a few also sent such rarities to him, and thus became a source of botanical and horticultural information for a recognized expert.[19]

As in the case of Clusius's male correspondents, by no means all of the women among Clusius's acquaintances were equally expert: some merely liked rare flowers in their garden; others were passionate gardeners who experimented and could select and distinguish. That kind of expertise seems to have bridged the barriers between the sexes and their respective roles, much as religious and political boundaries were to some extent overcome by this shared passion. In fact, neither men nor women appear to have regarded either the involvement or the expertise of women in these domains as exceptional at all. But there were also differences. Women rarely seem to have made – or sent – 'plant lists' or catalogues of their gardens, and it is rather unusual to find detailed and lengthy descriptions of plants in their letters, as some men did for purposes of comparison and identification. It would have been tempting to explain the absence of long descriptions by pointing out that the plants which reached women already had been named and identified, if Jean Boisot had not stated so explicitly that several rarities were seen for the first time in his region in his niece's garden. And it seems more likely that a lack of catalogues and detailed plant descriptions in their letters reflects practical circumstances rather than their interests. Another difference is that there are no references to women taking part in field trips, botanizing in hills and mountains or actually going to the ports in order to obtain rare plants. These elite women probably could not go far beyond the boundaries of the home, garden, court and city to search actively for new rare plants. But they had servants, male relatives and friends. They did know how to make certain plants come to them, and it is unlikely that they were less interested than the men in creating botanical collections. Finally, publishing about plants seems to have belonged exclusively to the male domain, even if writing about them did not. But even in published texts women still manifested themselves, just like those male correspondents of Clusius who did not themselves publish. They sent botanical information to Clusius, who regarded it as serious enough to mention them by name – thus giving them equal status in this respect with the men and turning himself into one of the few sixteenth-century scholars to acknowledge his female sources.[20]

Lastly, there is no doubt that these ladies were as passionately involved in 'our activity', as Boisot called it, as the men. Jacqueline Viess, for instance, wrote in 1572 from Ghent to Clusius thanking him for his gift of plants: 'also for letting me share in your treasure which you call small but I find very great, and you could not have given me a greater pleasure, I have planted them all today' (Viess, 6 August 1572). The response of Cathryna Quadt, Lady of Lantzkron near Cologne, to a present of rare plants from Clusius shows as much enjoyment and appreciation. The man who delivered several bulbs and the leaf of the ficus indicus (opuntia cactus) to her reported to Clusius on the same day in 1592 that 'she was so very happy about this present that she will have great trouble sleeping tonight' (Schollier, 10 October 1592). And the Italian court composer Filippo di Monte wrote to Clusius about a lady in Augsburg who took a great pleasure in plants and gardening:

> I received the box for the Lady of Augsburg, and know that it will be as dear to her as anything in the world, as she is exceedingly fond of such things and has a garden in Augsburg that is so well organized and full of every delight that it is a genuine pleasure to see. (di Monte, 13 July 1586)

That passions for plants could also run high in a different sense, emerges if we take a closer look at Clusius's two principal female correspondents, Marie de Brimeu and Anna Maria von Heusenstain, and their respective interests in gardens and plants.

4 FEMALE EXPERTS: ELEGANCE AND RIVALRY

Beauty and Garden Design: Marie de Brimeu

'I remember seeing a parrot at the late Marie de Brimeu's house before she left Holland which was unequalled in the colours, variety and elegance of its feathers', Clusius wrote. Almost all of the feathers covering its body were reddish, including its tail, which was partly red and partly blue. The feathers on the back and the wings were yellow, red and green, intermingled with blue. It had a white ring around the eyes marked out by wavy black lines. Clusius could not recall ever having seen a description of such a species. This bird was so fond of Anna de Hyllen, a noblewoman and relative of Marie de Brimeu, 'that it followed her wherever she went through her room and could not see her touching any piece of clothing without giving it a nip, so jealous was it'.[1]

Princess Marie de Brimeu was not only the owner of this rare and beautiful parrot but also an enthusiastic grower of tulips and other rare flowers and shrubs.[2] In the course of her life she created several superb gardens in the Southern and Northern Netherlands, and she was in touch with a number of plant lovers and gardeners in this part of the world. Her letters show her to have been not only an intelligent and resolute, sometimes imperious, melancholy and probably quite headstrong woman, but also a cultured one who wrote a beautiful and elegant French. She was born around 1550 into a Southern Netherlandish aristocratic family, spent much time during her youth in the Bruges and Malines area, inherited the county of Megen in 1572 from an uncle, and married twice. By the time her first husband died, in 1578, she may have already been thinking of changing political and religious allegiances. Her husband, Lancelot de Barlaymont, had been daring in the military campaigns of the Spanish against the Dutch rebels, but clumsy in political matters. By the time Marie de Brimeu remarried (1580), with Charles de Croÿ, Duke of Aerschot and Prince of Chimay – whom we have met earlier in his role as collector and patron of Jacques Plateau – she had become a staunch supporter of Protestantism and the Dutch Revolt. She converted her husband, who was some ten years younger, and the couple moved to Dutch territory. Croÿ's conversion did not last, however. There were political conflicts with William, Prince of Orange, and Croÿ returned in 1584 to Catholicism and the Habsburg regime. Marie regarded him as a traitor, left him and moved further north. From 1584 until 1600 she lived alone – that is, accompanied by ladies in waiting and female relatives, and surrounded by a wide circle of friends – as a wealthy and independent woman. From 1590 onwards she was based in Leiden, in a house with a garden adjoining the plot where the Leiden *hortus* would be established a few years later. Among her friends were several aristocratic and learned ladies, such as Princess Louise de Coligny (widow of William of Orange), Madame de Brederode, Madame de Matenesse, Madame DeFresnes, Marie's sister Bermont de Brimeu, and especially Anne de Lalaing, widow of Willem de Hertaing, Seigneur de Marquette. All of these women were personally known to Clusius, some of them also corresponded with him, and nearly all seem to have had an interest in gardening.

Marie de Brimeu's contacts in Leiden included the rich patrician from Zealand Johan van Hoghelande, a close friend of Clusius, who was known as one of the most expert gardeners in that part of the world. Both Marie de Brimeu and Hoghelande were influential in the circles of the newly created university of Leiden. Their joint efforts were instrumental in convincing the university governors to invite Clusius to come to Leiden, and in persuading Clusius to accept that invitation.[3] Although Hoghelande and Brimeu both took gardening very seriously and seem to have visited each other's gardens in and just outside Leiden regularly, there were practical jokes too. By employing sulphur and making a clever use of the fading light at dusk, Marie convinced Hoghelande that she had some rare peonies which could change colour from white to brown.[4]

Shortly after Clusius arrived in Leiden in October 1593, Marie de Brimeu was invited to move from Leiden to The Hague by the Dutch Estates General (who paid her rent) in order to be closer to the court: as a consequence, the correspondence between Clusius and Brimeu continued, while they could also visit each other. From at least 1594 onwards attempts were made to arrange a reconciliation of Marie de Brimeu with her husband, Charles de Croÿ, without her having to give up her Protestant faith. At the same time she was also involved in diplomatic efforts to arrange a peace between the Southern and Northern Netherlands. By 1600 some sort of truce had at least been concluded with her husband. Marie de Brimeu left for Liège, where she was occasionally visited by her husband, but their relations were not close. Her correspondence and plant exchanges with Clusius continued after her move back to the Southern Netherlands. She asked him in 1602 for help in stocking her new garden with plants from Holland. After her return to the Southern Netherlands Marie de Brimeu mainly spent her time planning new gardens and trying to improve her health by visits to spas. In April 1605 she died.

While Marie de Brimeu may have been exceptional to the extent that she generally operated as an independent woman, involved herself in high politics and fought innumerable legal battles concerning her control over her domains and income in the Southern Netherlands, her interest in flowers was shared by many women in her surroundings. The roots of the friendship between Marie de Brimeu and Clusius may, in fact, have lain in the Southern Netherlands in the circle around Jean de Brancion, and have gone back as far as the late 1560s, when she must have been about twenty years old.[5] Even at this early date Marie already referred to her garden, the plants she was going to send to Brancion, the seeds sent to her by Clusius and the gratitude she felt to him for offering to help her restore her garden, which had suffered from bad weather (Brimeu, 23 February 1571).

Many things about gardening can be gathered from Marie de Brimeu's letters – such as the persistent thefts from both her own and Clusius's gardens, the sending of seeds, rare plants and plant portraits back and forth between them, the location of the Leiden garden she put at his disposal, and the importance she attached to gardening. In one of her last letters from Liège, she even stated that she had moved house because of the continuous thefts from her garden, and had chosen another one mainly for its better protected garden.[6] What emerges very clearly from all of them is that she took an enormous pleasure in gardening, and during all of her life tried to become more expert in the art of gardening. Expertise, as she stated, grew with practical experience. She was interested, for instance, in colour changes from one generation of tulips or other bulbs to another, and hoped that experience over time would teach her more about this phenomenon: 'but it seems to me that there are some that have changed colour into another kind than last year and I did not know at all that this happened, but experience teaches many things' (Brimeu, 24 May 1593). Another way of becoming more expert was to learn from those who knew more about such matters, first and foremost Clusius himself. Marie de Brimeu shared the high esteem for Clusius as propagator of special flowers with their mutual friend Justus Lipsius. Upon the receipt of a letter from Clusius accompanied by a large basket full of bulbs, seeds and roots, she wrote:

> I will receive them certainly with complete affection, as much because of their rarity as for the opportunity that you say you have had to send them to me, knowing how much I enjoy gardening. That pleasure, I confess to you, is certainly great but is as yet more enjoyment than science [*de joli et non encores de science*], because I have neither many hours to spend on it and become more expert, as I would want, nor a good instructor who can introduce and assist me in this subject, like you who – as I have understood from monsieur Lipsius – are the father of all the beautiful gardens in this country, on account of both your knowledge of simples and the liberality which you have shown towards many and now towards me. (Brimeu, 18 September *c.* 1590–3)

Marie de Brimeu thus clearly distinguished her way of gardening, which was then still mainly a pleasurable activity, from something closer to a science or serious expertise, the level which she wanted to attain under the guidance of Clusius. She also indicated in which respects she wanted further instruction. After reporting in July 1592, for instance, to Clusius how the rare plants (such as double red ranunculus, various tulips and anemones, and a crown imperial) that he had given her were doing, she asked him: 'I therefore thought to reorganise my garden according to your counsel and advice and to comprehend how bulbs should be kept and in short with the help of your skill become a better gardener than I am' (Brimeu, 9 July 1592). Clusius was not her only advisor. Lobel sent her a catalogue with the plants in the garden of an English friend so that she could mark those plants which she wanted to have. Some women did therefore use plant catalogues even if they did not draw them up themselves.[7]

Of course Marie de Brimeu was flattering Clusius and presenting herself as a modest pupil, but that does not make the distinction between a 'hobby' and a serious field of expertise or a 'science' any less relevant. In the same letter, Brimeu went on to talk about garden design, and compared the garden with rich tapestries made of gold and silk:

> but also on account of your promise to add even more [i.e. gifts of plants], which I would certainly be ashamed to ask for, if I had not experienced your liberality before and understood from the riches of your tapestries, which now and then really surpass those made of gold and silk, that nature surpasses arts. (Brimeu, 24 January 1592)

At one moment she must actually have thought of modelling the design of her garden in terms of colour variety upon a tapestry sent to her by Clusius.

> and you promise again to continue [sending plants] in order to populate and increasingly embellish a garden which has been newly created, and which, in the course of time, I could make resemble the design of the beautiful tapestry you have sent me since, on top of everything. I would estimate and love it in place of those beautiful tapestries that I have lost in these troubles. But I know well enough how many difficulties I will encounter in this undertaking, because of both the rarity of the flowers and colours and the great numbers of them that would be needed. (Brimeu, 24 January 1592)

Anyone who has seen the famous Flemish tapestries of the sixteenth century with plants and animals that were often directly modelled after the illustrations in the learned works on nature of the period, will appreciate the intricate mirror game of reflection back and forth between garden, tapestry, nature, science and art that Brimeu was playing here.[8]

Marie de Brimeu was more involved in, or perhaps simply more explicit about garden design and the overall visual effects of her garden than the other women we have heard about. In several letters to Clusius she included painted portraits of her most beautiful tulips.[9] And she even considered having her Leiden garden painted, so that Clusius – then still in Frankfurt – could have a better idea and give her more specific advice:

> I almost wanted to have my garden, as new as it is, painted and send this painting to you, in order to receive better instructions concerning the arrangement which I should follow to embellish it. But it seemed better to wait until those flowers that I have will have come out, so that you can make a better judgement of their disposition. (Brimeu, 24 January 1592)

Some of the things that still 'disgusted and displeased' her about her garden were that she had given it a round shape and that the beds were too narrow. But she did not dare to restructure it in a major way without Clusius's advice, in particular about the order which she could follow in planting, since it involved transplanting bulbs that flowered in different seasons. A year later, in 1593, however, she was proud of the results. Her garden was even beginning to get a reputation to the extent that some fellow gardeners – indicated as 'those of this profession' – were becoming envious.[10]

Horticultural and Botanical Expertise: Anna Maria von Heusenstain

In 1592 Clusius sent seeds of the double delphinium, which he had received from Anna Maria von Heusenstain in Vienna, to Marie de Brimeu in Leiden. Writing to thank Clusius, Brimeu promised that she would not be as stingy with her rare plants as 'the lady from Vienna', about whom Clusius had clearly been complaining (Brimeu, 24 January 1592, 9 July 1592). In the community of sixteenth-century naturalists liberality and generosity were as important for women as for men. But rivalry and envy were equally strong. Marie de Brimeu alluded to national differences in liberality (or the lack of it), complaining that the Dutch were tight-fisted with their plants: 'in a country where those who have something are stingy keepers instead of amiable gardeners who help new garden lovers' (Brimeu, 18 September *c.* 1590–3) – punning on the similarity in French between keepers (*gardiens*) and gardeners (*jardiniers*). Anna Maria von Heusenstain's letters inform us in more personal detail about rivalry. But they also demonstrate very considerable female botanical and horticultural expertise, the important role of women in the distribution of exotic plants from Constantinople, and the professional-botanical rather than aesthetic use of plant portraits.[11]

Anna Maria von Heusenstain belonged to an aristocratic family which owned houses and gardens in both the city of Vienna and the nearby countryside. Her husband had a long career at the Emperor's court. They were members of a Viennese circle of prominent garden owners and plant lovers connected with the court, government or university. Clusius must have come in contact with them when he arrived in Vienna in order to advise Emperor Maximilian

II on the creation of a special garden.[12] They frequently asked Clusius for advice and for bulbs or seeds, and at the same time provided Clusius with easy access to rarities from Turkey and the Levant. In the same circle we find, for instance, Eva Ungnadin, the wife of David Ungnad, Baron de Sonneck, imperial legate in Constantinople. The latter sent various rare plants from Turkey to Clusius in Vienna.[13] Further members were Anna Aicholtz Starzerin (d. *c.* 1596–7) and her husband Johann Aicholtz (1520–88), a highly esteemed physician in Vienna as well as an anatomist and botanist, a professor in and (four times) dean of the faculty of medicine, and rector of this university during 1574–5. Clusius lived in the Aicholtz house during much of his stay in Vienna, and undertook many botanizing trips in the Alps together with Aicholtz. After the latter's death Anna Aicholtz continued to take care of their famous garden, which involved much more than simple maintenance: in 1591, for instance, she wrote to Clusius 'I am now building a pavilion and gazebo in the garden; in summer I want to be there, outdoors all day' (A. Aicholtz, 2 April 1591).[14] Anna Aicholtz remarried, to Zacharias Starzer, a magistrate of Vienna, who also took an active interest in gardening. Her brother, 'Monsieur Unversagt', was a high government official about whom one of Clusius's correspondents wrote in 1599: 'he governs almost the whole country, being at present the principal of the Secret Council' (H. Bloeme, 4 April 1599).[15] Thus, in Vienna we find ourselves in much the same kind of circles as earlier in the Southern Netherlands: landed and court aristocracy, high government officials of a generally aristocratic background, and some medical specialists who linked court, city and university circles. All of these persons were in close contact with each other and with Clusius during the latter's period in Vienna. Those who continued to correspond with him when he moved to Frankfurt (1588) and later Leiden were mainly the women: Anna Maria von Heusenstain, Anna Aicholtz and Eva Ungnadin.

We cannot discuss the Viennese-Austrian elite and their interest in plants here as we have that of the Southern Netherlands in earlier chapters, but their interests seem to have been very similar, and comprised both exotic and rare plants, horticultural innovation, the growing of rare bulbs, propagation and colour variety, and plant collection.[16] Madame von Heusenstain took an interest in all these aspects. Her letters to Clusius show her as a less cultured person than Marie de Brimeu – she wrote in Viennese dialect and her spelling is idiosyncratic to say the least – but she was certainly as sincerely passionate about plants and her gardens. In fact, she must have been almost obsessed with them. In all of her twenty-five extant, detailed letters she mentions very few personal matters, refers in one line to her husband who almost died of a serious illness, and mentions her children hardly ever but in the context of gardening: her young daughter Polyxena already had her own garden and hoped that Clusius would send her a rare plant. Madame von Heusenstain only briefly asked after Clusius's health, and referred to his possible transfer from Frankfurt to Leiden only in the sense that it might make her access to Clusius as provider of special plants more difficult. She also showed herself much more concerned about the thefts from his garden than about anything else that might have happened to him. It took the capture of her son by the Turks, general plundering by both foreign and 'friendly' soldiers, the destruction of her Viennese garden by 'Walloon' troops and the death of her cattle – in the course of 1605 and 1606 – for her to write briefly about her personal troubles.[17] But soon she was already busy again planning 'a nice little country estate' on the mountain at Starhemberg, 'where everything flowers earlier than in Vienna' (Heusenstain, 1 March 1606). The main things that seem to have interested her were the state of her gardens in Vienna and Starhemberg, obtaining rare plants – either via Clusius or through her various connections in Constantinople – growing and propagating these, and showing them off to rival garden owners.

Heusenstain prided herself publicly on her special relationship with Clusius and the access this gave her to rare plants. This did not improve her relations with fellow garden owners, as she herself noted: 'they are quite malicious to me; they say all the time I should let Sir Clusius give me something rare. I am bragging all the time of Sir, saying that Sir can send me something rare' (Heusenstain, 2 January 1590). The letters that went back and forth between Clusius and several other female garden owners in Vienna in which their mutual complaints were hinted at, give us some idea of the permanent pressure of such exchanges and the honour and reputation involved. In the summer of 1589, for instance, Anna Aicholtz wrote to Clusius that she had red Turkish lilies in her garden and that everyone was asking her for her colourful tulips, while complaining that Anna Maria von Heusenstain had enough of them in her garden, still wanted more and refused to share them with others (Aicholtz, 15 May 1589, 28 August 1589). The latter, in June 1591, in her turn complained that 'nobody receives anything' from Anna Aicholtz: 'My white oleander was gone, and I begged her for one; her husband had promised me one a long time ago, but she does not give me one' (Heusenstain, 17 June 1591). By August 1591 Madame von Heusenstain was apparently worried that Clusius might think badly of her on account of her reputed stinginess:

> I have real sorrow because I think that Sir is unfriendly towards me because he does not want anything at all from me and writes me in the last but one letter that I have anyhow so much more than Sir and do not give anything to anyone. I just believe that there are some people who want to create enmity between me and Sir, so that he will not send me anything any more. But I hope Sir will not do this and will not believe that I give things to just anyone; I do not do that and it would not serve my garden. (Heusenstain, 15 August 1591)

Heusenstain's interest, passion and expertise were very much those of the expert collector of rare plants, and she must have had several gardeners working for her, given the fact that she had a town garden as well as an estate with a garden at Starhemberg. At least one of these gardeners was a woman, a 'Gärtnerin'.[18] Heusenstain's regular reports to Clusius about her garden covered a whole range of events and activities: from mishaps and plants that had suffered or died on account of bad weather to the appearance of new colour varieties, her experience with grafting or sowing, the best ways to plant certain bushes and her desire for particular colour varieties or plants that were lacking in her collection or had died. In June 1591, for instance, she wrote that two of her double ranunculus had flowered, but many of her double delphiniums had been flattened during a thunderstorm. Her small myrtle bush, a present from Clusius, had died, while a bed with tulips had flowered beautifully. She wondered, though, whether she or her female gardener had done something wrong when sowing tulip seeds, because the flowers that these tulips produced were all either red or yellow and not variegated like those of one of her relatives. What she really missed in her garden were an orange coloured 'Betonie' (probably woundwort, Stachis officinalis L.). One of her three Spanish lilies had flowered and its flower had both white and yellow petals. She had planted the Spanish lilies and the double 'Märzblume' (daisy) all together in a pot. A green anemone had flowered, but a single white one which had been a present from Clusius had rotted. There were various white roses as well, and she had already successfully grafted one (Heusenstain, 17 June 1591).

Unlike quite a few other male and female correspondents of Clusius, Madame von Heusenstain was extremely specific in her requests for plants: they had to fill gaps in her collection, add something to what she already had or be of a specific colour. She asked Clusius several times for 'a flesh coloured' and a double 'Betonie', brown tulips, seeds of chequered tulips, double red anemones and:

> Let Sir not forget the yellow lily and the double daisy, I am all the time dreaming about them ... I ask Sir for a sprout of the white hyacinth. I have grown it from seed, but none of them flowers yet. But the yellow saffron that I have grown from seed really flowered yesterday. (Heusenstain, 17 June 1591)

Later, she hoped that her double daisy would flower, 'which has never happened yet in this country. I would be overjoyed' (Heusenstain, 15 August 1590). And she was in tears when one of her favourite rarities died or when she found out, upon opening a package from Clusius with rare plants, that everything inside was as dry as dust after the long journey: 'and I have, as God knows, cried on account of the red saponaria. I had so much looked forward to it' (Heusenstain, 19 May 1592).

Anna Maria von Heusenstain was by no means only on the demanding and receiving side, however. She regularly provided Clusius with plants, many of them rare and some of them exotic. He only had to let her know, for instance, whether he wanted another opuntia cactus, originally from America (Heusenstain, 9 January 1589). The then still rare horse chestnut, native to mountainous parts of the Balkans, is another example.[19] In July 1557 the Flemish physician Willem Quackelbeen had written from Constantinople to the Italian botanist Mattioli (then living in Prague) about twenty-seven different plants which were for the most part unknown in Europe at that time. The horse chestnut was one of them. A horse chestnut tree may have already been growing in Bologna during the 1560s, and there are references to the tree in the correspondence of the same period between the botanists Giacomo Antonio Cortuso in Padua and Ulisse Aldrovandi in Bologna, but it was still extremely rare there in the 1570s, and only marginally less so in the 1580s and 1590s. Clusius saw its flowers for the first time on a coloured drawing of a branch with flowers that Madame von Heusenstain sent him in 1603 (Figure 4).[20]

Clusius's repeated efforts to obtain horse chestnut trees for his successive gardens, and the difficulties and expenses faced by both Clusius and Anna Maria von Heusenstain in the early 1590s and again in 1603–4 in order to transport a young chestnut tree to Clusius, show how highly it was regarded at the time. In 1576, while in Vienna, Clusius had received his first horse chestnut – a small tree, not the nut – from David Ungnad in Constantinople, and in 1581 he received some horse chestnuts (the nuts) from Constantinople.[21] Their distribution followed a familiar pattern. Clusius certainly sent one or more nuts to Jean Boisot in Brussels, together with advice about how best to grow them. There was, as usual, some delay in their transport. Boisot complained in May 1582 that he had received neither the rare fruit of the laurocerasus (cherry laurel), which had likewise been first introduced by Busbecq and Quackelbeen, nor the 'castanea equina' via Plantin, who was apparently acting as an intermediary. But the seed or plant eventually arrived, for a few years later, in May 1586, it was growing in Boisot's garden in Brussels. Boisot had had his horse chestnut grafted on 'an ordinary chestnut', as Clusius had advised him (J. Boisot, 7 May 1582, 9 May 1586). Another horse chestnut from the 1581 consignment in all likelihood ended up in the garden of Anna Maria von Heusenstain. Ten years later she had a big chestnut tree, about which she wrote in 1591 to Clusius (who wanted a new horse chestnut for his garden in Frankfurt) that her 'Türkischer Kastanienbaum' had flowered beautifully, but that all the nuts had been blown off by the terrible weather. By 1603 Clusius again wanted a horse chestnut, this time for the garden in Leiden. Madame von Heusenstain tried to send him a small exemplar, 'a rather beautiful little chestnut tree', but the transport was no easy matter. The small tree was sent back and forth between Vienna and Frankfurt several times for lack of secure further transportation, and eventually may have been returned to Austria for safe keeping during the winter season. That same year a coloured drawing of a chestnut branch with flowers sent by Heusenstain did manage to reach Clusius, who found that its flowers (whitish with little purple spots) were very different from Mattioli's description.[22] By 1604 Madame von Heusenstain's own horse chestnut at Starhemberg was so big that it bore several hundreds of flowers: 'Sir will not believe how beautiful my tree was today' (Heusenstain, 10 June 1604).

Anna Maria von Heusenstain received dozens of rare plants directly or indirectly from Constantinople – from her own couriers and via her husband or other

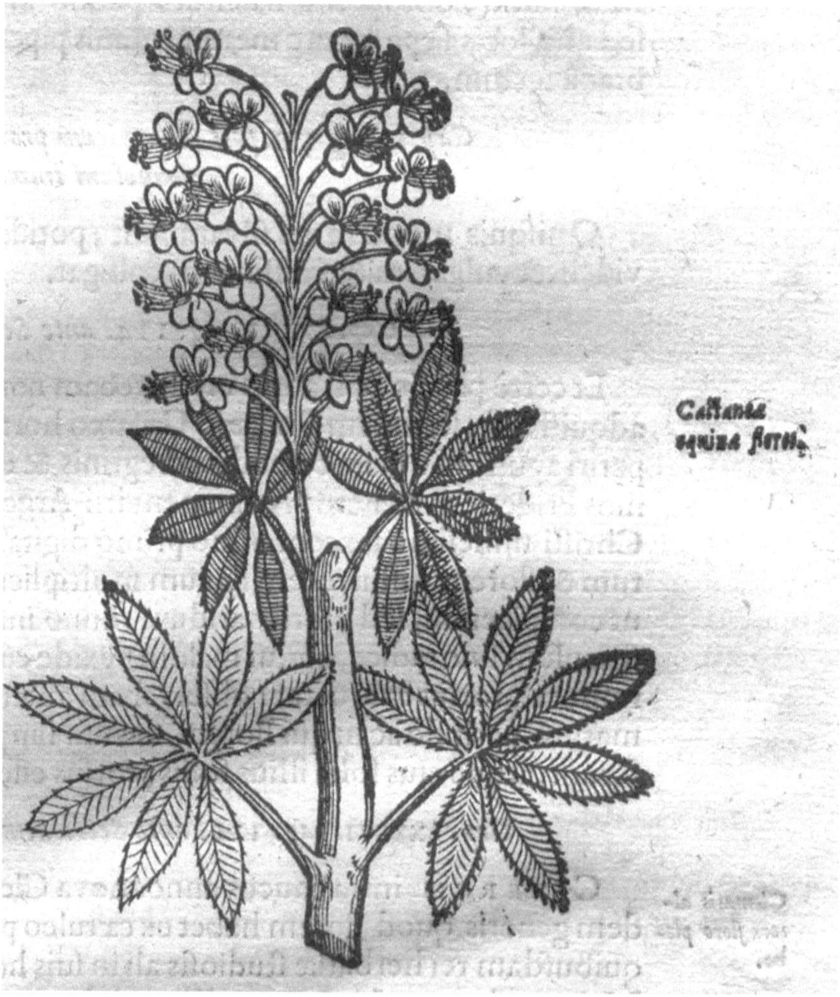

Figure 4. Horse chestnut branch with flower sent to Clusius by Madame de Heusenstain. From unpaginated Appendix to Clusius's *Rariorum* in his *Exoticorum libri decem* (Leiden, 1605). Collection Leiden University Library, UBLWGW_THYSIA 2202.

high-ranking officials. She generally grew them first in her own garden before passing on new exemplars, seeds, cuttings or bulbs to Clusius. The latter mentions a Byzantine colchicum, ranunculus asiaticus, and many varieties of anemones growing in her garden, all of which had come from Turkey.[23] Couriers seem to have gone back and forth between Vienna and Constantinople, perhaps even with the single purpose of conveying plants or letters about them:

> Herewith I am sending Sir two small flowers that someone from Constantinople has sent me per courier. There should also have been double *Schmalzblümchen* [a kind of ranunculus] ... But I have immediately written to him per courier that he should send me two or three that are good. But if he does not trust them he should send me them with flower and all. (Heusenstain, 1 November 1592)

Madame von Heusenstain never mentions plant portraits in connection with garden design, as did Marie de Brimeu. Clusius and Madame von Heusenstain both used portraits for precision and identification. Either the one or the other had a portrait painted of the desired plant, which Heusenstain then sent on from Vienna – by courier – to Constantinople in order to facilitate the choice of the right plant and prevent mistakes:

> Concerning the painting, Sir may really believe that I would like to show Sir the painter who has painted it. He has made it a little bit too red. The [other] painting that Sir had given me I gave to my servant to take with him when he went to Constantinople, but he has not brought it back to me. (Heusenstain, 1 November 1592)

In the communication among themselves both used plant portraits in order to make sure that they were talking about the same plant, or to convey the colour of flowers or the particular shape of plants – and thus for identification and comparison. Uniquely among Clusius's female correspondents, moreover, Heusenstain showed an interest in the wild plants of her own region and country. Clusius saw an ornithogalum in her garden which had come from Tirol, and she helped him obtain wild plants from her part of the world – both upon his request and on her own initiative.[24] During his years in Vienna Clusius himself sometimes bought wild plants that grew in the nearby mountains from female root cutters ('rhizotomae mulierculae'), who brought the plants to the market in Vienna. He asked them for the local names of these plants: some wild cyclamen of which he bought the tubers were called 'Sawbrot' (pig's bread) or also 'Erdeapfel' (ground apple) and in Hungarian 'dizno repa' ('porcinum rapum' or pig's turnip).[25] The hepatica trifolia (a type of anemone) too was brought regularly from the mountains to the Viennese market by these local women. But in 1588 Clusius heard from Heusenstain's husband that it also grew

> not far from the castle of the Heusenstains at Starhemberg, in those woods which extend from this castle to the snowy mountain and there too it has a double blue

> flower, as Heusenstain's wife wrote to me in Frankfurt, and she sent me one of its flowers wrapped in paper and enclosed in her letter, in case I doubted this. And she would have sent me half of the plant (of which she had only one), if war with the Turks had not broken out in that province.[26]

This probably does not mean that Anna Maria von Heusenstain herself went botanizing, although she must certainly have explored the extensive grounds of her estate. She, like Clusius himself in his Viennese days, was in regular contact with local women who collected and sold wild flowers and herbs from the mountains: 'Concerning the things that I should send Sir, I send them herewith. Only not the small white *Hahnenfuss*, because the one that the women had is completely spoilt and they have not been able to find one recently' (Heusenstain, 15 August 1590). In May 1591 these women were selling beautiful wild flowers all the time: 'the women brought me today some of the early flowers, those which have leaves like violets' (Heusenstain, 7 May 1591).

These exchanges, which are so different in both style and contents from those between Clusius and Marie de Brimeu, highlight how great the individual differences between Clusius's female correspondents and their types of interest in plants were. They also go a long way towards explaining why Anna Maria von Heusenstain is *the* most frequently mentioned female source (15 times) of botanical information in Clusius's printed works.[27] Clusius saw special plants in her garden and estates; he accepted seeds, dried flowers, roots, bulbs and portraits of plants from her; and he received information from her about the characteristics of plants, the period when they flowered and the exotic provenance of some. Clusius also explicitly mentions her contacts with Constantinople, and the fact that she included local wild flowers in her 'cultissimo horto'. Above all, Brimeu was only rarely a source of information for Clusius and much more often a recipient, while Madame de Heusenstain gave Clusius probably as much as she received.

No *Hortus conclusus*

Women, first and foremost the Virgin Mary, have traditionally been associated with and depicted in the walled or enclosed garden, the *hortus conclusus*, a space removed from the public domain. That image has little or nothing to do with the female plant lovers and gardeners who belonged to the sixteenth-century elite. Gardens might to a limited extent fulfil the role of a haven from the dangers of politics, warfare and religious strife, but they were by no means a domain outside society. For female garden owners gardening meant exchange with men and women, locally, regionally and internationally. It required – and resulted in – experience and expertise, involved them in a passion, was a source of emotions and a highly social activity: it entailed showing off these living collections, designing the garden and having constructions built in it; it involved the garden owners in rivalries and friendships, and provided one of the few spheres in which women could be in control, operate almost as independently as men and even be publicly recognized as such. Clusius discussed plants and gardening with many of his female correspondents on the same footing as with his male ones. His printed works also show that the expertise of women in the sixteenth century could still find its way into what was becoming a scientific discipline that for several centuries to come would be closed to them.

What these women contributed to the domain of plant knowledge and gardening was a potent mixture of practice-based knowledge concerning both gardening techniques and the cultivation and identification of plants, stylistic interest in garden design, informal but sometimes very high education, wealth and thus economic power and patronage or at least practical assistance to experts such as Clusius, and wide-ranging international connections. The traditional connection of noblewomen with estate gardens and the dispensing of plant-based medicines may have facilitated their role as garden owners and designers as well as male acceptance of their expertise concerning plants. In terms of the contents of their interest in plants this tradition seems, however, to have been totally irrelevant to the female garden owners discussed here. Not one of the women among Clusius's correspondents showed any interest in medicinal simples or vegetables. Their love of plants and gardens – like that of the male garden owners – was directly linked to the spreading cult of collecting and nature. They were actively involved in a creative and fashionable competition which stimulated the demand for rare plants, and they pursued, as Blunt and Stearn have put it, 'the beautiful rather than the useful'.[28]

Blunt and Stearn have also been among the few to point out the important role of flower-loving ladies at Vienna in the introduction of exotic plants from the Middle East, and we may well assume that it was not only the Southern Netherlandish elite in general but the women in particular who introduced the passion for flowers and the fashion of gardening into similar circles in the Northern Netherlands.[29] As creators of fashion and agents in the spreading of fashions in court and related elite circles in Europe, these women should be regarded as an important driving factor in the spreading of a cult that – in its slipstream – entailed the need for more specialized, more professional expertise and expert consultants: in other words, for men like Lobel and Clusius, who thanks to the patronage of male and female elite garden owners could turn themselves into what we now call botanists.

III ITALY

5 GROWING EXPERTISE: GARDENERS, COLLECTORS, NATURALISTS

Italy and the Botanical Renaissance

By Clusius's age Italy boasted a long horticultural tradition which went back, via the formal gardens of the Renaissance and the monastic gardens in which fruit, vegetables and medicinal plants were grown, to the pleasure and utility gardens of the ancient Romans. The early start and the long horticultural as well as academic and medicinal tradition should not, however, be taken to suggest that every single development of the Botanical Renaissance happened first in Italy and only later, under Italian influence, in the rest of Europe.

Together with Spain, Italy had been the country to receive the very earliest information about nature in the New World shortly after its discovery, but by the last quarter of the late sixteenth century many exotic novelties arrived first in Northern Europe, since the overseas routes to Asia via the African coasts were by then beginning to be controlled by the British and the Dutch.[1] While Italian university botanical gardens (Pisa, 1544; Padua, 1545) and chairs for the teaching of medicinal botany (Rome, 1513/14 in the form of a separate lectureship; 1539, in that of a chair; Padua, 1533; and Bologna, 1534)[2] were indeed the first in Europe, early printed and illustrated encyclopedic works on natural history appeared no later in the German-speaking world than in Italy: Mattioli's influential comments on Dioscorides first appeared in 1544, but publications by Brunfels, Bock and Fuchs came out in the 1530s and early 1540s, and Gessner's works on natural history were published from 1541 onwards. As we have seen earlier, moreover, the tradition of the realistic representation of living nature in art seems to have emerged more or less contemporaneously in the Burgundian territories and in Italy.

Even with respect to the history of great gardens the chronology is not unequivocal, while mutual influences in Europe are intricate. Acidini Luchinat, for instance, has documented the history of a number of Medici gardens in Tuscany, and argues that formal gardens started appearing in Italy as of the thirteenth and fourteenth centuries.[3] The particularly interesting estate of Poggioreale, created in the late 1480s, comprised a royal residence and park just outside the city walls of Naples. It has been argued that the brief invasion of Italy by the French King Charles VIII and his visits in 1495 to Poggioreale were of decisive importance for the introduction of Italian garden models in France.[4] And perhaps they were trasmitted from there to other parts of Northern Europe. Yet, if we think of the Southern Netherlands during the fifteenth century and of the Coudenberg palace at Brussels with its great park and the menagerie that was added to it during Philip the Good's reign (d. 1467), it seems that there was a different, perhaps even independent, tradition in the Burgundian area. Like Poggioreale, those northern estates united pleasure with utility functions, hunting with the keeping of domestic animals, and woods, orchards and kitchen or medicinal gardens with pleasure gardens.

The chronology and geography of great aristocratic gardens in Clusius's own time is no less intriguing. Some of the most famous Italian villas and gardens (Villa d'Este at Tivoli, *c.* 1560–70; Villa Medici at Pratolino, and Villa Lante at Bagnaia, 1570s–80s) originated almost contemporaneously with some of the most prestigious Elizabethan gardens in England (Kenilworth in Warwickshire, 1563–75; Holdenby in Northamptonshire, 1574–83; Theobalds in Hertfortshire, 1575–85; and Nonsuch in Surrey, 1579 and onwards). Charles de Saint Omer's estate near Bruges and Brancion's garden at Malines of the 1560s fit in the early decades of this same chronology. And the vast estates of the Count of Arenberg and the Duke of Croÿ of the late sixteenth century were probably inspired by Italian models and can only be compared with the top collections and estates of the Italian nobility.[5] Whatever the precise chronological intricacies, there is little doubt that Clusius's age saw the creation of a large number of great pleasure gardens, which set the fashion for many smaller ones.

At roughly the same time (*c.* 1550–60) as garden owners in and near Vienna, Italian garden owners were delighted by the major influx from the Middle East of plants – especially bulbs, tubers and corms – that were becoming available for the first time as a direct consequence of the high frequency of diplomatic interaction between the Italian states and the Habsburg Empire, on the one hand, and Constantinople, on the other. Diplomats brought back rare bulbs. In Italy, as in the rest of Europe, the fascinating new plants from the Middle East stimulated a passion for exotics and the fashion of rarity.[6] The new plants were also immediately used in the competition for status. It was not considered worthwhile to spend a fortune on garden design and then fill the garden with common plants. Exotics enhanced the prestige of a garden and its owner, and a nobleman's garden should therefore contain more rare *naturalia* than that of a man of lower social status. Exotic *naturalia* found their way rapidly and without distinction into both university botanical gardens and the private gardens of the aristocracy: the horse chestnut, for instance, made its way as quickly into the Pisa *hortus* as into the Medici gardens at Pratolino.[7] The new bulbs and tubers were less easy to deal with, however, than the perennials that had been the favourites until then, and their introduction triggered the growth of new kinds of horticultural and botanical expertise. Different methods of propagation and techniques of bedding out had to be learned. Experiments were made to discover the most suitable soil types.[8] The garden was changing, therefore, and horticultural and botanical knowledge were changing with it. Clusius himself, at the age of eighty-three, was well aware of the changes that had taken place in Italy during his lifetime: 'Because the flower lovers have so much increased in numbers in Italy they will undoubtedly be desirous to have plants brought to them from various places which they can buy'.[9]

Our exploration of who these flower-lovers were, what this search for novelty entailed and to what extent special types of expertise concerning living nature were developed in this context, will touch upon the phenomenon of collecting in Italy, the formation of a largely academically trained and publishing Italian community of naturalists, and the history of gardens and garden owners. Against the background sketched in the major works of Olmi and Findlen on natural history and collections, and on the history of Italian gardens and plant collecting by Masson, Lazzaro, Coffin, Tongiorgi Tomasi and Zalum Cardon, we will look in particular at plant specialists in Italy and their interest in new *naturalia*. Who exactly were Clusius's exchange partners in Italy?

First Contact

The fact that Clusius never visited Italy and personally met only a very few of the Italians with whom he corresponded is no drawback to such an exploration. In fact, that contact at a distance offers some decided advantages given the length, character and quality of Clusius's exchanges with Italian friends.[10] These lasted for at least forty years and are documented from 1566 until shortly before his death in 1609. They were not concentrated on one city or territory, besides, but covered the most relevant parts of the Italian peninsula. Through his correspondence Clusius may, indeed, have had access to a larger part of Italy than did some of his Italian friends themselves. Already by the mid-1560s Clusius had sufficient if not always grammatical Italian to ensure that there were no language barriers during these crucial early contacts. From the start Clusius's Italian contacts were therefore not automatically limited to the learned elite.[11] Apart from the inevitable postal problems and the pervasive problems of political and religious controversy and censorship, there were few practical barriers, consequently, to prevent Clusius from getting in touch with whoever he wanted to befriend in Italy. That makes his selection of Italian friends all the more significant. We may assume that the men with whom he maintained contacts at such a distance were, to him, some of the most interesting experts on natural history in Italy.

Clusius was lucky enough to find two key figures who opened the world of Italian naturalists to him in the persons of Jean de Brancion in Malines and the aristocrat Giovanni Vincenzo Pinelli (1535–1601) in Padua. Both were instrumental in shaping Clusius's network in Italy. Already in the mid-1560s Clusius could assist Brancion in the latter's exchanges of plants and seeds with some important Italian collectors, *naturalia* lovers and *curiosi* of that period, such as Giacomo Antonio Cortuso in Padua, Alfonso Pancio in Ferrara, the Contarini family in Venice, and perhaps even the papal physician Ippolito Salviani in Rome.[12] It was indeed Brancion who first introduced Clusius in 1565 or 1566 by letter to Italian collectors and naturalists, and set the tone for decades to come. The geographical focus of Brancion's own contacts (north-eastern Italy) always remained the core area of Clusius's contacts.

It can hardly be a coincidence that the correspondence between Clusius and the second key figure, Pinelli, commenced in 1575, the year of Brancion's death. It ended only with Pinelli's death in 1601, and from the 82 extant letters from Pinelli to Clusius, all of them written in Italian, we can tell that the frequency of their correspondence was relatively high from at least 1589 onwards, with 3–4 up to even 14 letters from Pinelli to Clusius per year.[13] Pinelli was a humanist with a strong interest in natural history, a collector of books and ancient manuscripts with one of the largest and best private libraries in Italy as well as his own botanical garden. His collection also comprised globes, maps, fossils, metals and scientific instruments, particularly in the domain of astronomy, optics and mathematics. The wealthy Pinelli, who was born in Naples to a family of Genoese background, studied law in Padua and remained there for the rest of his life, attracted by the stimulus of the university circle, the botanical garden and the vicinity of Venice as the centre of the book trade in the Mediterranean. Pinelli acted as patron, mentor and intermediary for many Italian scholars and scientists, including the young Galileo. He was personally in touch with more or less everyone in Italy who counted as a humanist, scientist or naturalist, and he put all of his collections at the disposal of what was no less than an informal private academy – a centre of erudition based at his home – which he coupled with a European network of exchange.[14]

Clusius's Italian scope widened enormously thanks to Pinelli, who brought many new and potentially relevant Italian figures to his attention, and for decades acted as intermediary. However much Pinelli knew about natural history and gardening, and even though he exchanged plants with several other Italian collectors and regularly asked Clusius for seeds of plants that he found hard to get in Italy or for information about a particular exotic rarity, plants were not his main field of expertise. That also seems to have been his own opinion – which forms an important indication that fine distinctions were being made at the time between the expertise of the wide-ranging connoisseur, which he undoubtedly was, and that of the more specialized experts who were focused on the study of living nature. When Pinelli told Clusius that a narcissus alexandrinus loved dry and stony soil and would not produce flowers under different conditions, he added: 'according to our gardeners' (Pinelli, 31 March 1598). On another occasion Pinelli deferred the question of whether some bulbs had been taken out of the soil at the right time to 'intendenti della professione', that is 'those who have a professional understanding of such matters' (Pinelli, 31 December 1598). Neither Clusius nor most of his naturalist friends would ever have made such remarks; they themselves were the experts and relied on their own observations and experience.

Via Pinelli gifts of rare plants reached Clusius, such as bulbs from Cortuso in Padua, some of which produced up to fourteen flowers per stalk, 'leontopetali' and narcissi from Pinelli's secretary Paolo Gualdo, and some thirty speckled narcissi from Cardinal Federico Borromeo (Pinelli, 27 July 1593).[15] Clusius also received tributes of a different kind, such as the invitation from the Grand Duke of Tuscany to make use of his garden (the Pisan *hortus*), while it was hinted that the Duke of Ferrara might be interested in his services.[16] Yet, his contacts with ruling princes and other members of the very highest aristocracy in Italy remained indirect, even if they were the owners of great gardens. The very limited role of his contacts with some of the better-known Italian naturalists is striking as well. With the most famous (and famously cantankerous) Italian botanist of his age, Pietro Andrea Mattioli (1501–77), Clusius never seems to have maintained any contacts at all.[17] The relative insignificance of his exchanges with Ulisse Aldrovandi (1522–1605) in Bologna is far more surprising given the latter's collection, exchanges with many naturalists, teaching of natural sciences (since 1561) at the university of Bologna and first directorship of its botanical garden, not to mention the fact that he was almost exactly Clusius's contemporary and that the two men shared so many interests.[18] The extant letters between Aldrovandi and Clusius show that their contacts remained incidental and very intermittent, albeit friendly in tone, during a period of three decades (1569–96), and are rarely remarkable from the perspective of expertise formation.[19]

Private Botanical Gardens of the Aristocracy: The North

Neither the very top of the Italian elite nor that of the academic world of naturalists in Italy was, therefore, among Clusius most relevant contacts. A closer look at the north-eastern part of Italy will give us some idea of those men whom he (or Pinelli and he) regarded as most important to him as correspondents, and the setting in which they operated.

As seen from Europe north of the Alps, Italy may have looked like one country, but in Clusius's age it was anything but united. In terms of political entities, the emphasis of Clusius's contacts was on the Republic of Venice, which included the great port of Venice and the hinterland of the Veneto with the university city of Padua and the smaller towns of Vicenza and Verona, and on the adjacent territory of Ferrara. The supreme power in the Veneto lay not with an individual prince but with the body of the Venetian Senate, which also had a decisive voice in the appointment of professors – including those of 'botany' – at the university of Padua and of the prefects of its botanical garden. Venice was crucially important in the importation of spices, herbs as well as exotic plants from the Middle and even the Far East. Prominent men from wealthy families in the whole of this region were interested in obtaining rare plants for their city gardens or summer estates, and so were the prefects of the Padua university garden and many collectors-naturalists and apothecaries.

In Padua, Verona and even more so in Venice, where space for gardens was especially at a premium, it was common for elite families to have estates and gardens outside town besides smaller town gardens. Some of these estates, although less well known than the Medici, Este or Farnese ones much further south, were real private botanical gardens where the owners cultivated both rare native plants and the exotic *naturalia* they obtained via exchanges with friends. Some of them were probably as well stocked as the university garden in Padua.[20] Members of the influential Venetian family of the Contarini were, for instance, known for their special gardens. One of these was located at Dolo, conveniently halfway between Venice and Padua, another in the environs of Verona. Giacomo Contarini (1536–96) was a leading member of Venetian and Paduan cultural circles, owner of a renowned private museum, collector of books, manuscripts and scientific instruments, patron and friend of Galileo and Palladio. Such interests and friendships immediately remind us of Pinelli, with whose circle he was indeed closely connected and with whom he exchanged information about rare plants, such as the ornithogalum from Constantinople that looked like a 'Persian plume' (Pinelli, 2 July 1591).[21] Another member of the same family – Nicolò Contarini (1553–1631), who was elected doge of Venice in 1630 – possessed 'vast' gardens in the environs of Verona. His datura was the model for an illustration in a work by the apothecary and collector Giovanni Pona from Verona, and Pona also saw the 'very new plant ... *bambage Indiano*' (probably Ceiba pentandra L., also known as kapok or silk-cotton tree) in Contarini's garden.[22] Yet another impressive private botanical garden in this area belonged to Cesare Nichesola (1556–1612), canon of the cathedral of Verona and a collector of both plants and antique inscriptions, the latter of which were displayed in his garden.[23]

There were also city gardens in this area, however, and some of these too were experimental gardens and botanical collections as much as pleasure gardens. The most famous of these must have been the garden of the nobleman Pietro Antonio Michiel (*c.* 1510–76) at San Basegio (parish of San Gervasio) in Venice. Michiel reminds us very much of his exact contemporaries Charles de Saint Omer and Jean de Brancion in the Southern Netherlands. His botanical and horticultural expertise was such, indeed, that he acted (on the request of the Republic of Venice) as supervisor for the Padua university garden and as advisor to its first prefect, Luigi Anguillara, in 1551–6. Michiel's own garden was a living collection where he grew indigenous as well as exotic species and an experimental garden. Like Saint Omer, Michiel recorded the rare plants that he grew as well as those about which he received information in the form of drawings or dried parts of plants from friends in Italy and abroad. From 1555 until his death in 1576 he had more than a thousand plants depicted in various albums, in which they were roughly grouped by the shape of the root or stem.[24] In Padua some of the private botanical gardens were at least at interesting as the university one. In 1587, for instance, Seseli tortuosum L., Pistacia terebinthus L. or turpentine tree, Lithospermum officinale L., columbines, many types of tulips and iris, double malva, various oleanders, clematis and acanthus mollis were growing there in private gardens, as well as large numbers of imported plants from Crete and Constantinople. The presumed presence of the coffee plant in a private Paduan garden in 1587 is especially interesting since its description of 1591 by Alpino is supposed to be the first one by a European writer.[25]

One of the best city gardens in Padua was the Bembo garden, created by Monsignor Torquato Bembo, son of the humanist and cardinal Pietro Bembo (1470–1547). Even compared with other gardens in northern Italy it must have been a notable collection, since a publication of the mid-sixteenth century which discussed the best gardens of that age described it as 'miracoloso'.[26] In 1590–1, the Bembo garden hosted a number of exotic plants that belonged to a young man whom Clusius regarded as one of the most promising experts on plants in Germany: Joachim Jungermann (1561–91), the nephew of Clusius's close friend Joachim Camerarius II and the presumed artist of the famous *Camerarius Florilegium.*[27] Jungermann – who died shortly afterwards on his way to Crete for a botanizing trip in Greece and the Levant – saw exotic potatoes growing in Torquato Bembo's garden and described him as a man with an 'admirable passion and eagerness in acquiring exotic plants' (Jungermann, 10 January 1591).

And in 1595 Pinelli put Clusius in touch with Orazio Bembo, who was by then taking care of this garden, so that Clusius could obtain seeds of rare plants from it for the Leiden *hortus*. The two men exchanged plant lists, and Bembo had no trouble in finding twenty-four species in 'our garden' of the list with Clusius's desiderata (Bembo, 13 November 1595).[28]

Such north–south exchanges of plant material were relatively common in the late sixteenth century and could involve large quantities of plant material. Around the turn of the century the nobleman Leone Lazzaro Haller von Hallerstein, commander of the castle of Piacenza, created some very impressive gardens in Piacenza and was familiar enough with the Low Countries to order plants from there for his garden, while he also had many of his flowers painted by a Flemish artist. Von Hallerstein's contacts with the Southern Netherlands seem to have dated back to the period in which he served in the campaigns of Alessandro Farnese, Duke of Parma and Piacenza. From 1592 he was in the service of the latter's son Ranuccio I Farnese (1569–1622). Von Hallerstein, although 'still a beginner and not yet an expert in botanical matters', showed 'a vivid and ardent desire both to practise and to be of service to you', as the *semplicista* Fra Gregorio da Reggio, who provided him regularly with plants, told Clusius.[29] Von Hallerstein was a man of great gardening skills and delighted in growing new things in his garden:

> He has the most wonderful and abundant garden of our days in so far as flowers are concerned; and if you had seen the great number of bulbs you would have been amazed, since the number of bulbous plants alone already exceeds 8,000, and they are the most rare and exquisite which can be found in the world. He has spared neither expense nor effort to have them brought from various countries. In fact, I do not know where he obtained a certain jonquil with a big double calyx, of which he had some 25 plants or rather bulbs. I have never yet seen it described by any author. I leave this to your judgement, because there were hundreds of each kind of rare bulbous plant, not to mention those with other types of roots which were likewise the most beautiful ones that can be seen.[30]

The personal involvement of Ranuccio I Farnese with gardens and gardening was just as impressive. He personally created a garden dedicated to his wife, Margherita Aldobrandini, with the plants given to him by Haller von Hallerstein, and 'it was he who transported everything and wanted to plant it with his own hands in that garden'.[31] Like Haller, Ranuccio took part in large-scale plant exchanges with the Southern Netherlands. In 1608, for instance, the Duke of Arenberg at Enghien had been informed that the Duke of Parma was sending a mule train with plants to the Netherlands, which also carried some orange and lemon plants as well as jasmines of Valencia. Arenberg told Matteo Caccini in Florence that the latter could add some of his own plants to the transport, while

Arenberg himself would send some 300 jonquil bulbs back with the Duke of Parma's men on their return journey.[32]

Clusius heard about almost every one of the above-mentioned gardens or their owners via the letters of Pinelli and his other contacts in Italy, and he was offered seeds from several of these gardens.[33] But only with Orazio Bembo and Haller von Hallerstein did he have brief exchanges. Given the impressive plant collections of both these men, Fra Gregorio's praise of Haller von Hallerstein, and the fact that Bembo was also hosting some of Jungermann's exotic plants, we must conclude that Clusius was only interested in direct exchanges with wealthy garden owners if these were real plant experts themselves and had access to extremely special plants.

Nature Study in the Veneto

Clusius's most interesting direct contacts in the Veneto generally belonged to rather less elevated social spheres, even if they shared the fascination with collecting and a passionate interest in rare plants with the highest circles of the aristocracy. Giacomo Antonio Cortuso (1513–1603) in Padua links the categories of (aristocratic) collector and professional. Cortuso was a patrician and garden owner who may have trained as a physician and certainly had a great interest in matters of medicine. He operated as a naturalist with an international network of exchange, but seems to have had no official connection with university life in Padua until he was appointed prefect of the Padua botanical garden (1590–1603) at the end of his life by personal order of the Venetian Doge. He was an excellent field botanist and procurer of plants which he sent to other naturalists as well as to elite garden owners.[34] Cortuso had his own garden in or near Padua and already in this period had his own international network of exchange in which some of the foremost naturalists of his age took part, such as Mattioli and Aldrovandi, Lobel (who in 1564 or 1565 did some field research in northern Italy and admired Cortuso's garden) and the Swiss naturalists Conrad Gessner and Jean Bauhin.[35] With all of them Cortuso exchanged plants, fruits, distillations, seeds and drawings. During the 1560s and early 1570s Cortuso was also one of the Italians who sent rare bulbs to Northern Europe. Brancion accepted, for instance, a dipcadi and a 'bulbus eriophorus' from Cortuso.[36]

After their introduction by Brancion in the 1560s, it took Cortuso and Clusius no time at all to set up a regular exchange of *naturalia* and their contacts eventually lasted for almost three decades (1566–92).[37] In 1567, for instance, Clusius sent Cortuso a selection of plants or seeds about which Cortuso wanted to know more: were they umbelliferous, annual or perennial, and what were their medicinal properties? Cortuso in turn sent many seeds, plant drawings (including one of a green branch of the Lebanon cedar with its fruit), dried plants, and some fossils ('petrified animals' or 'petrified shells and fish') found 'inside large and very hard rocks in our highest mountains of the areas near Padua and Vicenza' (Cortuso, 27 August 1567).[38] He had collected much of this material during field trips, which extended to various mountainous parts of Italy and to Crete (1568). An Alpine plant belonging to the primula family which he sent to both Mattioli and Clusius was named after him and is still called Cortusa Matthioli L.[39]

The packages sent to Brancion and Clusius in Malines sometimes contained significant quantities of seeds, bulbs and dried plant species; the latter can only have served research purposes. In the spring of 1569, for instance, Cortuso sent off a package with 75 dried plant species, well preserved between sheets of paper. That list occasionally mentions the medicinal uses of the plants as well as their folk names, and Cortuso's letter too indicates that he took a great interest in herbal medicine. A small part of it runs as follows:

> – the coronary forest rosemary which is always fragrant
> – sweet marjoram ...
> – the lunaria of the Arab healers
> – hippophaes of some spiny plant which produces a milk that I do not like
> – Levantine helicrysum
> – capers
> – Pliny's thinnus wood laurel
> – the juniper cedar, that is the Phoenician cedar as written by Mattioli
> – Italian forest chick peas. (Cortuso, 27 March 1569)

The list contains – indiscriminately mixed together – herbs and edible plants growing in Italy, wild plants growing in Italy without a known medicinal use, and exotic plants. Among the latter we find, for instance, a plant of 'which the tubers or the hairs are used in the Indies for very soft beds or pillows', seeds of the 'fico d'inferno' (Mexican poppy), and nasturtium indicum (Cortuso, 27 March 1569).[40] But the majority of names on the list belong, as far as we can tell, to the Mediterranean wild flora. Few them can be classified as decorative rarities suitable for aristocratic show gardens. Some thirty years later, when Cortuso had become prefect of the botanical garden in Padua, his interests still seem to have been extremely wide-ranging, even if the indigenous flora of Italy and the eastern side of the Adriatic continued to be one of his favourite topics. And even when Cortuso was a relatively old man, in 1592, he was still planning a research trip to 'Schiavonia' (Slovenia) to stock up the Padovan garden.[41] Cortuso was therefore an all-round naturalist rather than a specialized one. Clusius's frequent references to Cortuso show that he regarded the latter's information about Italian plants and other *naturalia*, much of which was based on personal observation in field and garden, as extremely reliable.[42] In his printed works Clusius included, for instance, a picture and the lengthy description of a marine plant based on Cortuso's information and on the dried parts of the plant which Cortuso sent him (Figure 5). Cortuso called it myriophyllum pelagium and also gave the folk names of Neptune's palm or Plume of the Nymphs. It grew in deeper parts of the sea, attached to the sand or mud of the sea bottom by its root and many hair-like fibres. Its tips contained many small 'nuts', as Cortuso informed Clusius, that did not contain seeds but very young plants, ready formed and similar to their 'mother', which fell off and sank towards the depths of the sea, to start growing from there. So long as it was green and attached, the stalk of this plant remained flexible, but cut off and dry it became 'as fragile as glass or coral'.[43]

Respect for high-quality information based on experience – especially out in the wild and in the museum – likewise formed the foundation of the relationship between Clusius and the Veronese apothecary Giovanni Pona. While Clusius's contacts with Pona's colleagues, the Calzolari family, remained limited to a brief exchange in 1596 concerning the inventory of the famous museum created by Francesco Calzolari (1521–1600), who was by then old, almost blind and retired to his country house where he 'takes a pleasure in cultivating his herb garden', his exchanges with Pona were frequent and intensive.[44] Pona too had created an impressive museum, which he called his 'repository', in Verona, and was involved in exchanges with a large number of naturalists and garden owners in Italy (such as Belli, Contarini, Fra Gregorio da Reggio, Imperato and Alpino), as well as some in the Netherlands (Lobel and Paauw) and Switzerland (Bauhin).[45] Pona and Clusius corresponded mainly about exotic and often dried *naturalia* and spices that were potential objects for the museum: 'particular items that are so much to my taste, exotic and singular' (Pona, 24 January 1606). Pona also hoped that in Amsterdam Clusius would manage to find 'a good choice of exotic things for me, because my repository hopes to be enriched from those parts by means of my Sr. Clusius' (Pona, 15 July 1603). Many of these items had a medicinal use, such as white cinnamon, nutmeg, costus indicus, cloves from the Moluccas or Winteranus cortex from Patagonia. The emphasis on the exotic in their correspondence and the fact that nearly all of the references to Pona in Clusius's works likewise concern exotic *naturalia* should not prevent us, however, from recognizing that Pona also had great expertise on the wild flora of the area surrounding Verona.[46] He had the favourite botanizing location in the whole of Italy within easy reach: Monte Baldo, a 2,218-m mountain overlooking Lake Garda, famous for its enormously varied vegetation. Clusius recognized the value of Pona's observations on those wild plants to such an extent that he included a Latin version of Pona's description of a field trip from Verona to the top of the Monte Baldo as an appendix in his own *Rariorum* of 1601.[47] In the following years Pona's original text expanded enormously: he inserted the findings of new field trips to the same mountain, and added further descriptions of plants, new illustrations and asides to plants that had been given to him in the meantime, while still maintaining the structure of his story as if it were the

Figure 5. Cortuso's marine plant Myriophyllon pelagium, also known as Neptune's palm or Plume of the Nymphs. From Clusius, *Rariorum plantarum historia* (Antwerp, 1601), p. ccli. Collection Leiden University Library, UBLWGW_CCLI.

recording of one journey to the Monte Baldo. In the Italian edition of 1617 the report of the trip alone takes up some 250 pages.[48]

Pona was by no means the first to go botanizing on Monte Baldo – a tradition of botanizing there went back at least half a century, and by the end of the sixteenth century Monte Baldo was visited almost every year by small groups of naturalists, both Italians and foreigners – nor was he the first to publish on it. In his much shorter account of 1566 Francesco Calzolari was the first to describe the plants of this mountain in a narrative that follows the itinerary up the mountain and describes the plants as they grew in their specific location on the Monte Baldo.[49] Pona took that mode of describing observations *in situ* further, paying even more attention to habitat and the riches of plant species, and turning his work in essence into a long report on plants in their places. Interestingly, Pona's work starts immediately outside Verona, and therefore also tells us about the cultivated fields and the foreign (in particular Cretan) plants grown in the private botanical gardens in this area. Continuing his journey up the slopes, Pona describes the individual species, nearly always mentioning their flowering period, whether they were evergreen or not, annual or perennial, and what their taste was like, while also giving long plant lists per location. In order to avoid repetition he presents plants in the places where they are first encountered on the way up the mountain or where they were most noticeable. For the highest summits, therefore, only those plants are mentioned that had not been already been encountered lower down:

> [on] the highest tops which are sterile to any other plant except for our smaller rock Trachelio which hanging from those almost falling rocks, grows there from a woody, frizzy, scaly and pale root, from which many leaves sprout, of which each and every one hangs from a petiole with a length of a palm and a half.[50]

Pona was extremely sensitive to place and location on the mountain, therefore, and perfectly aware of the great differences made by altitude and by a position on the sunny or the shaded sides and spots of the mountain. In many cases he also pays attention to humidity, exposure to wind or frost and snow. More than two centuries before Alexander von Humboldt and Aimé Bonpland published their work on plant geography in the tropics (1807) with its famous map in which vegetation zones were combined with altitude, both Calzolari and Pona thus sketched similar, if less systematic images in words.[51]

The Veneto and Crete

Whereas Cortuso, Calzolari and Pona were all involved in relatively short field trips, which must have ranged from a few days to several weeks, the physician Onorio Belli (1550–1604) from Vicenza concentrated on the systematic and long-term regional survey.[52] Born in Vicenza to a family of architects, physicians and apothecaries who were interested in collecting and closely connected with the Palladio family, Onorio Belli studied medicine in Padua during the 1570s, and maintained lifelong connections with Cortuso and the Pinelli circle. It was indeed Pinelli who suggested that Belli contact Clusius. Belli left in 1583 for Crete – which was then ruled by Venice and had a large Venetian colony – as physician in the service of the Venetian governor general (Provveditore generale) of Crete. Even after his patron's return to Italy he stayed on in Crete as town physician (*c.* 1595–9) of Chania. He eventually returned to Vicenza and Padua.[53]

Both Belli and his patron were seriously interested in antiquities and *naturalia*. Already from the very first two years of his stay on Crete Belli undertook an ambitious twin project: to investigate and chart as much of the flora and the ancient Greek monuments and inscriptions on Crete as he could manage, and to publish them.[54] Crete had by then been one of the favourite goals of early modern naturalists for decades – a botanical paradise on account of the huge variety of plant species that could be found there on a relatively small surface as well as its large number of endemic and relict species and a perfect research location for what was one of the primary goals of field research for most sixteenth-century naturalists: the identification or rediscovery of the (medicinal) plants of antiquity, in particular those described in Dioscorides's *De Materia medica*.[55] Belli too was 'searching Crete for the genuine plants of the Ancients so that I would be useful to the Respublica and would communicate my findings to friends' (Belli, 24 September 1593). None of the other sixteenth-century naturalists spent more than a limited period of time on Crete, however, or undertook a systematic investigation of its flora, and Belli's field research soon took him beyond classical *materia medica* into an even more wide-ranging project of regional exploration, identification and description. As Belli put it, looking back on his Cretan period, 'we combed the whole of Crete and various other regions to find out about plants' (Belli, 28 March 1602). Had Belli been more fortunate, his botanical work would have resulted in the first regional flora of Crete, just as Clusius's own field trips resulted in the first Spanish and Pannonian ones. Belli never published his work on Cretan botany, however, and we only know parts of his writings on this topic via his correspondence and the works by Alpino, Pona and Clusius. Belli did finish a work on Cretan antiquities, but this too is lost.[56]

Belli was a truly innovative 'archaeologist': the first to identify Candia as the old Heraklion and to describe Knossos; the first to study the geography of Crete in some detail and to make careful drawings of the ruins. His research methods were the same, and his investigations of the local flora were as thorough, observant and careful as his exploration of the ancient Cretan ruins. Belli relied above all on first-hand observation and asked the local population for information about both ancient ruins and the local vegetation. Of the *c.* 400 plant species known for Crete Belli discovered some seventy, and some of these, such as a cannabis-like plant, as he wrote, had certainly not been known to the ancients.[57] Belli's incredibly detailed descriptions of wild plants first of all served the purpose of correct identification. He spared no trouble in this respect, resorted frequently to plant comparison, was extremely critical – combining common sense with local information and a philological approach – with respect to both classical and contemporary authors, and also referred to plant characteristics (such as scent) which are not that often mentioned by other authors.[58] Discussing, for instance, the plant called Aspalato II by Dioscorides, Belli explained that it could be found all over Greece and Crete and was known there as 'aspalato'. He had gone to great lengths to find out whether there were scented types:

> while the woody roots and the bark are completely odourless, and I have had more than two thousand plants cut off or dug up in diverse parts of the island but have never found one that has the least bit of scent, nor can anyone be found in Candia who has ever seen a scented one.[59]

Belli's letters often include some information about the local use of plants. The chamaeleon albus, for instance, is

> gummi or mastix producing, which grows abundantly; it has a very big and strong smelling root. In June and July a droplet of gummi emerges which the children collect and of which they chew a little bit, afterwards rubbing it with their fingers and smoothing out; folded together in the shape of a bubble they push it over their other hand with force, so that the bubble bursts because of the blow and produces a noise; in which kind of game the children take an enormous pleasure. (Belli, 24 September 1593)[60]

While Belli surveyed the indigenous plants of Crete in the field, he experimented with exotic plants in his garden, growing them from seeds which he received from Egypt, the Middle East and East Africa. In this respect Crete was an ideal location as well: a stop-over on the shipping and trading route to Venice of spices and many other types of precious merchandise from the Middle and Far East. Experimenting in the garden led to Belli's descriptions of some curious phenomena. The first concerned the baobab (or abavo) tree, of which he had received seeds in 1595 via a port in Eritrea. Within a year, the seeds had grown to plants of 1.5 feet, which in summertime could be put out in their pots in the garden. There, Belli observed that each of the four plants which had grown from the seeds of a single fruit produced a different type of leaves (Belli, 15 February 1596, 15 August 1596). Even more interesting was his observation of a particular form of heliotropism in a plant which he called 'Bean from Jemen' or 'Abrus according to Alpin' (probably the Abrus precatorius).[61]

> And they turn so that they face the sun: for the leaves which are pinnate, that is, with a long peduncle joined to the opposite petiole and always with an even number, are very closely pressed together by sunset on the part that is turned away [i.e. from the sun], so that the points of the leaves face the ground, and the long peduncles to which the leaves are attached bend towards the earth during the night, so that they just touch the stalk, and remain like this all night; and when the sun rises they open up and the peduncles turn themselves on their spot almost at right angles from the stalk. By midday the leaves gradually start to face the sky, and during the later part of the afternoon they in their turn close very tightly together, so tightly that the one touches the other, and the tips [of the leaves] face the sky in exactly the opposite way to what they did during the night; afterwards they open up when the sun goes down. This periodical turning around of the leaves to the sun I regard as wonderful and worthy of observation, and on that account I thought it of particular significance for you. (Belli, 15 August 1596)

Clusius did not only regard this particular account as worthy of attention, but found Belli's observations on both the wild plants of Crete and the exotic *naturalia* grown in his garden so valuable that he published a whole group of Belli's letters integrally as an appendix to his own *Rariorum*, just as he did with Pona's writings on the Monte Baldo.[62]

Although Belli's work on Cretan botany was never published, the results of his botanical investigations thus did emerge in other ways. Some ended up as text in Clusius's printed works. Physically, the plant material sent by Belli took root in the richest gardens of the Veneto – such as those of Pinelli and Bembo in Padua, Pona in Verona, the Contarini outside Verona – and in the gardens of his contacts abroad: Clusius in Holland and the Bauhins and Platters in Switzerland.[63]

Conclusion

Northern Italy combined easy access to rare *naturalia* from abroad (Crete, the Middle East, parts of Africa and the Far East) and good postal connections to Northern Europe via the port of Venice with the presence of an influential university tradition and a great number of wealthy and erudite noblemen with a passion for collecting. Many of the latter were the owners and patrons of gardens of the living collection type and highly appraised rare *naturalia* and other curiosities. Some also collected antiquities, maps, scientific instruments, books and rare manuscripts. Most often the wealthy garden owners relied on others to find rare *naturalia* for them in Italy and abroad, organize their gardens or act as consultants. Only a few of them became horticultural and botanical experts themselves, thus uniting the roles of owner, patron and expert. These were the persons who interested Clusius most, on account of their first-hand access to wild plants and their expertise on *naturalia*. Clusius's selection of relevant correspondents – attaching most importance to who was closest to the original plants in their setting – was thus consistent with every other aspect of his methodology. As in the Southern Netherlands, the private botanical gardens and collections of the period which generally belonged to wealthy aristocrats played a crucial part in the shaping of botanical expertise: as collections such gardens (or rather their owners) were in constant need of novelties, while the successful growing of the newly introduced plants required new types of expertise that were partly botanical and partly horticultural.

A small number of men among Clusius's correspondents in northern Italy – all of them bourgeois professionals connected with medicine – appear to have had no explicit connection with the gardens of the aristocracy. Several of them created specialized collections of *naturalia* themselves and all were engaged in making surveys of the plants they investigated in the wild. What united them with the persons who acted as horticultural or botanical consultants for aristocratic garden owners was their first-hand knowledge of nature. Their expertise was always connected with various types of botanical field work and with a talent for observation, a great interest in identification and a desire for novelty.

6 NATURE IN THE GARDEN: THE RISE OF THE EXPERT NATURALIST

In no other part of Italy were Clusius's contacts so concentrated, intensive or long-lived as in the Veneto – in the sense of both direct exchanges and indirect information about other garden owners, collectors and naturalists. In the rest of Italy his contacts were fewer and farther between in a geographical sense, and for the most part concentrated in the 1590s and 1600s.[1] That, of course, was primarily a result of Pinelli's role and the way in which Clusius's network functioned. Clusius's other Italian contacts were as interesting, however, in terms of their expertise and specific interests.

Botanical Advisors to the Ruling Families

Members of various religious orders – and the Capuchins in particular, as Olmi has shown – played an important part in research concerning living nature in Italy.[2] Two of them were especially relevant to Clusius, and acted as horticultural and botanical advisors to some of the most important ruling families of their time. Evangelista Quattrami (b. 1527), an Augustinian friar from Gubbio in Umbria with a doctorate in theology, acted as chief herbalist and distiller for several members of the Este family from about 1569, both in Ferrara and on their estates in Rome and environs. The gardens with which Quattrami was involved were, indeed, among the most prestigious locations in the whole of Italy. During the late 1570s, he acted as superintendent of the Este garden on Monte Cavallo (Quirinal) in Rome. Shortly before, he had directed (or advised on) the planting of the Villa d'Este in Tivoli near Rome, which was then being constructed on the orders of Cardinal Ippolito II d'Este.[3] During the mid-1590s he organized yet another Este garden, where he 'has so well ordered the plants, and ordered them by species that it will be a consolation to see such an organized garden and with so many things in it'.[4] He was unfortunate in his service of the Estes, however. In 1598, after the end of the rule of the Estes over Ferrara, Quattrami told Clusius that he was forced to return to a monastery in his home region after so many decades of service to the family. Moreover, not a trace was left of the ducal 'giardino dei semplici' when soldiers were quartered in the fortress where it was located.[5]

Very much in line with the experts we have found in the Veneto, Quattrami was both a supervisor of gardens and an expert field botanist and explorer with a first-hand knowledge of many mountainous zones in Italy and even some abroad. The rare wild plants that he found ended up in the Este gardens but were also given to the many friends and naturalists (such as Aldrovandi) with whom he was involved in exchanges. Quattrami is said to have visited Constantinople and certainly explored many mountainous zones in Italy, the Dalmatian mountains and Andalusia to find new botanical species.[6] During the 1560s, for instance, some sixty years before members of the Accademia dei Lincei went herborizing on the Umbrian plateaus of Norcia and Castelluccio, Quattrami had already been part of small plant-collecting expedition in that area together with Luigi Anguillara (then prefect of the botanical gardens of Padua) and several other herbalists.[7] And in 1596 Quattrami came back from explorations in the Apennines and the Garfagnana with 300–400 alpine plants. Once the intense summer heat was over, he planned to give them a permanent place in the Este garden 'following their order'.[8] Quattrami was very interested in the correct identification of wild indigenous plants and serpents, but also developed considerable expertise concerning exotic plants, roots, fruits and spices (especially various kind of peppers) and their uses.[9]

The field of operations of the Capuchin friar and *semplicista* Fra Gregorio da Reggio (d. *c.* 1618) was even larger than that of Quattrami. Fra Gregorio was originally from Reggio Emilia and for many years took care of the garden, pharmacy and infirmary of the monastery Monte Calvario at Bologna, while he also spent shorter periods in other Capuchin monasteries and in Piacenza. The learned Fra Gregorio's botanical interests ranged far beyond medicinal plants and his field trips extended to nearly all parts of Italy, and even beyond. He used to go off almost every year into the Alps to collect the best plants available, and was very familiar with the flora of the mountains near Brescia and many parts of the Apennines.[10] While Fra Gregorio's field trips may have originated in his work as *spezialista* for the Capuchins, their goal was often to find rare plants for the gardens of noble patrons. Unlike Quattrami, however, whose almost lifelong patrons were the Estes, Fra Gregorio worked for several different noble families, almost as a kind of freelance procurer of rare plants cum horticultural consultant. Among his patrons were Haller von Hallerstein in Piacenza, the ruling Gonzaga family of Mantua, and Cardinal Farnese, who was stocking the illustrious Farnese gardens on the Palatine hill in Rome. In July 1608, for instance, Fra Gregorio wrote from Innsbruck to Vincenzo Gonzaga, 'I have been busy for two weeks in these Alpine mountains to discover something rare for your garden'; some ten months later, after another trip to Germany, he returned to Mantua, but was most upset to see that the plants which he had discovered with great effort 'on the harshest mountains' had been badly neglected.[11] In Fra Gregorio's case his fieldwork for noble patrons was accompanied by plant hunting for sheer pleasure and the interest in discovering new *naturalia*. Some of his plants were, in fact, collected on purpose to be sent to Clusius and published by him.

> The past year I have been in the mountains of Brescia where I have discovered rare plants which do not grow in my Apennines. Now I am hoping to go next season to Sardinia, a truly most famous island, where I believe that I will find without any doubt a great many plants, both marine ones and others, which have not yet been described by any Author, because I have seen no one who discussed them, nor has anyone made these journeys and described everything that can be found on the island. (Fra Gregorio, 9 March 1606)

The suggestion to go to Sardinia had come from his fellow Capuchins, who had told Fra Gregorio that coral was fished there and that fishermen occasionally found 'exquisite objects in the mountains that are at the bottom of the sea' and saw 'various plants and trees' there (Fra Gregorio, 9 March 1606).

Fra Gregorio made detailed and accurate observations both in his garden and out in the wild.[12] He was, in fact, probably one of the very best field botanists in Italy and certainly one of Clusius's most intelligent and observant Italian correspondents, who moreover created a personal herbarium which must have mainly served his own research. Among the rare plants which he grew in his garden, Fra Gregorio seems to have had a special fondness for a particular group of plants from the New World: he had some 25 different kinds of capsicum or 'Indian pepper' in the garden of his monastery at Bologna,

> each of them different only in the fruits which I have observed for several years in our garden of Monte Calvario at Bologna, and a marvel to see such a great variety of silique, and they create such a beautiful harmony for the eye when all of them are ripe that it is stupendous to see such a sweet sight ... I know of no one but me who has observed such a variety of Indian peppers. (Fra Gregorio, 9 March 1606)

The capsicums were not the only American plants there. During the dog days of summer, Fra Gregorio used to go into the garden after dinner to get a bit of fresh air. Every evening after the Ave Maria he saw some

> big butterflies with a long beak which they inserted now in one flower then in another, and I believe they continued doing so for a large part of the night, for I have observed them even until two o'clock in the morning. The flowers which they visited were the *stramonia*, *nicociana*, *noce metelle*, and *meraviglie di Spagna*, and I believe that they took their food more from these flowers than from others. Although I tried to catch them many times, I did not succeed in seizing any for fifteen consecutive evenings; finally, with a kind of net, I have captured five or six of them which I am sending to you, and I have given them the name Corpulent Night Butterfly.[13]

The names of the plants visited by these moths give us a further glimpse of the special plant collection that was once housed in Fra Gregorio's monastery garden. *Stramonia* must be Datura stramonium, whose flowers remain closed during the daytime, open completely at night and emanate a penetrating smell that attracts moths, which pollinate them. *Nicociana* may have been real tobacco or more probably one of the purely decorative plants belonging to this genus such as Nicotiana alata: its flowers too open during the evening and night and are known to attract moths. *Noce metella* (Datura metel or downy thorn apple), a close relative of the Datura stramonium, is highly poisonous. All three, like the capsicums, the potato and tomato, belong to the family of the Solanaceae, of which many members are poisonous but may also be used in herbal medicine. The last of the four plants mentioned, the *Meraviglia di Spagna*, was the Mirabilis jalapa, also known as the Four o'clock flower or Marvel of Peru, which emanates a strong scent from late afternoon until sunrise and is often pollinated by moths. The prominent presence of attractive – though often poisonous – plants from the New World in Fra Gregorio's garden and his obvious attention to them suggest that he regarded his garden as a combined source of beauty, healing, food and knowledge on nature, a locus of observation and experiment as much as a living collection.

Fra Gregorio was, in fact, so interested in the capsicums that he had some of them drawn from life and wrote a treatise about their appearance, the variety of types, best ways of growing them and how to prepare them, in which he also refers to authors like Bauhin, Mattioli, Lobel, Cordus, Dodoens and Fuchs. Clusius, to whom he sent his treatise in 1609, must have been impressed in particular by Fra Gregorio's detailed observation and meticulous description of the appearance of the plants and the inner structure of the siliqua or seedpods. As in the case of Pona, Clusius translated the Italian treatise into Latin; together with thirteen accompanying illustrations the text was eventually included in Clusius's posthumous *Curae Posteriores*.[14]

Fra Gregorio's personal contacts and exchanges were not limited to Aldrovandi and Clusius, to whom he donated many *naturalia*, including dried plants, and with whom he shared numerous observations on plants and insects as well as tentative identifications.[15] Fra Gregorio also exchanged information and dried plants with other Italian naturalists such as Belli, Alpino, Quattrami, Corvino, Pona, Colonna and the prefect of the botanical garden in Pisa, Francesco Malocchi, and he even briefly mentions an exchange with a fellow apothecary in Antwerp, by the name of Adrianus Beyerlinck, who in 1602 offered to send him a complete catalogue of his city garden. The major part of Fra Gregorio's dried plants ended up, however, in his own vast herbarium, which he called his 'giardino di cella' (cell garden).[16]

Fra Gregorio and Fra Evangelista Quattrami had long-term working relationships with aristocratic families based in the northern half of Italy, but their field of botanical operations was much wider. The same was true of one of Clusius's interesting contacts in the Grand Duchy of Tuscany, Giuseppe Casabona (*c.* 1535–95), who served the ruling Medici family. In Tuscany, Florence and its environs were the focus of Medici court life and the location of some of the most splendid aristocratic estates in Italy, such as the Boboli gardens and the Medici villa at Pratolino, while Pisa had its own university and *hortus botanicus* as of 1544. The latter town had, moreover, for centuries been the centre of the trade in drugs.[17] In terms of both aristocratic gardens and the academic teaching of botany Tuscany was, therefore, quite as important as the Veneto. Yet Clusius's contacts in Tuscany remained far less extensive than in the Veneto until the very last three years of his life, when he found a new friend and exchange partner in the young nobleman, lover of rare plants and plant dealer Matteo Caccini (1573–1640) in Florence.[18] During the last decade of the sixteenth century and the first years of the seventeenth century Clusius mainly exchanged information with two prefects of the Pisan *hortus botanicus*: Casabona, prefect from 1591 until his death, and the Friar Minor Francesco Malocchi, prefect in 1596–1613. With the latter, the exchanges mainly concerned the plant lists of their respective botanical gardens in Pisa and Leiden, while Malocchi also promised Clusius a catalogue of the Pisan museum which was located right in the *hortus botanicus*.[19] The exchanges with Casabona were more intensive.

Casabona had been working directly for the Medici in Florence and Pisa since the early 1570s. Just like Quattrami or Fra Gregorio, Casabona procured rare plants for his patrons and made field trips for these purposes in Tuscany itself and in the foothills of the Alps, the area of Rome, the Apuan Alps, along the Ligurian coast and on Corsica and Crete (1590–1). He also donated the seeds of rare plants (many from Crete) to Clusius and the Leiden *hortus*, which was in the phase of planning and early construction in 1593–4, and to Aldrovandi and various garden owners and plant experts in Europe with whom Clusius too was in touch.[20] Those contacts included three very influential figures indeed: the naturalist Joachim Camerarius II, the prince-practitioner Wilhelm IV of Hesse-Kassel – to whom Casabona sent elegant and exotic plants – and the Brussels nobleman Jean Boisot, who was very well connected with a large part of the Habsburg governing aristocracy.[21] It was Casabona's great botanical expertise and his ability in procuring rare plants which put him in contact with Boisot, Wilhelm IV and Clusius in the first place, and made him such a suitable director of the *hortus* in Pisa which, as he wrote in the spring of 1592, was being (re)constructed on a new location (Casabona, 20 March 1592).

Casabona thus took the exercise of plant procuring one step further; literally, by also working for 'customers' (or potential patrons) abroad; and in terms of the developing character of the role of botanical consultant, by almost, but not quite, turning himself into a commercial plant procurer. That much we may deduce from the conflict that arose about the issue whether Casabona had or had not been compensated financially by Wilhelm IV for his services. Casabona stated that he had sent plants to Wilhelm for some seven years with almost no compensation, whereas Wilhelm (or his staff) maintained that Casabona had been paid, but that plants had not (always) been received. Perhaps, as Casabona suspected, Wilhelm's *semplicista* Joachim (Gille) had embezzled a sum meant as payment.[22] However, the interesting aspect of the conflict is rather that it points to this 'confusion' of idioms between a commercial bond – in which payment would have been demanded and made long before the seven years were over – and the bond of patronage, in which such long-standing 'debts' were not at all unusual, but which rarely functioned over such distances, while Casabona was, moreover, also in the service of the Medici.

In Tuscany the Medici were the ruling family. In Rome they were only one of the powerful aristocratic families involved in the establishment of prestigious villas during this period. After the sack of Rome in 1527, a major phase of (re)construction of villas and gardens started in the city and its environs around the middle of the sixteenth century. In what were then its outskirts, villas, summer houses, parks and gardens were created on almost exactly the same spots as those of the classical period, since locations were chosen on the basis of much the same criteria as in ancient times: availability of water, an attractive position, convenient connections with the city centre, a cool breeze during the hottest months of the year. Among the most famous villas of the sixteenth century are the one created for Pope Julius III around the middle of the century (now Villa Giulia) on the Via Flaminia; the villa of the Cesi family from Acquasparta (between the Pincio and the Quirinal, in the area of the Roman *horti* Sallustiani); the Villa Medici on the Pincio, on the location of the ancient gardens of Lucullus; and several villas on the Esquiline hill, where in former times the gardens of Maecenas and various other Roman patricians could be found. The effect was that 'by the end of the century the panorama of the city was not very different from that of late Republican and Imperial Rome: the places where once the *horti* could be found now housed the gardens and sumptuous residences of the Roman patriciate'.[23] Meanwhile, the gardens of the Vatican and several other papal gardens in Rome were restocked and expanded as well, and a new *Giardino dei semplici* was created at the Vatican in 1561, which by the late 1560s or early 1570s under the physician-botanist Michele Mercati (1541–93) had become a real botanical garden with some 470 different kinds of plants.[24] In 1581 Montaigne visited Rome, and he named as the most beautiful gardens first of all those of Cardinal d'Este on the Monte Cavallo (Quirinal) and the Farnese gardens on the Palatine.[25] By the turn of the century the latter were already known as one of the best places in Rome to see exotic plants.[26] It cannot be coincidence that two of Clusius's correspondents, who acted as botanical advisors to the principal ruling families of Italy, were closely connected with these outstanding gardens: Quattrami (1570s) as superintendent of both the Este garden on the Quirinal in Rome and the Villa d'Este in Tivoli; and Fra Gregorio as provider of rare plants to the Farnese gardens on the Palatine. And given his decades of service to the Medici, it also seems plausible that Casabona procured rare plants for the Villa Medici in Rome. While Clusius may well have heard about the big garden projects in and close to Rome via these three men, his direct contacts with Rome remained minimal.

Practice and Experiment: Urban Botany

In Rome the Dutch apothecary Enrico Corvino, the physician Joannes Heckius, Prince Federico Cesi and the Neapolitan nobleman and plant lover Fabio Colonna (1567–1640), who was moving back and forth between Rome, Naples and various villages in their surroundings, all formed part of a link that might have resulted in a formal connection between Clusius and the Accademia dei Lincei, but never did.[27] Instead, Clusius preferred frequent exchanges with the wealthy apothecary, collector and naturalist Ferrante Imperato (*c.* 1525–*c.* 1615) in Naples.[28] With its *c.* 300,000 inhabitants by the third quarter of the sixteenth century Naples was the principal port of the whole of the Mediterranean as well as one of the biggest cities in all of Europe in this period. It was also the capital of the Kingdom of Naples which comprised the whole southern half of mainland Italy and together with Sicily and Sardinia formed part of a large territory controlled by Spain and ruled by a viceroy. For much of northern Italy Naples was, moreover, a major source of delicate plants.[29]

In 1586 Pinelli, himself originally from Naples, introduced Imperato to Clusius with the following recommendation: 'I want to say that Sr Ferrante Imperato, a renowned apothecary at Naples with great expertise concerning simples, will be more than ready and most active in serving you if you would wish for something from there' (Pinelli, 16 May 1586).[30] A first direct contact followed in October 1588, when Imperato wrote to Clusius in order to offer his services and ask to be included among his friends. Appropriately, he followed up his letter with the gift of plants

> which seemed to me more worthy because they have thus far not been described by others in so far as I know, which is not such an easy thing to do in the ocean of so many new plants that you have explained in your most learned volumes. (Imperato, 10 June 1589)

Typically for Imperato these gifts mainly concerned marine plants (seeds of a 'convolvolo marino', the 'palma marina' of Theophrastus, and a sertularia, 'named thus by me'), a subject in which not that many naturalists in this age were interested. Imperato also added pictures of the same plants of which he sent seeds or dried parts, for he remarked: 'on the picture the plants as a whole can be seen; they produce neither flowers nor fruits as is typical for marine plants' (Imperato, 10 June 1589).[31]

Imperato is now primarily remembered for his impressive museum in Naples with its collection of stones and minerals, animals, and living and dried plants, the latter of which filled many herbarium volumes. The collection was kept in a large room of his house, which like his apothecary's shop was located at Piazza Santa Chiara in the heart of the city, while the living plants could be found on the terrace and probably in a larger garden outside the town centre.[32] Among the animals were crabs, serpents, chameleons, crocodiles, salamanders and lizards, an otter, seahorses, numerous fish and shells, as well as peculiar or particularly interesting parts of animals, such as the foot of an elephant, teeth of a hippopotamus and the saw of a sawfish. The strange or monstrous forms included a starfish with eight arms, an oyster with several as yet incomplete pearls, a pygmy, some bones and teeth of a giant, and a moonfish. Among the vegetable items were algae, various types of wood, many marine plants and several kinds of nuts, papyrus and seeds of an American melon. And there were great numbers of minerals, stones and fossils. A live Egyptian ichneumon walked around in his house, and in 1648 an English tourist observed a live tortoise in the museum (of which Imperato's son Francesco was in control by then) which was said to be more than a century old.[33]

Like the collections of his fellow apothecaries Pona and Calzolari in Verona, Imperato's museum could be visited. It drew considerable numbers of visitors both from Italy and abroad, first of all physicians, apothecaries and naturalists with a professional interest, such as Peiresc and members of the Accademia dei Lincei like Federico Cesi, Johannes Faber and Fabio Colonna. But many visitors simply came for the sense of wonder and the curiosity of it all. Everyone in Europe was talking about Imperato's museum, wrote Jacques Plateau (Clusius's collector friend in Tournai) to Clusius in 1604: 'it is as famous as the Arsenal in Venice' (Plateau, 8 February 1604). The stunning effect on many visitors should not blind us, however, to the function of the museum for Imperato himself: it was a study and laboratory at the same time, a research location.[34] Imperato was both collector and an investigator of nature, a man who went out on numerous field trips in various parts of Southern Italy in order to observe *in situ*, who experimented and who analysed a large range of natural phenomena. He published his observations and ideas in his *Dell' Historia Naturale* (1599), in which plants and animals form a relatively minor item in comparison with the sections on physical geography, geology, meteorology, optics, soil types and their uses, fire, minerals, the wind, metals, the colour of water and many other topics.[35]

Perhaps because Imperato's published work concentrates on other topics than living nature, while nearly all of his enormous herbarium which probably comprised more than eighty volumes is lost and nothing remains of his garden, his botanical expertise is still underestimated. Imperato is said to have been extremely expert in drying plants: sixty years after its creation, by the mid-seventeenth century, the plants in his herbarium were reported to have kept their appearance, characteristics and even flavour.[36] This herbarium must have been one of his research instruments, much as his garden formed a location where he could try out and study wild plants collected during field trips. Throughout the 1590s and often with the assistance of Pinelli, Imperato sent Clusius many seeds, bulbs and tubers which he grew in his garden, had found in the wild or had been able to obtain from friends. In 1593, for instance, Imperato wrote about the variety of bulbous and tuberous plants in his possession: many different colours of iris bulbosa, four roots of the true apios of Dioscorides, narcissi and various types of hyacinths 'that grow here spontaneously', the white-flowering lilium marinum or hemerocallis valentina, three types of ornithogalum and dentaria, arisarum with broad and with narrow leaves, colchicums, the moly of Mattioli, a hyacinth with white flowers, a yellow narcissus with leaves like a rush, two types of leucojum, and so on.[37] Some of these had come from friends, but to obtain others he had sent a man specifically for that purpose all the way to Montevergine in Campania, while yet others had come from Puglia. In 1598 and 1600, in parcels with plants sent by Imperato to Clusius, bulbous plants formed once more the dominant category.[38]

Imperato was involved in various exchanges with two fellow naturalists in Naples, Bartolomeo Maranta (d. 1571) and Giovanni Battista della Porta (1535–1615), but also developed long-distance exchanges of information and *naturalia* with fellow naturalists, apothecaries and *curiosi* such as Pona, Calzolari, Guilandino, Cortuso, Aldrovandi and Mattioli in northern Italy, Gaspar Bauhin in Switzerland and Joachim Camerarius in Nuremberg. In short, he operated just like Clusius and many other sixteenth-century naturalists, and looked for contacts with men who were as inquisitive and curious about natural phenomena as he was himself. It is no coincidence, therefore, that his published work, like that of Clusius, presents research on natural history as a collaborative effort and that many colleagues are thanked in it.[39] That recognition was explicitly expressed already around the middle of the sixteenth century by Francesco Calzolari – like Imperato an apothecary and a major collector and field botanist:

> I may not yet be worthy, but I am really interested and would like to reach a certain degree of perfection in these matters ... and here and there be on friendly terms with

> all groups, because nowadays when one person raises a question the other learns ... but as I said above, without the help of one's friends one cannot do anything properly.[40]

None of these men was physically surrounded by a research group on the spot. To a considerable extent their networks, like that of Clusius, constituted 'virtual' research groups, which were Europe-wide and mainly linked by correspondence.

While fitting into international, European models of research organization in this respect, Imperato's research practice also belonged to a Neapolitan tradition of artisanal science. This was epitomized by the Accademia dei Segreti of Girolamo Ruscelli during the mid-sixteenth century with its goal of experimental research, union of scholars and craftsmen, and strong participation of artisans, in particular apothecaries, herbalists, alchemists and gardeners. Of course, not everyone in Naples followed that tradition, as the differences between Imperato's practice-oriented and practice-based expertise and Della Porta's more philosophically oriented one make clear; nor did this situation persist far into the seventeenth century.[41] But in Naples knowledge of plants was predicated on the workshops of the apothecaries, many of which originated in or were still connected with the religious communities, and flourished thanks to the city's trade. That situation stimulated a special combination of theoretical and practical knowledge, in which the latter was very much based on the manipulation of plants and spices during the preparations of herbal medicines and on a great familiarity with the colour, smell and taste of leaves, flowers, berries, bulbs and roots.[42] It looks, therefore, as if the botanical knowledge in Naples represented by Imperato was shaped by a pragmatic, intellectual tradition which was sponsored neither by the state or governing prince nor by a university, but was rooted in artisanal practices and greatly advanced by the excellent access to natural novelties via trade. That tradition was very different from the academic ones of Padua and Pisa, the elite settings of private academies like that of Pinelli or the Accademia dei Lincei, and from the court settings in which many Italian naturalists operated whom we have met earlier.[43] But it shows many similarities with the one to which Pona and Calzolari in the Veneto belonged: those two men and Imperato were apothecaries of the erudite kind whose expertise was rooted in an artisanal tradition, based in close proximity to the major ports of their age, and therefore within easy reach of exotic and new *naturalia*, minerals and spices.

Unlike Della Porta, Imperato was never asked to join the Accademia dei Lincei, although its members knew him and his museum very well and received very considerable gifts of *naturalia* from him (in 1610, for instance, he sent 150 different kinds of seeds to Johannes Faber, and some to Federico Cesi and Johann Schreck), and although he was one of the most respected naturalists in Naples.[44]

As Olmi has argued, this was probably connected with reasons of class and politics. Whether it was also connected with a difference in scientific approach between the Lincei and Imperato – did the Lincei think that Imperato was too much focused on the practical aspects of research on natural history rather than on the systematic organization of knowledge, as Stendardo has suggested? – is still open for debate.[45] There is no doubt, however, that Imperato explicitly recognized the similarities between his own pragmatic, experimentalist and non-philosophical approach to the knowledge of nature and that of Clusius, and indeed called Clusius the foremost among his mentors in the natural sciences (Imperato, 28 June 1600). Interestingly, Clusius did not become a member of the Accademia dei Lincei either: he was invited to join the Accademia, but refused, perhaps because he regarded the Lincei as not empirical enough.[46]

The Rise of the Botanical Expert

Given the fact that Clusius had the language, time and contacts at his disposal to establish exchanges with those persons in Italy whose knowledge concerning living nature he found most relevant, his selection of correspondents gives us a significant slant on Italian natural history of the late sixteenth and very early seventeenth centuries. All over Italy, but with concentrations in the north-east, Tuscany and in and around Rome, the highest strata of the aristocracy created private botanical gardens, which often formed part of larger estates and competed in riches and variety of rare plants with each other. By no means all of the owners of such great gardens were botanical experts themselves, even if many of them were connoisseurs and collectors in a more general sense. Rather than with the aristocratic owners of those gardens Clusius generally preferred exchanges with men who had first-hand, practice-based expertise concerning living nature as well as book learning. Socially, those men belonged to two different but overlapping spheres. The first comprised the experts to whom the aristocratic garden owners turned for botanical and horticultural advice and the procurement of rare plants. We find members of the lower aristocracy among these consultants as well as friars, physicians and an occasional apothecary, but all operated in the context of the aristocratic courts and patronage. The second sphere was that of bourgeois and mainly urban medical professionals: physicians and especially apothecaries. The latter in particular operated in the urban cultures of Verona and Naples, in or close to the major ports of Italy, and were less directly connected with the courts or with university life.[47]

Clusius by no means shunned contact with the Italian universities, as his exchanges with Aldrovandi and Pancio show. Yet, the role of university-based naturalists among his correspondents was much less important than might have been imagined and his contact with men who became prefects of university botanical gardens was almost never predicated on this function: it generally started many years before they were appointed to these functions, at a time when they were not connected with the universities but already renowned for their botanical knowledge. That Clusius's choice was not inspired by a negative view of book learning should almost be self-evident given his own great erudition and that of many of his friends. To become a botanical expert, however, much more was needed than book learning. At the time, that conviction lay at the root of the botanical research of the majority of Clusius's correspondents, out in the wild or in their gardens. Olmi has aptly summed up this 'deep-seated conviction', which itself was both a result of and a further incentive for the changing nature and status of knowledge concerning living nature:

> a thorough understanding of nature required not only the teachings of those who spoke from university lecterns or successfully practised medicine and spread the results of their observations via printed works, but also the help of nature lovers and above all of those ... who 'got their hands dirty' through the activity of gardening, catching birds, or herbal medicine, and who had great expertise and knowledge concerning animals, plants and minerals[48]

The contemporary Italian plant expert to state most explicitly as well as radically that plant knowledge should be rooted in practice was the Venetian nobleman Pietro Antonio Michiel. Criticizing Mattioli's edition of 1554, and stating that many illustrations did not depict real plants but were inventions, he thought it unimaginable that a naturalist would try and picture a whole plant on the basis of only a small part. For a good understanding it was necessary, he said, to cultivate plants, observe their development from beginning to end, and to render this in both word and image.[49] In his emphasis on the practice of cultivation and the observation of change over time Michiel thus even went a step further than momentary observation, inserting the factor time.

While Clusius's connections with the sphere of the Italian universities were less prominent than expected, the similarities between the private botanical gardens of elite collectors and the public botanical gardens of the universities were, on the other hand, far greater (especially in terms of the quality and diversity of their collections) than their difference in status might suggest. In some respects it is helpful not to attach too much importance to the distinction between private and public. University botanical gardens are generally treated as official institutions of learning while the private gardens of the Italian aristocracy are primarily seen as prestigious courtly settings. Display and rivalry were by no means absent from the university gardens, however, nor were learning and research lacking from private gardens.[50] The Michiel, Bembo and Cortuso gardens, for instance, may well have played as important a role in botanical research as the university gardens. It has been pointed out that the creation of university gardens by opening private aristocratic gardens to the public should also be seen as part of the public image strategy of certain Italian rulers, and in other respects too the transition from private garden of a ruling family to public university garden was a gradual one. University gardens did have an important teaching function and were open to the public, but they remained to a considerable extent under the personal control of the rulers of the respective states.[51] Men like Casabona and Cortuso were appointed as prefects on the basis of their services as plant and gardening consultants to the ruling family in question as well as their expertise and personal networks, and not on that of a university career.

In so far as living nature was concerned, gardens – from Fra Gregorio's monastery garden or Imperato's *giardino pensile* to the living collections of Michiel or the stunning gardens of the Villa d'Este – were one major location of learning, observation, research and experimentation for Clusius's correspondents. The other was formed by the open countryside and especially by particular mountainous zones of Italy (such as Monte Baldo, the Norcia and Castelluccio plateaus, the Apennines), the Alps and Crete. Field research served a threefold purpose, whether it consisted of short and small-scale trips or of lengthy regional surveys: procuring new rare plants for elite gardens; the investigation and identification of medicinal plants and the sorting out of classical *materia medica*; and the discovery of new plants purely for the sake of scientific curiosity. The expertise of virtually all of Clusius's Italian correspondents was shaped in these locations. And it was research *in situ* which must have inspired this generation of naturalists to pay attention for the first time to the context or habitat of plants – in other words to look at how plants fitted in their places. Whatever their social background or precise function, in terms of interest and expertise, Clusius's Italian correspondents shared many characteristics with him: practice-based expertise on *naturalia*, great curiosity, experience in field botany and an excellent knowledge of wild plants, a fascination with novelty, interest in the possible use of plants but an even greater and critical interest in identification of *naturalia* on the basis of correct and detailed observation, experience in growing and propagating plants, and an aptitude for horticulture. As experts on nature we may therefore call them Clusius's colleagues and exchange partners, though not necessarily or in every respect his equals.

While the interest in nature and gardening was age-old and had a particularly respectable tradition in Italy, the function of botanical and horticultural consultant and plant procurer was a new phenomenon there during the sixteenth century. That may be inferred, at least, if we combine the information about the various men discussed above with evidence concerning the staff of the papal gardens. Records of payments to various types of gardeners and garden supervisors go back long before the sixteenth century, but only from the papacy of Pope Pius V (1566–72) 'the appointment of a botanist on the papal *ruolo di famiglia* became standard'.[52] In the Quirinal gardens the new position of botanist (*semplicista*) emerged in 1569, and was filled a few months after its creation by Fra Evangelista Quattrami. Several of Clusius's friends thus formed the very first generation of a new genre of specialists concerning nature: the plant expert or botanist. The emergence of a new professional role and category in itself indicates the growing awareness of the relevance of specialized knowledge and the increasing need (socially, financially and in terms of the quality of its contents) to have it recognized as such. Examples like the appointment of Jacques Plateau as the expert keeper of the *naturalia* collections of Charles de Croÿ in the Southern Netherlands in 1604 suggest that this process was taking place at much the same time all over Europe. Indeed, Clusius himself was in no different position for most of his life. Neither the search for new plants on the part of *semplicisti* or *spezialisti* nor the emergence of this new role itself can thus be understood without the rivalry between the aristocratic owners of the great gardens, in Italy or elsewhere. The grandeur of gardens and the presence of large numbers of rare flowers in papal, princely, ducal or university botanical gardens was an integral part of the conspicuous display of wealth and power. And the expert *semplicista* became an asset and instrument to aristocrats (and universities) who aspired to honour and prestige, joining the ranks of the artists, composers, mathematicians and other scientists patronized by princes.

Sometimes that rivalry was fought out over access to rarity and to information about rare *naturalia* – in flagrant contrast with the code of liberality and freely sharing to which Clusius and so many of his friends subscribed (or at least said they did). Rivalry between Padua and Pisa as representatives of the Veneto and Tuscany, and possibly even the more fundamental conflict between the obligations felt by a botanist-consultant towards his aristocratic patron versus this code of free sharing, lay at the bottom of a tell-tale controversy between Pinelli and Giuseppe Casabona, botanical consultant to the Medici. Casabona returned to Italy after a plant-hunting trip on Crete 'rich in noble plants and seeds' (Pinelli, 28 August 1591). But he left Venice so quickly, to Pinelli's surprise, that

> we at Padua learned at the same time of his arrival in Venice and his departure for Florence, so that we know nothing about the results of this journey, as the Duke of Candia had written to several persons that he would send great things with Casabona. But he [i.e. Casabona] may have reckoned that he could in this way ingratiate himself with his patron, without communicating anything to his friends for the present, although he had many times promised to do so. (Pinelli, 28 November 1591)

Casabona continued in his refusal to share his finds on Crete, speculating, according to Pinelli, that his patron's desire for exclusive rights to these *naturalia* would drive up his own remuneration – a comportment that was hardly suited to a man of honour, but certainly seems to have been effective.[53] Shortly after his return from Crete, Casabona was appointed to supervise the reconstruction the Pisan *hortus*, and in 1592 he became prefect of the new garden.

Such inherent conflicts of loyalty may also have stimulated certain plant lovers of high social standing to become involved in the selling of plants – an activity in which, it seems, mainly itinerant salesmen were involved until, more or less, the turn of the century. The conflict concerning the payment or compensation for plants sent by Casabona to Wilhelm IV of Hesse-Kassel suggests that he had almost begun to act as a commercial plant seller. It was another friend of Clusius, the young nobleman Matteo Caccini in Florence, who in the very first years of the seventeenth century actually took that step, though without renouncing his role as noble plant lover. Caccini was probably the first nobleman in Europe not only to own a special garden with a wealth of rare plants, but also to act as a garden designer and professional plant seller, who even sent catalogues and price lists of available plants to customers.[54]

The fact that a vast amount of new plant material was becoming available to be studied in the sixteenth century was an outcome not simply of the investigative activities of researchers with an interest in natural sciences (with or without connection with the universities), but also of the desire of gardeners-collectors for new items for their living collections and the concomitant prestige. In their role as botanical and horticultural consultants, an important category of naturalists indeed owed their jobs and their very existence as a professional category to these aristocratic patrons. Put differently, there were, therefore, reasons non-intrinsic to the natural sciences that had major effects on both the quantity of material studied, the professional roles of those studying it, and their methods. The success of natural history as an emerging discipline did therefore not merely 'rest on its social success at court'.[55] It seems much more likely that the vast interest in natural history emerging *from* the courts – and pushed by the powerful incentives of fashion and the need for display and honour – gave a decisive impulse to the fusion of plant knowledge generated in that court context, with knowledge originating in the circles of academically trained physicians and practice-trained apothecaries.

IV FRANCE

7 FIELDWORK IN FRANCE: EXPLORING THE INDIGENOUS FLORA

On 7 November 1600 Canon Joachim Levenier at Bordeaux sent Clusius a box with bulbs and other plants that he had collected during botanizing trips in the Pyrenees and the Bordeaux region. The box was shipped but the ship was wrecked and the box with precious plants lost. A separate letter, which arrived in Leiden on 19 January 1601, tells us, however, what the box contained and how rare some of these plants were in Levenier's opinion:

> Crocus with long, hair-like leaves and a white flower, 2. They are very rare and I have found only three.
> Crocus with long, hair-like leaves and an almost purple flower, several bulbs. The two white ones are apart on the other side of the box.
> Fritillary with a white flower, also apart
> Mountain fritillary with an almost black flower, 5
> Gentian mayor, a big plant
> Lily-hyacinth (*hyacinth liliac*), 2
> Hemerocallis with yellow flowers, 30 & some small ones
> Hemerocallis with yellow flowers without any spots, 1, it is wrapped up in cotton in order to distinguish it
> Sea narcissus, which is commonly called pancratium with a white flower, 6. I have not had the red one in flower yet, as soon as it flowers I will send it to you
> Pseudonarcissus Aquitanicus with a campanula-shaped yellow flower surrounded by white petals, I send you 6
> Mountain pseudonarcissus with pale-coloured flowers, 4. (Levenier, 7 November 1600)

Levenier hoped that Clusius would 'liberally share' the 30 yellow hemerocallises 'with whom he wished'.

Less than two weeks before Levenier's dispatch, Jean Robin Jr likewise wrote to Clusius. Levenier and this young and promising botanist, the son of the Parisian court apothecary and royal herborist Jean Robin, knew each other personally and by reputation. In October 1600, young Robin had just arrived in Middelburg after a journey devoted entirely to the gathering of rare plants which had kept him away from Paris for several years. In the course of his travels Robin had visited Spain, from where he sent bulbs to his father in Paris in 1598–9, and even Guinea on the west coast of Africa (Robin Sr, 10 April 1599). Robin Jr had not planned to arrive in Middelburg or expected to find Clusius in Holland. Otherwise

> he would have brought back many more and other things, such as wild animals and seeds attached to their branches and twigs, but since I did not think of you at all nor of any [other] friend, I have not burdened myself except with such things that are more easily carried such as seeds. (Robin Jr, 26 October 1600)

Some intriguing and paradoxical aspects of these two examples raise questions that are central to the present chapters which focus on France. Although Levenier was based in an important French port which could give him access to exotics from overseas, he appears mainly to have been fascinated by indigenous wild flora, going out on field trips to obtain rare native plants. In contrast, the prime interest of the Paris-based and court-connected apothecary's son and herborist was exotic *naturalia*, and he travelled to faraway places in order to collect them. Does this suggest that 'place' in the sense of the location, base or home of naturalists was not so relevant after all to the shaping of their interests? Following in the footsteps of Levenier and of a few contemporary French naturalists and contacts of Clusius we will first explore the character of their interest in nature. In the following chapter we will look more closely at the interest in both rare indigenous and exotic *naturalia* among a circle of bourgeois collectors in Levenier's home region (south-west France) and in the Parisian court circles to which the Robins belonged.

The Plant Collector Canon Levenier at Bordeaux

Joachim Levenier (also Venerius; *c.* 1565–1619) – a man whose name was completely absent from the history of botany until 2005 due to the fact that he never published – was, in fact, a botanical expert and the owner of what must have been one of the best private botanical gardens of Western Europe around the turn of the sixteenth century.[1] Levenier came from a Venetian family that had moved to France some two centuries earlier. He was born at Bordeaux and chose a career in the Church. He became canon of the Bordeaux cathedral (*c.* 1586), and eventually vicar general (1607–15). It seems likely that a common friend put Levenier and Clusius in touch with each other, and that the first extant letter of December 1597 from Levenier to Clusius was indeed one of the first in their exchanges.[2] Whereas Levenier's three letters of 1597–8 are in Latin and contain many elaborate compliments to Clusius as the foremost expert in matters of plants, relations subsequently became less formal. Levenier changed to French, and his tone became less deferential to Clusius and more critical of the expertise of others. Certain plant sellers in particular came in for some harsh remarks:

> Monsieur, it has recently happened that certain *herboristes* passed by who had come from Mount Tabo near Tarascon in the Pyrenees; who told me among other things that they had brought ... red hemerocallis. This, as I insisted to them, is false since I have visited these mountains and know well what can be found there, and also they [i.e. these plants] had not yet been in flower when they took them from the mountains. They also said that they had white martagons, with which I disagreed for the same reasons, adding that there are no such flowers in the Pyrenees and that the only person who has them am I myself and those with whom I have shared them. They also had white autumn flowering crocus, but how could they distinguish these in the month of June from the violet ones! ... And they were finally forced to say that they could not guarantee any of this, since they could see very well that I am not a man who is deceived easily. (Levenier, 6 July 1605)

By 1603 Levenier did not even hesitate to point out to Clusius that the latter had been fooled by some other fraudulent Parisian plant sellers in Holland: 'You have indubitably been deceived concerning the martagon with the white flower, the hyacinth with leaves like a lily and a yellow flower, and the white iris bulbosa, as experience will show you' (Levenier, 12 March 1603). Worse, Levenier wrote that he knew these particular plant sellers, and had little respect for their knowledge. They apparently visited him on the way to and from the Pyrenees, often ate at his table, and in order to obtain rare plants from the Pyrenees made use of a man whom Levenier himself had actually taught to recognize such plants. Otherwise they would never have found anything, he wrote (Levenier, 12 March 1603). It may not have been Levenier's intention, but the message was still that the famous Clusius, by then recognized as top botanist of Europe, had been fooled by ignorant men whose scant knowledge was but a pale reflection of Levenier's own expertise.

Clusius took no offence. He refers many times to Levenier in his works and repeatedly calls him 'a most learned man'. The frequency of their correspondence too indicates that these two men who never met in person were seriously interested in the plants they could offer each other as well as each other's expertise.[3] Although their relationship commenced with Levenier asking for rarities, Clusius on the whole seems to have been at least as often on the asking and receiving side as Levenier. What Levenier wanted from Clusius, besides respect for his expertise, was clear from the start: rare plants, advice on how best to grow them and contacts with other plant lovers. What Levenier had to offer can be reconstructed from his letters and their interesting, often alphabetically ordered, appendices which list the plants, tubers and bulbs that he sent to Clusius, and from the posthumous inventory of Levenier's garden. That inventory listed only those plants that could be transplanted to a new location, and indicates that Levenier's great garden was dismantled soon after his death. It mentions the impressive number of some 2,000 plants and bulbs: among them pomegranates with double flowers, a peach tree with double flowers, a small apricot tree, orange trees, a dozen or so special roses (some with double flowers), oleanders with red flowers, lilacs, a jasmine with yellow flowers, a primula auricula (*oreille d'ours*) which he also sent to Clusius and 'a great number of cyclamen, martagons, *purprains*, narcissus and hyacinths put in the allee and borders of the garden'. Equally impressive is the quantity of 2,000 or more flat tiles, to delimit the platforms and sections of this garden for the conservation of the plants and flowering bulbs.[4] The garden was clearly organized in the usual style with compartments, and extremely rich in rare plants and the varieties with double flowers which were in fashion during this period.

What Levenier sent to Clusius – both on his own initiative and upon Clusius's request – naturally reflected only in part what grew in his garden. Anyone sending plants in this age knew that some travelled well, others not at all. Levenier mainly sent bulbs, tubers and seeds to Clusius: hemerocallis, several kinds of wild hyacinths and fritillaries, gentian, crocus, single and double narcissus, pseudonarcissus, colchicum vernus, crocus stellaris, moly, cyclamen with ivy-shaped leaves, dens caninus, a yellow wild orchid, pancratium, allium, many different types of iris, ornithogalum, some tulips and a peony, besides seeds of 'thymelaea' and the yellow aster.[5] Apart from the American nasturtium indicum (Tropaeolum spp.), which Levenier mentions in passing in a letter dealing with field botany, not a single one seems to have come from America.[6] Many of these plants were rare and new, however – even to most plant lovers of this particular region – and had been recovered from the wilder parts of south-west France. Two of the special roses (of which Levenier seems to have been particularly fond) mentioned in the inventory of his garden had come from the Northern Netherlands, however: the *rosa cinnamomea* with the double flowers had arrived in January 1600 from Leiden, and in April 1600 a 'rose with 300 petals' reached him from Middelburg.[7]

Such detailed information about the variety of plants in a sixteenth-century private garden is rather unusual, even if we take the better-documented gardens in Italy and the Southern Netherlands into account. Letters (1611–12) from Levenier to Caccini in Florence tell us, moreover, which plants he did *not* have by that time but very much wanted to include in his garden – and thus give us an even better idea of the potential scope of his living collection.[8] Of digitaria, *aurea mala* (literally golden apples), the Cretan leucoium and the American 'granadilla, which is also called passion flower' he needed only the seeds, while for most of the tuberous or bulbous plants one exemplar was enough. In Levenier's long list of desiderata we further find seven types of anemone, each precisely indicated by the shape of the leaves and the colour of its flowers; two types of asphodel; a 'golden apple' from China; Jove's beard (houseleek), a plant native to the French and Italian Alps; the American canna indica; cineraria; a clematis with double blue flowers; various cyclamens; 'gelsominum from Guinea, or syringa arabica'; helleborus niger; nine types of hyacinths; seven types of iris; leontopetalum, leucoium and muscari from Crete; a red Byzantine lily; two types of lychnis; two varieties of martagon lily; an unspecified plant from the Moluccas; seventeen types of narcissi, again clearly specified by colour and form or size of flower; a peony; two kinds of primula; a tulip with three red and three yellow petals, and three different double roses.[9] By this time Levenier's interest had therefore begun to extend to a few exotic plants, but the main emphasis was still on European plants and on bulbous and tuberous plants that originally had come from the Middle East. Given the fact that the manuscript treatise *Agricoltura teorica* (1595–6) by A. del Riccio mentions peonies, hyacinths, narcissus, musk roses, cowslips, tulips, daffodils and anemones as plants particularly suited for the garden of a king, Levenier's garden must have been at a princely level at least.[10]

Its valuable plants also made it vulnerable, however, and his garden, like those of Clusius and Marie de Brimeu, was continually threatened by thieves. Some of these were professionals, but others were fellow plant lovers and rival garden owners who came to visit by day and to steal by night – as Levenier remarked, with some broad hints at whom he had in mind.[11] The bulbs of his most beautiful yellow hemerocallis were stolen, for instance, 'which produced up to 36 or 40 flowers, which I had put aside to adorn my garden' (Levenier, 7 November 1600). This called for defensive measures, especially when Levenier was away from home. A loyal servant repelled attacks by 'thieves and bad folk' and made a valiant use of 'good harquebuses and munitions' to safeguard his master's garden.[12]

While Levenier obtained some of his rare plants by buying them, much of his plant material must have come from barter with other plant collectors. These exchanges were a regular source of conflict, and it must remain undecided whether Levenier always remained on the right side of the law in his efforts to obtain rare plants, especially given his emphatic statement 'My intention is just to obtain beautiful plants by every legitimate means open to me' (Levenier, 28 August 1601). His comments to Clusius on various conflicts and misunderstandings concerning exchanges of rare plants tell us much about his relations with fellow experts and rivals, and show both his real expertise and the way in which he wanted to present himself as an expert to Clusius. Levenier sketched some of the main rules of the unwritten code of behaviour which applied (or should apply according to him) to exchanges between honourable collectors. The first rule was that gift giving and exchanges which involved payment should not be confused. Both were possible, even between the same persons, but it should always be clear whether something was a free gift in the context of friendship and counter gifts, or a transaction in which some form of direct monetary or material compensation was involved:

> What I give I do give freely and what is given to me is more pleasant to me because of the love of the one who gives than because of the object itself, even if things are concerned of which I am extremely fond. (Levenier, 10 February 1601)

Levenier was particularly annoyed with Johan van Hoghelande, a Leiden garden owner and close friend of Clusius (see Chapter 10), who had complained that certain gifts of Levenier's had cost him much and had promised to send various rare bulbs in return but never done so. Offended, Levenier stated that he had never in his life sold anything and certainly no rare plants. Hoghelande, who was said to have some 2,000 tulips alone in his garden, had shown himself to be extremely stingy, moreover, and had not even sent Levenier the six tulips for which the latter had asked.[13] Liberality and sharing were the great virtues in this community of exchange. Promises of counter gifts should always be honoured.

Levenier the Plant Hunter

Although Levenier refers only rarely to the published works of naturalists (mainly those by Clusius, Lobel and Dalechamps), he frequently uses Latin names when describing plants, and must have been well read in such matters. The brief phrase 'which I cannot find at all in the books' (Levenier, 20 August 1602), shows that he used to look up plants in works on natural history. His predominant interests and sources of expertise, however, were the practices of field botany and gardening. The latter focused on the collection of rare plants – whether native to France or not – and his attempts to acclimatize the wild plants which he discovered during his field trips to the setting of the garden. The former should be understood in his case as the systematic exploration of the regional flora of south-west France and parts of the Pyrenees and northern Spain, with the specific purpose of discovering new and rare wild plants.

Levenier's knowledge of the flora in the regions he explored may well have been unsurpassed at the time. He proudly wrote: 'There is no year in which I do not discover something new, travelling on purpose to different places' (Levenier, 8 July 1601). Not far from Bordeaux he explored the Island of Oleron in the Bay of Biscay just south-west of La Rochelle in 1600, searching 'with my men' for pancratiums; during another trip to a nearby island he discovered a new pancratium. In a long and rhetorical passage (in which he countered a statement by a fellow plant lover which he interpreted as a slight), Levenier emphatically claimed the honour of the discovery of various wild plants and, thereby, his status as a naturalist:

> Is it not I who have discovered the yellow roses in the mountains of Vabres? Have I not discovered a forest full of red, white and variegated peonies? Have I not been the very first to discover white and green fritillaries? Has he [i.e. his detractor] ever seen two types of yellow fritillaries? Has he ever had them of an almost black colour? Has he ever seen martagons that are white with little purple spots inside and red outside? Was it he who discovered the totally white martagons? Was he the first to know of the yellow hemerocallis? Has he ever heard people speak in the Pyrenees of the small and low-growing red rose with the smell of the musk rose that grows there? Has he ever heard anyone else than me say that double daffodils grow near Blaije, some 10 to 12 leagues from here? Haven't I discovered on the Islands Est, Oleron and in Brouage the sea iris and the iris that you call Pannonian? (Levenier, 10 February 1601)

As a wealthy man of high social standing Levenier travelled and herborized accompanied by servants who must have done most of the carrying and digging. He regularly sent out his servants as scouts to explore the terrain and find out from local inhabitants where rare plants had been sighted – very much as one would do on a hunting expedition.[14] Almost every spring or early summer he went off to the Pyrenees for a number of weeks with a staff of assistants. Local knowledge, the expertise of the shepherds and villagers, was enormously important in this respect. Levenier's description of his trip in the summer of 1601 conveys very well how big the areas were that he explored and how he operated, using local information or the local inhabitants themselves in order to obtain the rarest plants. On this particular trip he dared not go too close to the Spanish side of the Pyrenees and the Spanish coast because of the many robbers and thieves. As a consequence he brought back less than hoped for, since the French side of the mountains was not as botanically interesting as the Spanish one:

> I have done everything I could do there, having normally five persons with me and quite often six or seven, and in spite of having made the shepherds take them all over the mountains, I have in the end found out by experience that what one finds on one mountain is also found on another one – or almost – and that the mountains on this side of Spain are very infertile in terms of rare and beautiful plants ... I have leafed through more than sixty mountains and have made the shepherds search all of them ...
>
> The purple [iris bulbosa] that I took is very rare there and I have never seen more than one bulb with its flower which a shepherd brought me at Banieres de Luchon on a Sunday when I was dining with monsieur Richier, the king's physician and professor of botany and medicine at Montpellier, who had come expressly to the Pyrenees with some of his students to see me. (Levenier, 20 August 1602)

Such a sign of respect from Pierre Richier de Belleval (*c.* 1564–1632) – who had been appointed to the new chair of anatomy and botany at Montpellier in 1593, and started creating the *hortus botanicus* there in the same year upon the request of the French King Henry IV – was an honour indeed.[15] By telling Clusius this story, Levenier emphasized his status as an expert, which he stressed even more by describing how his own plant identifications were actually more reliable than Richier's:

> I have also found a kind of phalangium with flowers like a lily, white, elegant and scented, *sieur* Richier thought that it was the *lilioasphodelus*, but that cannot be the case for certain reasons, and once he had listened to these he did not want to insist but preferred to acquiesce. (Levenier, 20 August 1602)

It might be tempting to regard Levenier – with his heavy-handed ways of emphasizing his own expertise, his derogatory remarks about others and his claims to be the prime expert and first discoverer of so many plants – as no more than a self-important plant hunter who was mainly interested in trophies and honour. The negative qualification would be unwarranted for more than one reason. Honour and primacy were (and are) significant motives for scientific or any other kind of discovery. In fact, many of Clusius's correspondents and friends aspired to a reference in the latter's published works as an expert or discoverer of a rare plant. Indeed Clusius himself was as interested in the scholarly honour of first publication as Levenier in that of first discovery.[16] The fact that several of Clusius's printed works contain appendices on topics which had nothing to do with that of the main body of the book but gave him the opportunity to publish quickly his latest novelties, bears out how important scoops were to him.[17] Levenier probably realized as much, for he wrote to Clusius: 'I had sent off this package to you before that to Signeur Hoghelande so that you would have the contentment of being served first', and 'When I obtain something new I always share it first with you, not only this year, but always in the future' (Levenier, 18 February 1600, 16 April 1600).

Nor is there a contradiction between plant hunting and developing great expertise in this field. The rare wild plants that Levenier discovered and subsequently tried to grow in his garden were indeed his trophies. As Levenier put it, 'I am absolutely eager to get such things' (Levenier, 18 December 1599). But in order to be able to recognize which plants were rare or even unknown, he had to have an extensive reservoir of factual knowledge concerning the local flora. That can only have been built up during many years of field trips in these regions which had hardly been explored in a botanical sense except by the rural population. The local inhabitants taught Levenier and his staff, but they themselves generally could not assess the rarity of these plants for lack of comparative material and information. The parallel between Levenier's explorations of the Pyrenees and European explorations of nature in other continents is thus unmistakable.[18]

Levenier's explorations of the Bordeaux region and the Pyrenees gave him detailed knowledge of the different flowering times of the plants in which he was most interested. And he was strong on identification; he was able to recognize many plants just by the leaves. In order to recognize and distinguish the varieties of the bulbous and tuberous plants that he liked so much, he paid close attention to the colour nuances of leaves and bulbs, and of course in particular to those of the flowers:

> I have also found a very beautiful violet iris of which the upper petals were as velvety and full of colour as the lower ones. There are also common ones which are covered with small and very dark or black spots on both the lower and upper petals, but those are not very clearly visible on the purple one. And I have also found one single iris of which the whole flower was spotted – both on the upper and lower petals – with purple and very dark red spots which were mixed and jumbled without any system. (Levenier, 20 August 1602)

Like Clusius and other plant collectors, Levenier was fascinated by the rare white varieties: he mentions a white fritillary, and white crocus vernus stellaris, 'gentianella', primula, campanula, hyacinthus liliaceus, small hyacinth, and mountain martagon, as well as a pulsatilla with big semi-double flowers which were blue outside and white inside, and an even more special one – of which he had only discovered two – which had twenty completely white petals. Levenier sent several of these plants or their dried flowers to Clusius, and promised him white and red rose bushes without thorns once the right season had come (Levenier, 28 August 1601). About the white iris Levenier (and Lobel) were almost lyrical: 'I have found the white iris bulbosa maior Pyrenea (which Lobelius calls the hyacinth of the poets) with a flower as white as snow, but only one among a hundred thousand millions of blue ones that I have seen' (Levenier, 28 August 1601).

Levenier – like Clusius and the wealthy garden owners in the Southern Netherlands and Italy – was in no doubt that the black fritillary was related to the ones with different colours, and recognized that family resemblances linked the various pansies mentioned above. But in this age the possibility could not be excluded that a fritillary, anemone, iris or pansy with an unusual colour might, in fact, be a new type of plant, and therefore a potential trophy to the plant collector. The great deal of attention paid to minute details of colour was therefore intimately linked with both the quest for rarity and the desire for beauty, but itself stimulated observation and attention to certain details – in particular of the flowers – which would turn out to be essential to the understanding of plant sexuality and propagation, and the classification of plants in families. Given the importance of colour, it is surprising that coloured 'portraits' of plants played no part in the exchanges between Levenier and Clusius. Levenier relied completely on words in order to convey his impressions of wild and cultivated plants to Clusius. Clusius seems to generally have done the same – but once he sent Levenier a coloured ribbon to show him the beautiful colour of some 'Silesian carnations' (Levenier, 16 April 1600).[19]

Like his Italian counterparts, Levenier was very aware of plant geography and ecology, both on a large scale – pointing out that the different slopes of the Pyrenees had a very different vegetation – and on the level of individual plants. Concerning the yellow cyclamen, for instance, he informed Clusius that it had leaves no bigger than the end of one's thumb without spots or marks like other cyclamen but with many veins, a small tuber with tough fibres sprouting from its underside (in contrast to other cyclamen) and a flower stalk three thumbs high which carried several small yellow flowers; on the mountain 'Saravieille near the port of Plan' it grew between stones which were always kept moist by water running down the mountain (Levenier, 20 August 1602). Levenier returned to some spots in the mountains several years running, remarking on how the numbers of certain rare plants increased or diminished. And he often mentions the geographical locations where he had found the plants that he sent to Clusius: on Mount Soutou or Mount Pujol in Spain, near the bridge of Suert in Spain, on Mount Campsaure or Mount Gars in France, near the village Obiglé in Poitou.[20]

From experience Levenier knew that certain wild plants did not adapt well to gardens, that others demanded special humidity or a particular type of soil, and that transplantation from the wild to a garden could cause a change of colour of the flowers. He explained to Clusius for instance that the hyacinthus liliacus did not grow well in his gardens and would not do so either in Clusius's one: it was smaller and produced fewer flowers than in its 'place of birth'. The big gentian had also died in his garden, but he sent a root to Clusius, so that it could be tried out in a different soil. Not one of the 'woody and fibrous' plants that he had brought from the mountains had survived the transfer, even though he had planted them with their clod. A white pseudonarcissus produced a pale-coloured flower in his garden, and a pale-coloured one had become yellow the next year.[21] In other words, Levenier knew how to translate knowledge about the original growing conditions in the wild into practical knowledge concerning acclimatization in gardens. And he was as aware of the causal explanations for the adaptation problems of transplanted plants as of some possible solutions.[22] In all these respects Levenier's types of expertise and interest were no different from those of Clusius.

For lack of more detailed information about Levenier's early training, background and contacts, we do not know how his passion for the French wild flora or his practice of field botany originated. His interest in rare plants fits in with the fashion of collecting, but only very few gentleman-collectors became expert field botanists, while Levenier's garden appears to have been of outstanding quality and riches even compared with the princely and university gardens of northern Italy. His field trips too were distinguished by their scale – as full-fledged expeditions – their geographical range, and his complete lack of interest in the medicinal use of plants. Given the regularity of his field trips, his long-standing interest in the Pyrenees and the fact that he kept track over the years of the best locations to find rare wild plants and their numbers, it is hard to imagine that Levenier would have kept no notes of his findings, but he left no other texts than his letters. He never published anything or even planned to do so, as far as we know.

Three possibilities thus suggest themselves. Levenier may have discovered field botany for himself, urged on by his passion for rare plants, and extending the model – in terms of practical organization – of the elite hunting trip to the exploration for rare plants. Secondly, given Levenier's personal acquaintance by the late sixteenth century with Richier de Belleval, prefect of Montpellier's *hortus botanicus*, his interest in field botany and the wild local flora may have been inspired by the practice-oriented teaching of natural history at the university of Montpellier around the middle of that century: the time, in fact, when Clusius himself was studying there under Rondelet. Levenier may even have studied there himself. He was certainly acquainted with several gentleman-collectors who also lived in south-west France and had been connected with the university of Montpellier. We will meet several of them in the next chapter, and they may have formed the third source of inspiration for Levenier's field investigations.

The Native Flora, the Apothecary and the Nobleman

Two more examples may help us to gauge some further aspects of the interest in the native flora in France. While Levenier cared very little about the medicinal uses of plants, the interest of Noël Capperon (also Caperon), a Protestant apothecary from Orléans, in the flora of his region must have originated in his profession. He had no Latin and it is unlikely that he ever trained at a university. Capperon's expertise concerning the flora of the Loire valley ranged well beyond medicinal plants, however. In fact, by the late 1560s, more or less during the same period when field trips were becoming more common in both Italy and the circles of French medical students, Capperon had even become internationally known as an expert field botanist. The English naturalist Thomas Penny, for instance, visited Montpellier, botanized on Mallorca and then proceeded to Orléans (between 1566 and 1569) to study plant physiology with Capperon.[23] The latter was known even in Paris as an expert with an excellent knowledge of 'simples' (J. de Vulcob, 6 November 1584).

Contact between Capperon and Clusius must have begun during the late 1560s, and Capperon sent Clusius several detailed descriptions of plants growing in the Loire valley in which he never forgot to mention their habitat:

> which I have found along the river Loire in a sandy spot ... its root consists of three little bulbs or tubers, roughly the size of big peas, oblong and white in colour. From these bulbs a leaf grows, similar to that of ornithogalum Dodonaeus, but bigger and without a rib in the middle, with a little round stalk half a foot high and at its top three leaves which are shorter than the first one at the bottom. In between its leaves grow two or three little stalks, each with a flower which is yellow inside and green outside. (Capperon, 12 December 1571)

In 1571 Clusius asked Capperon for details of a 'variegated bulb' which Capperon called fritillaria, a name used by the locals as he explained, 'because the white and red marks or spots are arranged in a regular pattern, just like our chess boards or draughts boards' (Capperon, 12 December 1571). It has indeed been said that it was Capperon who first introduced the fritillary to England, in 1572, where it was also known as Caperon's narcissus. The name still exists, as Fritillaria (Caperon) L.[24]

Whereas Capperon's botanizing trips had very different cultural roots from those of Levenier, the outcome was similar: high-quality and detailed knowledge of the flora of a particular region or sub-region, even if that knowledge was never published or even put down in writing. Our final example of a French naturalist with an interest in native flora is the nobleman Nicolas Claude Fabri, Seigneur de Callas and later de Peiresc (1580–1637), who is best known for his enormous erudition, his contacts with, influence on and vast correspondence with key figures in European intellectual and artistic life from Galileo to Rubens, and his myriad interests which ranged from astronomy to natural history, and from antiquities and coins to optics, law and medicine. His knowledge too remained unpublished in his age. It can only be laboriously reconstructed through his vast correspondence.[25]

Some of these many contacts and activities still lay in the future during the years in which the young Peiresc and the old Clusius corresponded (1602–8), but Peiresc's interest in living nature was already marked.[26] At the beginning of this period the then twenty-two-year-old Peiresc was studying in Padua, and in close contact with Pinelli and the naturalists Pona in Verona, Alpino in Padua, Aldrovandi in Bologna, and Imperato and Della Porta in Naples. It was Pinelli who encouraged Peiresc to get in touch with Clusius, which resulted in a lively exchange of letters and *naturalia* right up to Clusius's death.[27] During those same years in which Peiresc corresponded with Clusius he was also in close contact with Pierre Richier de Belleval, and he sent Clusius detailed descriptions of Richier's botanical garden at Montpellier not long after Levenier met Richier in the Pyrenees in 1602.[28] Peiresc's extended trip to Northern Europe in the course of 1606 was marked by his interest in rare *naturalia* and special collections. He met Jean Robin Jr in Paris (who showed him some exotic fruits which had been erroneously described by Clusius and asked Peiresc to suggest corrections to Clusius), stayed with Charles de Croÿ near Brussels and admired his special collections (sending him pink flamingoes afterwards), visited Lobel in England, saw the vast collection of Bernardus Paludanus in Holland and visited Clusius in Leiden.[29] The exchanges between Clusius and Peiresc resulted, among other things, in donations from Peiresc to Clusius of various kinds of seeds, mushrooms, fossil shells and corals.[30]

Peiresc – like his contemporary and fellow nobleman Federico Cesi (1585–1630), who at exactly the same time and at the age of eighteen was busy founding the Accademia dei Lincei in Rome (1603) – was an intellectual omnivore. But unlike Cesi, Peiresc was not interested in creating a great collection, a body of information or even an academy. His goal has been summed up rather as an 'invisible, Europe-wide college of savants of which he was administrator and patron'.[31] No wonder that Peiresc got along with Clusius, who had been doing little else than creating a community of exchange in Europe for most of his life. In so far as *naturalia* were concerned, Peiresc was much less focused than Levenier or Capperon. He liked both the rare and the not so rare: wild indigenous plants, dried and living exotic ones, living ichneumons and flamingoes, horned hares, strange fish, folk names, medicinal uses of plants, their habitat, how to prepare and eat them, where to find them, how to grow them and so on. From the start of his correspondence with Clusius he put all of his knowledge as well as his already considerable network of friends and assistants at Clusius's service, in particular to procure plants and seeds from the wild flora of the Provence in which the latter was interested. In February 1604 he sent Clusius fresh seeds of 'our seseli from Marseille' and the roots of 'tragacantha' which grew nearby, and of a plant that an apothecary called 'centonica'. He also sent someone to cull a rare plant which the sailors of Marseille called 'tartonraire' and used for purging; it is remarkably effective, as Peiresc added, 'both above and below' (Peiresc, 25 February 1604). Each of these particular plants had a medicinal use.[32] In the same letter Peiresc asked Clusius to send him a list of the plants from the Provence in which the latter was interested. A very long list indeed must have arrived, and a year later Peiresc sent a box with a large number of the requested plants to Leiden. Most were sent in the form of seeds or bulbs, but Peiresc also included a picture of an interesting mushroom, a gum or resin, and several fossils of shells and plants with the location where they had been found. Interestingly, his list of seeds and bulbs was bilingual, with a Latin and a folk name (in French or Provençal dialect) for each plant, such as: 'Arbutus / vulgo arboussier'; 'lentiscus / lentiscle'; 'olea / olivier'; 'Rhus coriariorum, Sumac / Foüuil'; 'Thymelea / Bouffe-galine', et cetera.[33]

Quite a few of the plants and seeds that he sent Clusius were not collected by Peiresc himself. He ordered someone to get plants for him, as in the case of the 'tartonraire', enlisted the assistance of a number of physicians and apothecaries, and wrote to friends asking for help in obtaining the plants that Clusius had requested:

> I wrote immediately to Frejus, Toulon, Montpellier and Avignon and elsewhere, and asked my friends over there very insistently to collect all the seeds that you desire. I was also at Marseille at this time and gave the same orders to someone else, so that I hope that, between them, they will get everything you want. (Peiresc, 9 August 1605)

It seems that Peiresc also asked the assistance of an experienced 'botanist' during his field trips, who could recognize and identify plants in all seasons.[34] Clearly he was capable of quickly mobilizing friends in his part of the world, and knew whom to approach in order to obtain local plants. There was apparently no shortage of people with a sufficient knowledge of the local flora.

But Peiresc himself went out into the field as well to observe both archaeological remains and living nature on the spot and collect *naturalia*. In February 1605 Peiresc announced that he himself would go to the area of Marseille once the right moment had arrived, to get seeds of the tragacantha for Clusius, since he did not trust his informants, some apothecaries, who had assured him that the plant did not produce seeds, while Peiresc knew that it did flower (Peiresc, 15 February 1605). By August of that year the time was ripe. Peiresc let himself be carried, as he put it, by his curiosity to the place where the tragacanthus grew abundantly and collected its seeds (Peiresc, 25 August 1605). His botanical fieldwork in his home territory was much more intensive. It seems to have originated at least in part as a side-effect of his hunting trips. Peiresc was explicit about the combination of hunting for game and looking for interesting wild plants:

> I have often enjoyed myself while out on a hunting trip by having all those seeds culled that seemed to me the most extraordinary, in order to find the ones that you had marked [i.e. on the list], and in fact in this way I at last had all of those in my possession which the apothecaries had not been able to find. But there were also many others for which you had not asked that I have wanted to send you nonetheless, since they had been gathered, hoping that you may enjoy them. (Peiresc, 15 February 1605)

The list of plants from the Provence which he sent to Clusius on this occasion even contains a separate section headed: 'those which Peiresc has come across in the fields and while hunting either along the seashore or in the woods, and in particular in the territory of Beaugentier' (Peiresc, 15 February 1605).

Most of these seeds had been collected in the hilly and wooded area surrounding the small village Beaugentier (now Belgentier) between Toulon and the Massif de la Sainte-Baulme. Those names by themselves would perhaps have meant little to Clusius, but as Peiresc explained his family owned a property near Beaugentier, the place where he was born, and the mountain with the name 'Col d'Anis' above it was famous for the special plants which were collected there for medicinal purposes.[35] In this area, as he told Clusius, the rarest plants of the whole of the Provence could be found, such as styrax (Styrax officinalis L.), and various marvellous mushrooms in which Clusius would be particularly interested. Peiresc sent Clusius one of the dried mushrooms itself (a Clathrus cancellatus Fr.), which after soaking was the model for his illustration of the fungus coralloides cancellatus (Figure 6).[36] But Peiresc also sent Clusius one of his famously detailed and readable descriptions, which just like Clusius's own helped to set the standards of scientific description:[37]

> They originate on the ground with the form of a snow-white chicken egg, and when they ripen the egg grows gradually and from its inside something like an *ajour* purse emerges (if one can call it that), perforated like the [needle]work which French ladies call rasoir, with grid-like markings, and it seems as if the interlaced threads or branches that create the shape of this purse are so many branches of real coral, on account of both the colour, which is perfectly red, and the size, which has the proportion of branches of true coral. Further, in order to receive appropriate nourishment from the earth, nature has provided them with a very straggly root, divided into several small threads of the same white colour as the skin of the egg to which they are attached. (Peiresc, 15 February 1605)

In stark contrast with the admirable beauty of this 'inanimate creature' its smell was fetid and almost insupportable. Peiresc speculated that the stench came from the grey and viscous liquid that could be found inside the red 'branches'. In order to observe such minute and interior details, he must have dissected the mushroom and its thin filaments. Peiresc's observations did not stop at 'identification and on-the-spot autopsy'.[38] Still based on ocular inspection, they moved on to the habitat of this mushroom – he had found it most often in the Beaugentier territory between the reeds growing along the banks of a little river there – and its function in a particular ecological context:

> Nonetheless I have judged that it had to have some good quality in it and some well advertised hidden virtue, given the fact that immediately after it has completely opened an infinite number of small insects start eating it, which desert all other meat (in order to run to this food) and even several other types of mushrooms which we regard as very good to eat. (Peiresc, 15 February 1606)

Had he had a painter on the spot, Peiresc added in the same letter, he would have had the mushroom depicted life-size and with all its vivid colours. Eye for detail,

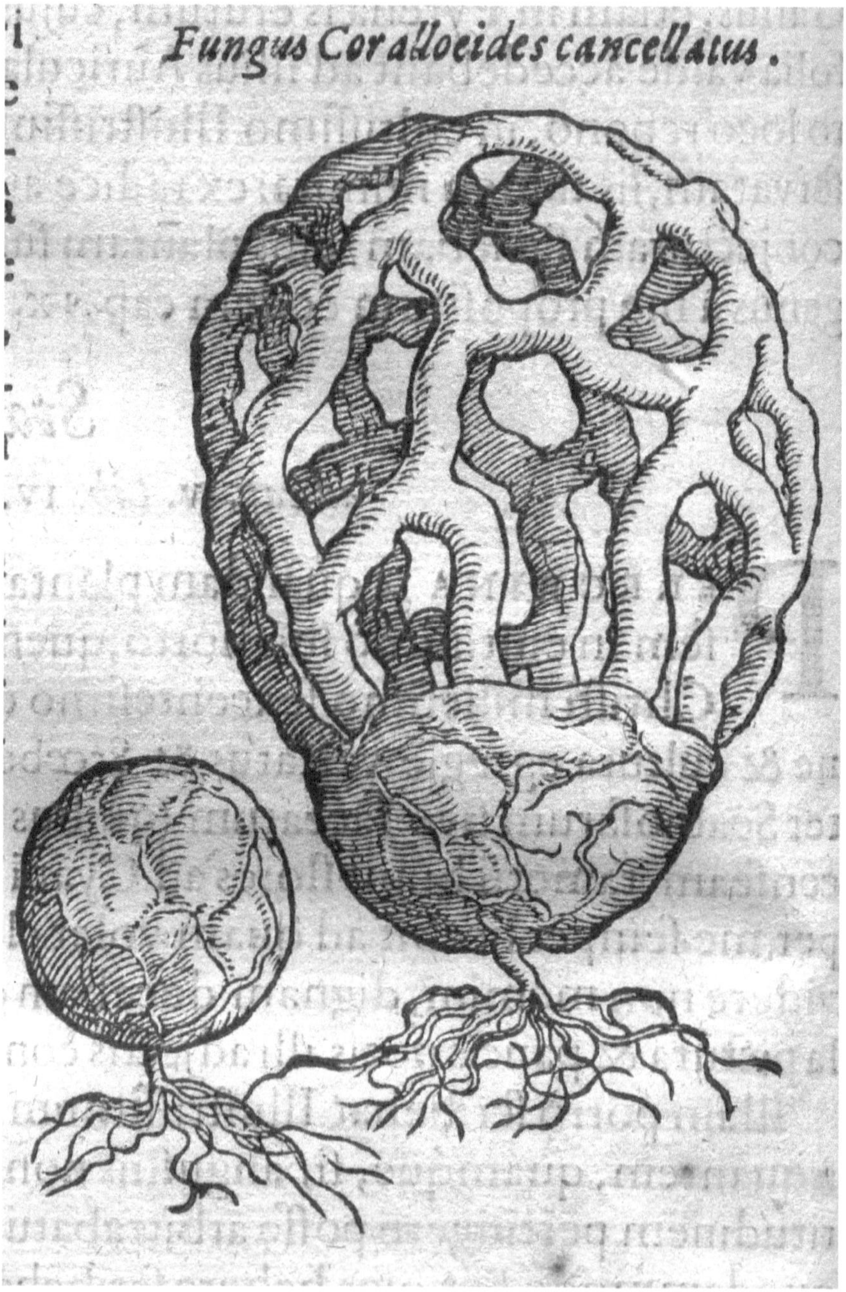

Figure 6. Peiresc's mushroom, Clathrus Cancellatus Fr. From unpaginated Appendix to Clusius's Rariorum in his *Exoticorum libri decem* (Leiden, 1605). Collection Leiden University Library, UBLWGW_THYSIA 2202.

the importance of autopsy, an interest in habitat and ecology, and a great talent for detailed description united Peiresc and Clusius, besides their common interest in creating a European community of information exchange. We have seen how Onorio Belli's methods of observation and analysis of Cretan antiquities and plants mutually reinforced each other, or more probably were part and parcel of a single method of observing and recording. In Peiresc's case, the same seems to have been true. Their example suggests that methods for acquiring expertise about nature were of great importance to the development of what would become known as connoisseurship in the domain of classical antiquities. Fieldwork, observation on the spot and local knowledge lay at the foundation of both.

In different degrees each of the three men discussed in the present chapter had made field botany his specialism. But the cultural roots and the aims of their practices differed. Levenier the plant hunter adapted the style of the aristocratic hunting expedition to his principal purpose as a collector of obtaining rare *naturalia*; he had hardly any interest in the medicinal uses of plants, and there is little evidence of the influence of a university tradition. Levenier never published anything because for him knowledge of the regional flora and of habitat was an instrument, not a goal in itself. Capperon's expertise concerning the regional flora of the Loire valley originated in the professional, non-academic tradition of pharmacy which was rooted in the local knowledge of simples. The interest in ecology and habitat here at first served the purposes of medicine, but soon went beyond those practical aims. For Peiresc, the aristocratic hunting trip appears to have provided a culturally acceptable form of field investigation, which served no other purpose than the collection of knowledge itself. Many other noblemen hunted, but only a few combined it with a serious interest in nature out in the wild and a great attention to precise, personal observation – which in his case were probably also moulded by his experience at the university of Padua. Three very different cultural traditions – non-academic folk medicine and pharmacy, fashionable and aristocratic hunting practices, and innovative academic approaches from Italy – thus joined during the late sixteenth century in shaping fieldwork in France and generating expertise concerning its native flora.

8 FRENCH COLLECTORS BETWEEN PORT AND COURT: THE DESIRE FOR RARITY

Levenier's correspondence with Clusius holds a surprise. After an almost complete absence, exotic plants emerge suddenly in a rather spectacular way in the very last of his eighteen long letters to Clusius written between 1597 and 1606. By June 1606 Levenier was expecting the return of his nephew Jean Bachelier, whom he had sent out some two and a half years earlier to 'Constantinople and other places in the Levant. He promises to bring me many beautiful and rare things which have not been seen before in Europe and to give me as much contentment as I can wish for upon his return' (Levenier, 13 June 1606). That emphasis on rare things which had not been seen before in Europe reveals the principal reason why Levenier had sent his relative to the eastern Mediterranean – much as Madame von Heusenstain had sent her own courier from Vienna to Constantinople. Levenier must have wanted *direct* access to exotic discoveries which were even more spectacular than his indigenous ones. A friend and fellow garden owner even told Clusius that Levenier had high hopes that 'one of his men', whom he had sent out to search 'all over the Orient and even the kingdom of China', would bring back 'excellent flowers which are completely unknown to us' (Vertunien, 25 February 1606). China was a bit of an exaggeration – we must suppose – but all the rest was true. The same letter of June 1606 in which Levenier told Clusius about his hopes for rarities from the Levant also reveals that he had direct contact by this time with the New World. Without further introduction or comment Levenier informed Clusius that he had recently received some *naturalia* directly from Mexico from an unnamed friend there, thus demonstrating that direct connections with Mexico were not a monopoly of Spain at the time. Nearly all items on the accompanying short list in Spanish, which he copied for Clusius, refer to Mexican fruits or trees, but it is unclear exactly which parts of them reached Levenier. For each item – such as 'Anonas de Guajaca', 'Cico çapot', 'elo suchil', 'cempual suchil' and 'elera de Havana' – the list gives the size, colour and flavour of the fruit or provides a short description of the tree or plant, and how it was used as medicine or food (Levenier, 13 June 1606).[1]

The earlier absence of references to exotica in Levenier's letters probably does not reflect a lack of interest on his side but simply a lack of access, while even the distinction itself between rare exotic and rare European plants may well have been quite irrelevant to Levenier because his prime interest was in rarity and first discovery, wherever a plant came from. That much can be inferred precisely from this venture of obtaining rare plants from the Levant via his personal agent. It suggests that this wealthy and well-connected man who lived in a port and knew how to make use of overseas transport, may have finally seen his chance to surpass even the excellent collections of his fellow enthusiasts and rivals in his home region by obtaining rare and exclusive novelties from the Levant. We know who some of those men were and can to some extent reconstruct what their interest in *naturalia* consisted of.

Bourgeois Collectors in South-West France

During the second half of the sixteenth century – the period of the wars of religion – the south-west of France was one of the core zones of French Protestantism. Its two ports, Bordeaux and La Rochelle, had two types of hinterland: nearby Poitiers, a provincial centre with a university and strong ties with the famous university of Montpellier; and distant Paris, capital and court residence. Poitiers was strategically located on the point where the main roads from the ports of Bordeaux and La Rochelle to Paris joined. While Bordeaux was mainly associated with trade and shipping between the northern and southern parts of Europe and in particular linked to Middelburg in the Netherlands via the wine trade, La Rochelle is better known for its transatlantic connections. From the late 1590s onwards, Clusius was in regular contact not only with Levenier, but also with a circle of collectors and garden owners in Bordeaux, Poitiers and La Rochelle. Levenier in Bordeaux made sure that packages with plants, bulbs or seeds from Clusius were transported to Poitiers and, vice versa, that letters or packages from the hinterland were shipped to Leiden via Middelburg. The four men with whom Clusius corresponded shared some of the bulbs and seeds or plants sent to them with other members of this circle. Levenier knew many (and perhaps all) of its members personally and must have been very well aware of the important role of exotic *naturalia* in their collections. Traces of the French contacts with America emerge in their family histories and collections.

The wealthy apothecary, poet and collector Paul Contant (1562–1629) in Poitiers is the best-known member of this circle thanks to his *Le Jardin, et cabinet poétique* (1609), a long poem that catalogues his collection of rarities ('singularitez') and garden.[2] Contant had an enormous library of some 4,000 books and a collection which reached the impressive size of some 4,500 objects: according to the inventory of 1628 it comprised some 100 bottles of essences or perfumes, an enormous number of small bronze statues, a 6-metre kayak, 150 shells, 150 'things from the earth' including fossils, more than 100 animals (or parts of them), a seahorse (found by himself in Venice) and four crocodiles. Clearly, many of these had no practical use to an apothecary. The collection had a strong focus on *naturalia* and reflected Contant's interest in rare and extraordinary things, monstrous forms, fossils and, especially, exotic *naturalia* and *ethnographica* from all parts of the world, such as ostrich eggs, the jaw of a lion, corals, 'Indian' feather decorations, armadillos, toucans, et cetera. A few objects must have reached him directly from North America, since Contant thanks Du Mont, governor of Nouvelle-France, in his *Le Jardin*. Contant's collection was organized to evoke wonder and highlight the spectacular. Both his collection and garden were famous and Contant showed them to French and foreign visitors whose high rank he rather pompously emphasizes in his poem. Yet, its monetary value was limited, since it mainly consisted of 'natural curiosities, of small market value'.[3]

Contant had visited Italy as a young man in 1582. He travelled via Montpellier, herborized in the Savoy and visited Rome, Venice and Padua. Italian examples may have given him the idea of combining a *naturalia* collection with a private botanical garden and a workshop or even *studiolo*. His garden was adjacent to his apothecary's workshop, but it is not clear whether it also flanked the space which housed his collection.[4] Just like Ferrante Imperato's collection in Naples, Contant's *naturalia* collection comprised an enormous herbarium, which he must have begun as a young man. It amounted to fifteen volumes by the time the inventory of his collection was made, containing some 3,000 indigenous and exotic plants. These volumes are depicted in one of the illustrations in his *Le Jardin*: an engraving of a large chest or cabinet with thirty-two small drawers in the top part and three bookshelves in the bottom part has an inscription saying that the drawers contained nature's *mirabilia*; on the shelves the fifteen volumes which contained the dried plants can be discerned (Figure 7).[5] Interestingly this picture shows us thirteen spines with the letters of the alphabet on them – suggesting that the dried plants were classified by name – and two spines which bear witness to another type of system, inscribed respectively: Bulbs, Vegetables.

While exotic *naturalia* and *ethnographica* occupied a prominent place in Contant's cabinet, his garden too comprised a number of non-European plants, but his poem indiscriminately mixes these and the European plants, mentioning the Lebanon cedar, cypress, pistachio tree, laurel, acacia, tulip, myrthe, aloe, opuntia cactus, crown imperial, double peony, anemone, ranunculus, hemerocallis, moly, ornithogalum, colchicum, narcissus, gentian, iris, columbine and martagon lily. The order in which Contant presents the plants in his poem – according to size and starting with the biggest, the trees – was probably inspired by the wish to impress the reader, just as his cabinet was meant to astound the visitor.[6] It is in

Figure 7. The Contant collection with herbarium volumes organized by alphabet or otherwise, in P. Contant, *Le Jardin, et cabinet poétique*, revised edn (Poitiers, 1628). Private collection.

the same context of rarity and attempts to impress that we should understand his emphasis on special tulips, the presence in his garden of shrubs and trees that were not often found at this latitude, and the unusual characteristics of plants that were less special. How the plants were actually distributed in his garden we do not know, but the distinction between exotic and indigenous does not seem to have played an important part in its organization, whereas Contant's strong emphasis on rarity fits in perfectly with Levenier's quest for it.

Contant's poem mentions Clusius both as author of scientific works and donor of plants. The references to twenty-three others from whom Contant received plants or other *naturalia* give us some idea of the social composition of this circle of collectors, *naturalia* and garden lovers. Among them we find two noblemen, four clerics, ten physicians, six apothecaries and one surgeon. Nearly all lived in south-west France (one in Bordeaux; nine in Poitiers and six in La Rochelle). Only two were based in Paris: one of these was Jean Robin Sr, the royal *herboriste*. There were some intriguing figures among these collectors and their further contacts in south-west France. The apothecary Samuel Veyrel, who owned a cabinet of curiosities at Saintes (between La Rochelle and Bordeaux), for instance, must have been a relative of Nicolas Veyrel, an apothecary in the same town, who in 1558 was accused of heresy together with Bernard Palissy (*c.* 1510–89), the French Protestant potter famous for his ceramics which represent and contain amphibians and other small animals. The poet-apothecary Paul Morisseau in La Rochelle, who gave Contant a 4-metre crocodile, came from a family with a long tradition of trade relations with the New World: they mainly sold weapons. And the Poitiers physician Jean le Roy, Seigneur de la Boissière, donated a monstrous lamb and a dog with eight feet to Contant's collection, but is better known as the painter of an album with drawings of rare plants which is still preserved in Paris and was dedicated to Thomas Garnier, yet another apothecary at Poitiers.[7] We find ourselves, therefore, in the circles of wealthy, town-based bourgeois professionals and a few noblemen. The prominence of men connected with medicine is striking, with almost equal proportions of apothecaries and physicians. According to Schnapper, the number of apothecaries in Poitiers during this period was, indeed, conspicuously high, and it looks as if the number of physicians was considerable as well. Nearly every inhabitant of Poitiers and La Rochelle who had some kind of medical training seems to have been affected by the passion for rare *naturalia*.[8]

Le Coq and Vertunien at Poitiers

Two men in this circle deserve some further attention. Contant's poem refers frequently to Pascal le Coq, Seigneur de Forges (1567–1632), alias Gallus, for both his donations of plants and exchanges with Clusius. It was probably he, in fact, who passed on some of Clusius's gifts of bulbs and seeds to Contant. Le Coq, moreover, made the drawing on which the illustration of Contant's cabinet with the fifteen herbarium volumes in *Le Jardin* was based: in fact, the text at the bottom of the picture of the cabinet says 'P. Le Cocq Med Doctor Et Decanus Pict'.[9] Le Coq had a distinguished and interesting career. He came from a wealthy family and was known as a learned man who had studied medicine at various European universities. He spent some nine years travelling all over Europe (and possibly even to India), during which period he explored their local flora and learned several modern languages besides Latin and Greek. After his return to Poitiers he became a fully qualified physician at the medical faculty in 1597, and eventually its dean. Plants and horticulture were therefore lifelong interests of his, and Le Coq is supposed to have acclimatized the Persian lily and anemones in Poitou. In 1621 he sold his garden to the French king, whereupon it became the official Jardin des Plantes of Poitiers.[10]

Le Coq's father-in-law was an equally learned physician at Poitiers. François de Saint-Vertunien, Seigneur de Lavau (*c.* 1540–1607), was a moderate Protestant and humanist, a member of the local elite and close friend for thirty years of the famous Scaliger. Vertunien la Vau – as he usually signed his letters to Clusius – also maintained friendly relations with such noted figures as the philologist and classical scholar Isaac Casaubon, the historian and magistrate Jacques Auguste de Thou, the poet and humanist Scévole de Sainte-Marthe and the humanist-jurist Claude Dupuy, a great friend of Pinelli's.[11] Like most physicians in Poitiers, Vertunien took his medical degree (1567–8) at the medical faculty of Montpellier rather than in Paris, and was taught there by Clusius's tutor Rondelet and Rondelet's successor Laurent Joubert. Vertunien regarded the latter in particular as his mentor, and dedicated his translation of Greek texts by Hippocrates into Latin to Joubert. Vertunien took up a medical practice at Poitiers, where the medical faculty granted him and three other physicians a kind of monopoly on the local medical market. After 1604 his health declined and he visited several spas trying to cure his gout.[12]

Neither Le Coq nor Vertunien is ever mentioned in Clusius's printed works, but their plant collecting practices show, like those of Levenier and Peiresc, how collecting and growing rare plants had become part of the culture of the educated high bourgeoisie and local nobility in provincial France by the late sixteenth century. Unlike Contant, who had neither Latin nor a university training, Vertunien and Le Coq formed part of the Republic of Letters given their formal training, wide-ranging experience, travels, erudition, linguistic abilities and personal contacts with the intellectual elite of their day. Vertunien's letters to Clusius show him to have been both up to date in his reading of the contemporary and classical botanical literature, and well versed in the practice of growing rare plants and plant identification. Referring back to his student days in Montpellier, Vertunien mentioned in 1599, for instance, that he was glad to see that Clusius had recognized an erroneous identification by a common friend, the late Monsieur Posthius in Avignon, who had mistaken a common plum tree for the myrobalan (Prunus cerasifera). And Vertunien could not help adding: 'by which he [Posthius] showed clearly that he was not well versed in this knowledge of simples'. Vertunien reminisced that even when both of them had studied under Joubert in Montpellier, Posthius had 'loved poeticizing more than herborizing' (Vertunien, 7 April 1599). The terminology used by Vertunien to describe this type of knowledge and his own and Le Coq's interest in *naturalia* is significant. In an explicit reference to the key notion of curiosity he calls himself and his son-in-law 'two very curious men'; and he further speaks of Clusius's great services to the Republic of Letters, in which he clearly included all those who studied nature (Vertunien, 18 October 1601). For Vertunien, therefore, who was first and foremost a medical man, plant identification formed an important part of the knowledge of simples – in the sense of plants in general and not just medicinal ones – which he saw as a serious branch of knowledge in which members of the Republic of Letters engaged.

Just like some other garden owners in the Southern Netherlands and Italy who regarded their gardens as living collections, Vertunien and Le Coq used Clusius's published works as a kind of checklist of desiderata for theirs. The two men assiduously studied Clusius's *Exoticorum*, which he had sent them:

> and we will hold it among the most precious treasures together with the first volume [i.e. the *Rariorum*], which we constantly have in our hands day and night: comparing it with the beautiful flowers and plants with which you have adorned our gardens and those of our friends. (Vertunien, 25 February 1606)

And Le Coq sent Clusius a list of the simples described in the latter's works that were not yet in their possession – just in case Clusius had some extra exemplars (Vertunien, 7 April 1599). As Vertunien's reference to herborizing during his student's days in Montpellier already indicated, his interest in plants (both medically useful and ornamental ones) was lifelong. Several decades before his exchanges with Clusius he already owned works by naturalists such as Pierre Belon and Garcia da Orta (Vertunien, 25 February 1606).

Vertunien and Le Coq had more than one garden in Poitiers. But their interest in gardening, talking about *naturalia*, reading and discussing Clusius's works was to such an extent a joint one that Vertunien speaks of 'my' garden, 'our gardens', Le Coq's garden, and sometimes of 'our garden'. Some of the letters are even signed by both men. We do not have a catalogue of their plants, but their letters evince a great interest in rarities, a fascination with colour variation and a predilection for rare white flowers reminiscent of Contant, Levenier and many other gardeners and plant collectors among Clusius's correspondents. The two rare plants from their part of the world which Le Coq and Vertunien offered to send Clusius were a cyclamen with ivy-shaped leaves and the white fritillary 'of which only two of three are found among a million others'; they finally managed to find one for Clusius (Vertunien, 18 October 1601).[13] Vertunien and Le Coq grew tulips from seed and Vertunien informed Clusius every year about the vicissitudes of the tulip bulbs which the latter had given them:

> This year we have had at least thirty different tulip flowers in our garden, thanks to your liberality, so that seeing this admirable beauty we do not cease to pray the Lord to give you a happy and long life ... Among others [there were problems with] one which three years ago produced a flower with a yellow base and white-edged purple petals, among the most beautiful that can be seen; a year earlier its stem was gnawed by a worm and it did not flower; having dug it up last year to plant it elsewhere, it has died to our great regret. But for one loss, ten are gained, which consoles us and is the same as what has often happened to you. (Vertunien, 4 October 1600)

In the spring of 1601 Vertunien told Clusius that many tulips were coming up, which made him hope that he would have hundreds of flowering ones (Vertunien, 10 April 1601). If Schnapper is right in stating that tulips were not widely distributed in France during the first quarter of the seventeenth century, could be mainly found in the botanical garden of Montpellier in 1598, and were on the whole a Flemish specialty, that would make Vertunien's garden with its large numbers of tulips all the more special.[14]

In many cabinets of curiosities of the period the category of the rare shaded into that of the monstrous and the anomalous.[15] They contained items like Contant's eight-legged dog, a deformed foetus, or objects that defied classification because they blurred and called into question the borderlines between *naturalia* and *artificialia*, animal and plant, or animal and mineral. It is unusual, however, to find an explicit example of the fascination with the monstrous in the context of gardens and living plants. Vertunien's garden produced such a monster. In April 1601 Vertunien reported that an orange crown imperial already had five flowers, but, he continued, there was also another

> monstrous one (which is the first one seen in Poitiers) with a paler colour than the ones you have sent us, which has grown two stems, one three finger-breadths wide, and the other two, which will have more than 50 flowers and in my opinion two or three tiers which however we cannot yet judge very well since it has just begun to grow. (Vertunien, 10 April 1601)

It was not the unusual, pale colour that made him use the term monster for this particular plant, but its excessive number of flowers – just as Contant's eight-legged dog was classified among the monstrous because of its abnormal number of limbs.

Perhaps Vertunien's explicit use of the term monstrous in this description was connected with his (and his son-in-law's) medical training. That same professional background may also have prompted these men to experiment with plants in ways that are reminiscent of the anatomical theatre. They cut in half the bulb of one of their crown imperials, and put the two halves back together in the soil in order to see what would happen. In the course of the summer the bulb only developed one flower stalk. When they dug up the bulb again, in the autumn, to check to what extent the two halves had grown together, they found that the halves were indeed rejoined and had produced six small bulbs the size of nuts, grouped around the big bulb and attached to it by long fibres (Vertunien, 10 April 1601). If their 'dissection' of this bulb was indeed inspired by their medical training in which ocular observation was of crucial importance, their description is one of the rare reports of botanical experiment by means of 'anatomical' methods.

Unexpectedly, given their easy access to the ports of Bordeaux and La Rochelle, Le Coq and Vertunien and, it seems, more indirectly the whole Poitiers community of naturalists, were mainly on the receiving end in the exchanges with Clusius. Few counter gifts apart from books, the white fritillary and the special cyclamen are mentioned. In this respect they were very different from Levenier. But what united all these collectors and plant lovers in south-west France is their great interest in the rare, monstrous, unusual, exceptional and marvellous. Exotic *naturalia* almost automatically belonged to this category, but to all of these collectors rarity and the capacity of evoking admiration for the owner-collector seem to have been much more relevant characteristics than a non-European provenance.[16]

French Court Connections and the Robins

We can make use of Levenier's personal network and Clusius's connections to move from south-west France to the French courts around the turn of the sixteenth century. The first of Levenier's 'court links' takes us some 100 km south-east of Bordeaux, to the small town of Nérac, another centre of French Protestantism during the second half of the sixteenth century. Nérac was one of the two seats of the court of the Albret family which ruled over much of Guyenne and Gascony in this period. With the formal conversion of Jeanne d'Albret to Protestantism in 1560 the region became a haven for many Huguenots. The surroundings of Nérac formed an attractive location for the construction of new country houses, and the Albret court and castle at Nérac became a focus for poets and writers. Jeanne d'Albret was the mother of Henry of Navarre (1553–1610), who would be King of France from 1589 to 1610 as Henry IV and order the creation of the first royal botanical garden in Paris. It was also the Albret family which created a large park and the Jardin du Roy at Nérac.[17]

In this context of parks, country houses, court culture and gardens we find Levenier's friend André de Nesmond, Seigneur de Chezac (1562–1616), president of the Parliament of Bordeaux, who lived in or near Bordeaux itself but was also connected with Nérac.[18] Nesmond seems to have been as much of a flower lover as Levenier. A common acquaintance, who knew both Levenier's garden and that of the Robins in Paris, even called Nesmond's garden 'the most beautiful in France for its abundance and variety' (du Casse, 2 February 1599).[19] We have almost no information about its contents, but it is clear that Nesmond had many unusual bulbs, both exotic and native wild ones, and was almost as interested as Levenier in rare indigenous plants although he was far less energetic in his search for them. Instead of undertaking a field trip himself, Nesmond sent out his servants to obtain rare wild plants from the spots where these had been discovered by Levenier – not in order to compete, as he was careful to explain, but in order not to deplete Levenier's stock. Nesmond also received bulbs from Venice, which probably arrived there from the Levant, so he must have used his own international contacts to enrich his garden (Levenier, 28 August 1601). Levenier often shared Clusius's gifts of bulbs (such as lilium susianum and lilium bizantinum miniatum) with Nesmond and, vice versa, Nesmond's gifts to himself with Clusius. Among the latter were bulbs of the white mountain martagon which Nesmond had discovered in the hills thanks to the indications of a local apothecary.[20] The fascination of both men with rare white flowers is manifest from Levenier's description of Nesmond's discovery:

> This plant, as I have heard on several occasions from the President, in no way differs from the common mountain martagon, with the same leaves and stalk, but fading in colour from green to white, with a white flower just like the columbine. It has a bud at the top which when it first opens is identical to the white lily in the white that is revealed, which then turns into white flowers without any spots or dots. (Levenier, 20 October 1598)

On behalf of Nesmond, Levenier sent bulbs of this white mountain martagon to the two persons whose botanical knowledge they respected most: four to Clusius and two to Robin in Paris.

After a long period of civil and religious wars Paris was reviving under the religiously tolerant King Henry IV. His initiative to establish a royal garden in Paris fits in well with his Nérac background and the Albret family tradition of creating parks and gardens. Henry IV may already in 1590 have been thinking of establishing such a garden, since Clusius was asked by his friend Marnix de Saint Aldegonde in the winter of 1590–1 whether he might be interested in working for the French King – presumably as a court botanist or creator of a new garden

– once the troubles were finally over. Marnix wrote: 'The king himself, who is a great lover of plants, desires nothing but peace after the furore of these civil wars ... so that he can restore botany in France' (Marnix, 8 November 1590).[21] Clusius was not interested, but a new garden was indeed created in Paris: by father and son Robin, who were also the two principal figures linking the Bordeaux-Poitiers-La Rochelle naturalists and the court of Henry IV and his Italian wife since 1600, Maria de' Medici. On 30 June 1597 Henry IV ordered Jean Robin Sr (1550–1629), court apothecary, and already since 1586 king's herborist, to create a new garden for the growing of simples for the medical faculty.[22]

By then Robin already had a famous private garden on the tip of the Île de la Cité, and it was probably this garden which contained a very fine collection of rare plants with an especially interesting number of exotica from the Orient, Africa, the West-Indies and a few from South America.[23] In April 1599 Robin Sr listed special bulbs which were then growing in his garden, such as an early flowering yellow African narcissus, a very small double white narcissus, a totally white big narcissus, a small totally white narcissus which carried ten to twelve flowers on one stem, a tulip with very green leaves and a purple flower mixed with yellow, ornithogalum arabicum, and a lilium susianum (Robin Sr, 10 April 1599). A month later a mutual friend wrote to Clusius about the many flowering narcissi and an especially beautiful yellow hemerocallis in the garden of the Robins (Combaud, 17 May 1599). Robin Sr himself planned to have this flower's portrait painted and called it a 'very beautiful and rare plant that has not been observed before by any other author', and offered Clusius a bulb (Robin Sr, 9 February 1598). According to the catalogue published in 1601 Robin's garden contained some 1,400 plants. Jean Robin Jr (1579–1662), also known as Vespasien, followed in his father's footsteps, and eventually was appointed lecturer in what by 1624 had become – on a new site in Paris – the Jardin Royal des Plantes.[24]

Among the exotic plants in the collection of the Robins the presence of North American plants is particularly significant. Warner has discussed the role of both Robins with respect to the introduction of North American and Canadian plants in Europe in some detail, and suggests that in 1601 two species could even be found in their garden that had probably come from the Cartier voyages to Canada of 1534–42: a thuja occidentalis and an actaea spicata (also known as snakeroot). When a new catalogue of the Robins' garden was printed in 1623 it comprised some 1,700 plants, including many of those already listed in 1601, but also many new ones, particularly from Canada.[25] Shortly before 1609 the spectacular Mexican narcissus jacobeus could be found in the Robins' garden, and when the American passion flower bloomed there for the first time at these latitudes in 1612 – exactly the year in which Levenier was asking Caccini for its seed – that special event was commemorated in various plates.[26]

In spite of the important presence of exotic and in particular American plants in the Robins' garden, the title of their 1601 catalogue explicitly indicates that they cultivated indigenous as well as exotic species: *Catalogus stirpium tam indigenarum quam exoticarum quae Lutetiae coluntur* (Paris, 1601). All flowers that were fashionable at the time increased spectacularly in numbers in their garden if we compare the lists of 1601 and 1623: anemones from 26 to 55; iris from 22 to 36, cyclamen from 3 to 16, hyacinths from 25 to 56, ranunculus from 19 to 32, narcissus from 26 to 103, and there were many tulips as well.[27] The particular mix of exotic plants with the fashionable flowers of the period, and a special interest in North American and Canadian plants characterized not only the plant collections of the Robins but also the (slightly later) Parisian gardens of the brothers Pierre and René Morin and their successors, prominent nurserymen of the 1620s to 1660s. Pierre Morin specialized in anemones, ranunculi, tulips and irises, and collected butterflies as well as shells, while René grew a wide variety of bulbs, besides herbaceous plants and shrubs. Pierre Morin's garden included some plants from North America and Canada, and several of the Morins' plants were introduced into England, among others via John Tradescant the Elder.[28]

Both Henry IV and his wife Maria de' Medici took an active interest in rare *naturalia* and gardening. The Queen in particular stimulated the fashion of nature at court. She personally visited Robin's private garden on the Île de la Cité, and promoted flower painting and flower embroidery – the latter had become popular among noblewomen in Britain and Europe, such as Mary Queen of Scots and Catharine de' Medici, in the second half of the sixteenth century – as well as gardening and the passion for live flowers. In fact, Pierre Vallet's *Le Jardin du Roy tres chrestien Henry IV* (Paris, 1608), in which some of the exotic flowers from Guinea, Canada and Virginia in Robin's garden are depicted, was dedicated to Maria de' Medici. Vallet was a gardener, court painter, and draftsman, and his book, which is now famous for its botanically accurate and naturalist illustrations, was actually meant as a pattern book for embroidery.[29] It was also Maria de' Medici who in 1603 recommended young Robin to her family in Florence, when he was sent on a plant-gathering trip to Italy:

> the son of a herborist and *simpliciste* of the King My Lord who has, upon his commandment, some years ago started a garden in Paris with various rare and sought after trees, plants and *simples* from far away provinces; he is now departing for Italy in order to see what kind of rare things he can find there, I beg of you to be pleased to let him see your gardens and to give him permission, if there are some simples there of which he could make a drawing or take seeds, to do so.[30]

Via his correspondents in Italy Clusius was kept informed about young Robin's Italian trip. Fra Gregorio da Reggio reported that Casabona had received two young Frenchmen (one the son of 'Giovanni Rubino')

> who had come here on purpose purely to search the whole of Italy for simples ... they stopped here with me in Bologna for two days and I believe that they are now in Rome, I don't know whether they will pass through Naples, they have promised me to return in the month of October. (Fra Gregorio, 8 July 1603)

Warner argues that the Paris garden of the Robins was important 'as a repository of plants that had been brought into France by the earliest exploration and settlement of Canada', and points out that Canadian plants were also sent by the Robins to the naturalists John Gerard and John Tradescant the Elder in England, Gaspard Bauhin in Switzerland and the Cardinals Farnese and Barberini in Rome. Vice versa, some Virginian plants reached the Robins from England.[31] Her emphasis is more on diffusion, however, than on the question of how the exotics actually reached the Robins or of where the rare non-exotic plants in their collection came from. We have too little information to trace their sources of rare plants in detail, but young Robin's plant collecting expeditions inside and outside Europe demonstrate that such trips were organized with this sole purpose, and that no *a priori* distinction was made between indigenous wild plants of Europe and exotic ones. As indications of early professionalization – but not specialization – young Robin's trips are as significant as Levenier's expert knowledge on plant geography and his proficiency in field botany. And it was, appropriately, Levenier who acted as host and guide to the young Robin in 1602 when the latter was on a trip to the Pyrenees and northern Spain.[32] From there Robin Jr sent packages with local rare plants to his father, but much of the credit for these discoveries must go to Levenier, whose guidance must have been of enormous value to the young Robin. Over the years Robin Jr greatly added to the Parisian garden, in particular plants from Italy, Spain, the Pyrenees and Guinea. He even had a catalogue printed of the plants he brought back from Guinea in 1603 under the title *Exoticae quaedam plantae a Johannes Robino juniore ex Guinea et Hispania delatae anno 1603*.[33] There is some evidence, moreover, that Robin Jr was an all-round naturalist whose interest extended to insects as well, and a collector of curiosities: the English flower and insect painter and collector Alexander Marshal noted next to his drawing of an African sword fly that he had passed on one of his two sword flies to the young Robin, 'a great Colecteur of Curiosities'.[34]

While the young Robin travelled abroad in order to find new plants, his father stayed closer to home. He undertook no strenuous field trips of the kind that Levenier or indeed Clusius were involved in. Yet, he too collected rare plants for the Parisian garden and contributed in a major way to its riches. In 1601 he made a long tour through the Southern Netherlands in order 'to see whether he can find something beautiful and rare for the King's garden' (Combaud, 28 August 1601). In 1601 Robin Sr visited all of the gardens in Plateau's home town Tournai, whether they belonged to apothecaries or men of the Church, and amassed a large number of common plants. Robin had gone through more or less the whole of the Southern Netherlands in this way, from Tournai to Ghent, then to Bruges, Antwerp, Malines, Brussels, Mons and Valenciennes, before returning via Cambray to Paris. He followed up this tour with written demands for rare plants to many garden owners, a practice which was appreciated all the less since he gave little in return:

> Since his return to Paris I have received two letters; the messenger told me that there were as many as eight or ten letters for Tournai alone ... in which he asked for new plants. I don't think he will have received many. From me he had not a single one. For more than 15 or 16 years I have not received a single plant from him. I also do not feel obliged to him, because he asks for such a large number of plants all at once that it makes one feel disgusted; if he still has one, he asks for 10 or 12. Seeing that he did not get from me what he desired, he has asked whether I wanted to sell him some. I have replied to him that they were not for sale. Since then I have had no news from him ... I have little trust in the French, it would be neither the first nor the second time that I have been deceived by them. (Plateau, 8 February 1602)

What at first sight might look like a division of labour between the son, who went out to obtain exotica, and the father, who stayed closer to home collecting indigenous flora, was in fact nothing of the kind. The young Robin simply undertook the more strenuous and dangerous trips, as befitted his age. Both collected indigenous and exotic plants. So long as the plants were new to the Robins and preferably rare, it was irrelevant whether they came from the gardens of the Italian nobility, the Pyrenees, the Guinean coast, the gardens of plant collectors in the Southern Netherlands or – indirectly – from the Levant.

Conclusion

Looking back on the chain of connections, from Bordeaux and La Rochelle to Poitiers and eventually Paris, and from Levenier, the Contant circle, Vertunien and Le Coq to the Robins, it is clear that the distinction port–hinterland did not coincide with that between a specialized interest in the exotic or the indigenous. The presence of exotic *naturalia* and *ethnographica* in the collections of Contant and his circle would be hard to understand without easy access via the nearby ports, but exotic *naturalia* could be admired as well in the court-linked gardens of the Robins in Paris. Indeed, every collector who could lay his hands on exotica – whether in the form of living or dried plants, seeds, roots, bulbs or of (parts of) animals – was proud to own them. The attempts by Levenier and Peiresc to obtain exotic plants and animals that had not been seen by others show as much. In so far as living exotic plants were concerned, the best place in France to see them in the period between about 1595 and 1630 must indeed have been the Robin (and Morin) gardens in Paris rather than any private garden in a port town like La Rochelle or Bordeaux. Relatively easy access in a purely logistic sense to exotic *naturalia* arriving by ship was, therefore, not in itself enough to stimulate the formation of great collections of exotic *naturalia* or that of high-quality expertise in this field.

In so far as French indigenous or more generally non-exotic plants were concerned, the Robins were leaning heavily on other experts: foremost among these were men like Levenier (and perhaps Nesmond and Vertunien) for the rare plants from south-west France and the Pyrenees, and the gardeners of the Southern Netherlands for indigenous plants of that region as well as novel plants from the Levant that were by then being acclimatized in Europe. The fact that Robin Sr combed through so many gardens of the Southern Netherlands incidentally confirms the high botanical and horticultural repute in which that country was held. It is likely, however, that the private gardens of Levenier and probably Nesmond in France and those of various plant collectors in the Southern Netherlands continued to equal if not surpass the collection of the Robins in so far as rare indigenous plants were concerned – and this may even have been true for bulbs that had originally come from the Levant.

For all of these French plant lovers, collectors and experts, rarity took priority over any distinction between exotic and indigenous – even though that distinction was indeed made at the time. Most exotic *naturalia* were rare by definition, and therefore regarded as relevant and fascinating. But the category of the rare also included most newly discovered and attractive indigenous *naturalia*, plants with flowers of an unusual colour, 'monstrous' varieties and other natural forms that induced wonder. Perhaps the clearest sign of the explicit awareness of the importance of novelty and rarity among both the bourgeois provincial and the aristocratic or court-connected plant lovers discussed in this section comes from Jacques Plateau. In 1603 he rather crossly wrote to Clusius that not only young Robin but half of Paris now went off to the Pyrenees in search of rarities: 'these Parisians now are flocking there as in a procession' (Plateau, 15 March 1603). His (rather exaggerated) point as a collector and connoisseur of rare *naturalia* was that rare indigenous plants were becoming accessible and thus threatened to lose their rarity. Robin Sr himself had complained that the French plant sellers who sold wild plants from the Pyrenees (and about whom we have heard Levenier's scathing remarks) 'went from town to town selling a great abundance of their plants so that they make all of them common' (Plateau, 15 March 1603).[35] That was also one of the reasons why young Robin was sent out farther afield in order to come up with real novelties – in his case not for commercial reasons but in order to maintain the high standards of rarity of the Robins' Parisian garden.

Given the priority of rarity over the distinction between exotic and indigenous, it is understandable that we have not found any signs of an intentional specialization in either indigenous plants or exotic ones. Whether a collector of *naturalia* emphasized exotic or indigenous plants seems to have depended mostly on practical factors such as access, wealth, the right connections and personal taste. Levenier's great expertise concerning the regional flora of south-west France was not merely the result of a choice for the indigenous but also of a (perhaps temporary) lack of direct access to non-European novelties. Capperon might well have wanted to know more about plants outside his Loire valley, but he probably never had the opportunity to study them. And Peiresc, who did indeed have the means and connections to obtain both indigenous and exotic *naturalia*, never chose the one or the other, but collected and studied both.

Whatever the intentions and aims of these sixteenth-century lovers of nature may have been, the effects in terms of the formation of expertise were indubitable. The resulting expertise concerning indigenous plants was of high quality indeed and Levenier, Capperon and Peiresc set standards of expertise that helped to shape the standards of professional botanical knowledge of the age even if none of them published. After all, Levenier's expertise was respected and his guidance accepted by both the Robins and Clusius; Capperon was sought out even by English naturalists for his specialized ecological knowledge; and Peiresc's accounts helped to set standards of descriptive precision in natural history. In so far as indigenous plants in France were concerned, the practices of the men discussed here thus contributed to the professionalization of botanical knowledge in particular with respect to field botany, ecological knowledge, habitat and plant geography; in this respect they were similar to the Italian 'field botanists' discussed earlier.

In terms of exotic *naturalia*, two different strands of expertise can be discerned. The first one consisted of a rather incidental kind of knowledge which is hard to pin down and which resulted from the relatively haphazard importation of *naturalia* from Africa and America that ended up in the provincial curiosity collections of south-west France. The second and more systematic one concerned exotic plants – in particular those from North America and Canada – which found a place in the Parisian, court-connected gardens of the Robins, and somewhat later in those of the Morins. Those plants were collected, catalogued, named and in most cases described and depicted. They formed part, moreover, of professional and international exchanges between plant collectors, botanists and garden lovers in France, England and Italy which promoted their introduction in various European countries and their insertion into the growing body of botanical knowledge.

V HOLLAND

9 DUTCH PORTS: CURIOSITY AND THE EXOTIC

In the appendices to his *Rariorum* of 1601 Clusius included a very lengthy letter by the Middelburg town physician Tobias Roels with marvellously detailed descriptions of manioc, yam, several palms and various other tropical plants. Roels noted in particular how they were grown, processed and eaten or otherwise used, and mentioned his native and European informants.[1]

> Our host ... showed us a thick, ¾ foot long root, blunt at both ends. At the bluntest end six or seven small thorny branches sprouted forth ... which should have carried many leaves; they consisted of 5 or 6 segments and were as long as the palm of a hand. But no leaves could be seen. The root has an almost black bark but is completely white inside. The single little branch that I send you was cut off from this root by me. As you see, it is woody, round, flexible, tough, and rough on account of the many very small thorns. I have asked him [i.e. his host] to send you the root ... which he took for the *inhame* [i.e. yam]; but I have heard differently from the Prefect of the Indies, who had added twelve black slaves from St Thomas [São Tomé] to his other servants. When he had fallen ill on account of the change of air and asked me to come and cure him ... I did my best to find out from them what was the case. He stated that this root was the one from which a flour is made that is called the flour of the *Manyouca* root, and which foreigners incorrectly call Mandioca. The *Manyouca* root produces neither flowers nor fruits, but is planted with its roots after it has been cut into pieces. Once these have been planted, they grow to be a thick and big root, usually after a year, sometimes after six months, depending on the fertility of the soil. As a result the roots sometimes grow as long as two times three quarters of a foot. The *Manyouca* was brought some years ago from Brazil to the island St Thomas which lies near the equator at 33 degrees longitude.
>
> The common name among the inhabitants of this island is *Manyouca*. But some of them declare that the Brazilians call it *Yuca*. Undoubtedly the name of the plant was unknown to the Ancients, since these western lands were hardly known to them.

> The root is for them an indispensable foodstuff. It contains a poisonous juice, but they nonetheless do not abstain from it. They know how to prepare it in such a way that the flour and the bread made with it can be eaten without harm. After the bark has been taken off, they cut the root into pieces and sticks, which they lay in the sun to dry, after which they pound them with wooden or stone utensils in very small pieces. Then they put them into sieves of tree bark or straw, and pound them with their feet until all the juice has been pressed out, which juice cannot be drunk without risking one's life. The afore-mentioned negroes told me that this juice cannot be turned into something that can be used, not even by boiling it, as has been stated by the learned Spanish physician Nicolaus de Monardes in his book about the simples of the New World ... I have also questioned others about the *Manyouca*, in particular that sailor who was a prisoner for three years on Hispaniola, an island in the New World, and the Portuguese surgeon who reached us some days ago from the town San Domenico [Santo Domingo] on the island of the same name. They assured me that this juice is most harmful and even lethal in its raw condition, but loses all its harmful effects when the fresh roots are cooked together with meat, as they had often observed. This, however, is unknown to many negroes. Possibly, another, harmless root has been taken for the *Yuca*. But how strongly that one differs from the root we are referring to here, I will discuss briefly when I will discuss the Inhame.[2]

Roels was not the first European to write about manioc or how its poisonous juices can be extracted, but Clusius may have valued his information all the more because it was also based on the personal experience and observations of African informants.[3] Roels's report continues for many more pages, and contains numerous other examples of his curiosity, precision, personal examinations and cautious reliance on the experience of the indigenous population of the areas where those plants grew or on the eyewitness testimony of persons who had visited those parts of the world. It also demonstrates his critical evaluation of evidence presented by various sources – comparing learned contemporary and classical knowledge with modern practical information from both indigenous and European informants. Unusually, Roels explains why he was so interested in tropical roots and fruits:

> In my description I have gone to such great lengths because those observations that concern the properties of foodstuffs are the most useful to medicine of all. Other means to repair our strength are not used continuously, but without food no life is possible either in health or in illness.[4]

Roels's investigations therefore served his professional interests as a physician and formed part of his investigations as a budding naturalist, while his report at the same time helped Clusius's research on exotic *naturalia*. Neither had commercial considerations.[5] Yet, if major trade routes had not been shifting in the late sixteenth century from the Mediterranean, Spain and Portugal to the North Sea ports, it is unlikely that yams, manioc, palms or novel evidence about their use would have reached Roels in Middelburg. The smaller ports of Middelburg and Enkhuizen are the first location to be examined in this section. Afterwards we will move inland to Leiden and its university, looking for the *couleur locale* of expertise.

Middelburg

Middelburg was one of the ports most immediately involved in the Dutch explorations of the East Indies, the creation of the Dutch East India Company (VOC) in 1602 and West India Company (1621), and the vastly increasing trade in exotic spices. In the course of the seventeenth century the town also became a centre of the slave trade. Within a few decades it was transformed from a relatively prosperous North Sea port into a booming port for the Atlantic and intercontinental trade. The initial phase of this transformation took place during the 1590s, the decade in which Clusius's contacts with Middelburg were most intensive.

A phase of major new prosperity and growth began for many Dutch towns with the closing of the port of Antwerp by the Dutch after the surrender of Antwerp to the Spanish armies in 1585. Almost two thirds of Antwerp's (pre-siege) 100,000 inhabitants left the city. Many wealthy émigrés from the Southern Netherlands – most famously the Moucherons – chose Middelburg as their new home town. Middelburg's wine trade and related business connections with and via Bordeaux, traditionally an important source of wealth, continued to flourish, and ties with Britain were strengthened by the establishment at Middelburg of the 'subsidiary court' of the Merchant Adventurers in 1582. New wealth stimulated a demand for art and luxury objects, catered to by many silversmiths, weavers, painters and glazers. And the population grew steadily, eventually reaching a peak of about 30,000 during the heyday of the West India Company and the slave trade.[6] New shipping companies were established in the 1590s and merged in 1600 to become the main shipping company of Zealand to the East Indies. Balthasar de Moucheron was one of its founders. He did not limit himself to the transatlantic trade, however, and in 1597 created a company to start trade with the Turkish empire, while he also sponsored Dutch expeditions (1593–4) to find a north-eastern passage to the Far East.[7]

One of Moucheron's principal business partners was Simon Jasperse Parduyn, a leading figure among Middelburg merchants who acted several times as burgomaster. Simon and his brother, the apothecary Willem Jasperse Parduyn (1550–1602), were among the richest inhabitants of Middelburg. Willem had a house and shop, the Golden Mortar, on the central market square. Both owned beautiful gardens.[8] Willem Parduyn was among the very first of Clusius's new Dutch friends to express his pleasure about Clusius's move to the Northern Netherlands. Only a month after the latter's arrival in Leiden he introduced himself in November 1593 by letter as a friend of a friend, and sent gifts, such as wine, oranges, lemons and pomegranates from Spain.[9] In the years to come

Parduyn used his shipping contacts to act as intermediaries for the transport of letters and boxes with seeds, plants, bulbs and other *naturalia* between Clusius and the Spanish naturalists-physicians Simón de Tovar and Juan de Castañeda in Seville, and between Clusius and several French plant lovers.[10] The world-wide shipping network of his brother and Moucheron also served to obtain rare *naturalia* from the North African coast, the East and the West Indies that he sent to Clusius, such as hyacinthus stellatus and mauretanicus (the latter from the area of Fez), the tropical root inhame or yam (Figure 8), palma saccifera, a booklet with paper from the paper tree, various tropical nuts, fruits, and beans, ricinus americanus (probably the plant that produces castor beans), a spectacular triangular fish from the Indian Ocean with information about its name among the Portuguese inhabitants of India, and a substance that looked like asphalt or bitumen and had come from Trinidad.[11] Several of these gifts thus came from parts of the world where merchants from Zealand were beginning to trade during this decade: Guinea, also known as the Gold Coast of West Africa, where

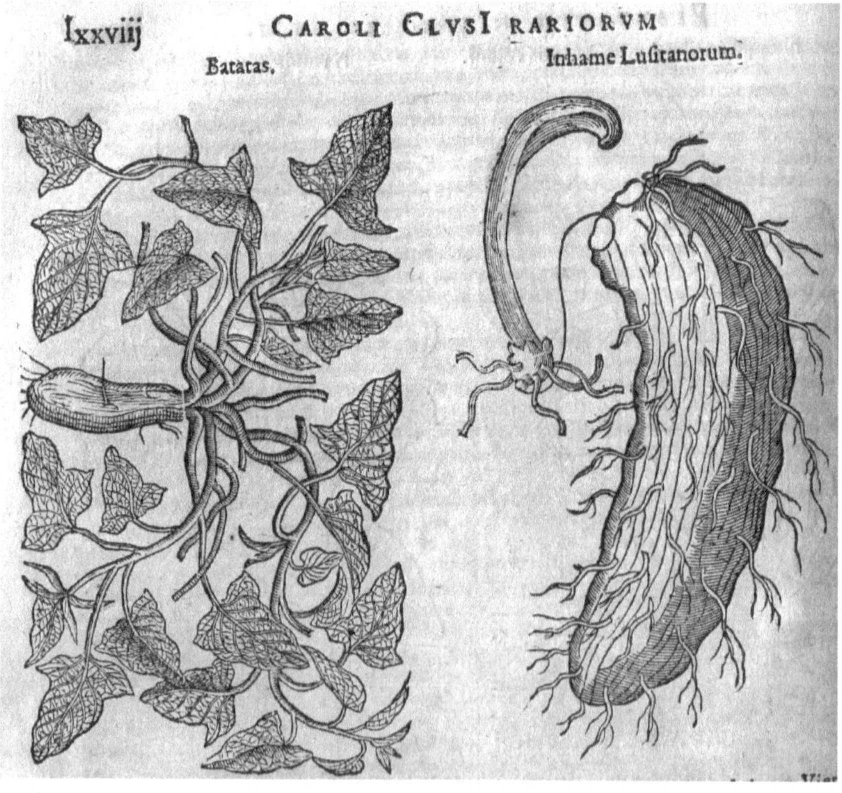

Figure 8. Sweet potato and inhame or yam. From Clusius, *Rariorum plantarum historia* (Antwerp, 1601), p. lxxviii. Collection Leiden University Library, UBLWGW_661 A 3.

many Dutch forts would be established in the course of the seventeenth century; and the 'Wild Coast' of the Guyanas, where the first presence of the Dutch is documented for precisely the mid-1590s, while the Dutch exploration and later colonization of this area was predominantly carried out by Zealanders.[12]

Parduyn showed himself both enterprising and courageous in obtaining exotic and strange *naturalia* or information about them for Clusius. In Middelburg he interrogated the crews of ships which returned from Guinea and the Far East about rare plants and animals. In this way Parduyn obtained for Clusius, for example, a fruit resembling a pineapple, fruits that were said to have come from date palms, and the beak of a bird whose body was said to be smaller than its beak (perhaps a toucan). He even went so far as to send Clusius 'creatures, both small and big, 6 big ones and 8 small ones, which I have had pulled from the bottom of the ships that had come from Guinea' (W. Parduyn 8 September 1599). And he did not hesitate to verify his information by the most direct and personal means at his disposal: eating. In the autumn of 1596 several Dutch ships transporting slaves from the island of São Tomé to Lisbon ended up in Zealand. Parduyn went on board one of the ships, which carried 'nothing special except many miserable Moors, both small and big ones, men, women and children' to a total of 90 persons:

> I have been on this ship to ask for something rare, but there is nothing except a root that these savages eat cooked or fried. I have eaten and tasted them fried; the savages call them *emiga*. I send Sir six roots, that is five of one kind and one other that is long and a bit knobbly of which they say that it is a different kind. I have wrapped one in grey paper because it began to sprout. I have planted two in pots to see what it will be. (W. Parduyn, 1 November 1596)

Willem Parduyn was therefore more than an extremely well-connected intermediary, and his friendship with Clusius was based on a shared passion for *naturalia*. Clusius respected Parduyn's knowledge and gardening expertise, while the latter was also a collector in his own right: he owned a cabinet with *naturalia*, especially shells – an item that later occupied a prominent place in many Dutch collections of the Golden Age.[13] His letters are full of explicit examples of his own love for plants, which went back quite a few years before his first contact with Clusius, and of the care he took of bulbs and seeds in his garden. He grew rare plants, tried to propagate them, and shared them with other *curiosi* in Middelburg. In his first letter to Clusius Parduyn explicitly introduced himself as wanting to become acquainted because of 'his love for plants of which he was a lover', and with the same letter he sent a catalogue of 'my flowers and simples', asking Clusius to choose any that he might like (W. Parduyn, 27 November 1593). Clusius reciprocated with gifts of highly appreciated bulbs and tubers. Many of these found their way into the garden of Parduyn's brother as well, and by 1596 the minister of Middelburg wrote to Clusius that the gardens of both Parduyns and that of Tobias Roels were 'children' of Clusius's own garden and fared all the better for it (De Jonghe, 14 May 1596).[14]

Roels's long report on tropical plants quoted above provides the most vivid evidence that an informal circle of local *curiosi* existed in Middelburg who liked discussing exotic and other rare *naturalia* and comparing the information which they gathered from their various contacts in and outside Europe.[15] 'Some months ago', the report starts, 'I had the pleasure to be invited for a meal by one of my friends, an apothecary'.[16] This was Willem Parduyn, who is described as 'host and head of table':

> There we tucked into the copious meal and the guests enjoyed the sight of living plants and all kinds of beautiful flowers ... and they entertained themselves with a sensible and useful conversation about flowers. Then the discourse turned to foreign and exotic plants [*peregrinas et exoticas*], and while talking during that busy meal about one thing and another I mentioned in passing ... your [i.e. Clusius's] letter in which you asked me to write something about the fruit of the caryoce.[17]

Although we cannot be certain that this was one of a series of meetings, it seems clear that the whole event was focused on talking about nature, while the guests must have been selected on the basis of this shared interest. Parduyn himself held forth about a root which he thought to be a yam, but which Roels later identified as manioc. Roels's phrase about their discussion of foreign and exotic *naturalia*, incidentally, also confirms that fine distinctions were made between these categories at the time, and it highlights what was one of the main interests in this circle of nature lovers.

Roels's report itself does not mention any of the participants in this meeting apart from Parduyn and a young relative of Roels, but the circle of *naturalia* lovers in Middelburg of the 1590s and early 1600s can to a certain extent be reconstructed. Middelburg was known at this time as a centre of lovers, collectors and painters of special *naturalia* who were generally in close contact with each other. Among them we find the local minister Johannes de Jonghe, who loved tulips and had one painted for Clusius; the apothecaries Johan Somer and Thomas de la Fosse; and the merchants Gerard Staels and Jacques Noirot.[18] The latter assisted Clusius with his shipping connections in Naples, Venice and Seville, and offered him seeds of West Indian flowers that had been personally picked there by a friend of his. He described himself in 1601 as 'a young and aspiring lover of flowers but not of medicinal herbs' (Noirot, 6 February 1601). Matthias de Lobel too could be intermittently found in Middelburg during the period 1595–1603; he lived and worked there as a physician, alternating with phases of work in England and various journeys.[19]

The most explicit confirmation that we are indeed dealing with a circle of local *naturalia* lovers comes from the evidence of tensions and rivalries. Minister Johannes de Jonghe complained to Clusius that he had planned two years earlier to send the latter bulbs of two types of early flowering martagon lilies from Constantinople ('Sultan Zambach') which were flowering in his garden. He had given them away to Roels, however, who had promised in his turn to share gifts from Clusius with the minister. But De Jonghe saw reciprocal gifts from Clusius arriving for Roels which the latter by no means shared with him: 'he acts neither justly nor faithfully because he keeps everything for himself' (De Jonghe, 14 May 1596).[20] The apothecary Johan Somer, son of the local bailiff, in particular seems to have irritated some of his fellow *curiosi*. He had travelled widely in various parts of Europe and, like Levenier's nephew, visited the Levant (1590–2). Somer's printed travel account describes many adventures and mentions the usual 'tourist attractions', such as the colossus of Rhodes, the Knossos labyrinth and the pyramids of Egypt, but his story does not indicate a more than average interest in *naturalia*. He mentions living coral being brought up from the Mediterranean, papyrus, parrots with pointed tails, vicious crocodiles in Egypt and wild cauliflowers in Cyprus. But his observations on *naturalia* are few and cursory, and his declaration that the great variety of flowers on Mount Ida (Crete) 'would have pleased a collector of simples' suggests that he did not regard himself as an expert naturalist.[21]

Somer's return to Middelburg with a mass of rare bulbs which his fellow citizens did not possess must have inspired him to claim a botanical fame that rather clashed with his lack of expertise. He wanted Clusius to publish his name next to four plants (Byzantine star hyacinth, small white tulips, a martagon lily with small or narrow leaves, and the yellow fritillary), several of which Clusius had received from Roels, Parduyn and De Jonghe, but which had originally come from Somer – at least according to the latter. And Somer asked Clusius to do the same for any plants that he would send him in the future (Somer, 8 May 1597). Earlier, both Roels and Parduyn had expressed some doubts about whether Somer had actually been to all the places that he claimed to have visited. They also informed Clusius that most of Somer's rare bulbs and plants had not come from the Levant at all, but from one of the Medici gardens in Florence where Giuseppe Casabona was the overseer.[22] And although Roels admitted that Somer did indeed have some rare items – 'certainly these are exceptional garden ornaments' – he also said that all of them had already been described by Clusius; he was openly critical of Somer's expertise:

> For the rest let me come to the point: since Johannes Somer is only an amateur when it comes to herbs and especially flowers (for he hardly knows any herb except the very few which he has) and since he has some which have never been seen by our root-cutters but which have been described by men who are well-trained in this study, he

> exaggerates their merit and holds them in the highest esteem, such as dens caninus, a spring crocus with a big white flower by some called wood crocus or mountain crocus ... a simple ranunculus. (Roels, 2 January 1594)

Envy may very well have played some part here, but Roels generally knew what he was talking about, and in the same letter named a number of plants in Somer's garden commenting on each whether it was rare or not, had actually produced the promised flower or not, and had already been described and published or not.

Like the Parduyns and Somer, Roels belonged to the local elite. Members of his family had been influential dignitaries at a local and provincial level both during the Habsburg period and after Middelburg had joined the Dutch Revolt.[23] He was also a highly educated, young, bourgeois professional. Roels, as town physician, speaks of Parduyn as 'our apothecary', and offered to have Parduyn send Clusius some flowers that Parduyn 'eagerly grows in his garden' (Roels, 1592, no further date). Such words imply close collaboration between these two men as well as a difference in professional status and training that is also evident from their letters: while Parduyn was almost certainly much wealthier than Roels, he had no Latin and no university training. Roels wrote fluent Latin. Some of his letters would indeed have looked like true humanist epistles if his eloquent praise of Clusius had not been so abruptly followed by purely matter of fact and technically phrased discussions of plants. And while Parduyn's network of commerce and exchange extended to French garden owners, the Spanish ports, Guinea and even the Caribbean, Roels had studied in Italy, travelled in Europe and met famous naturalists.

The first meeting between Roels and Clusius dated from 1589 and took place in Frankfurt. Roels, who was already familiar with Clusius's works, visited the latter on his way to Padua, where he planned to finish his medical studies. He carried a letter of recommendation from Johan van Hoghelande – a friend of Clusius born in Middelburg who had moved to Leiden. Clusius often assisted young men of good background, and thanks to him Roels could visit the famous garden of the Elector Palatine at Heidelberg. Roels's interest in plants therefore predated his acquaintance with Clusius, and while Roels's sojourn in Italy must have served his medical training, it was also an extended botanical research trip in which the medicinal qualities of plants seem to have played only a minor part. After Roels's return to Middelburg he married and was appointed town physician in 1592 with special care for the poor. One of his direct colleagues, as he was happy to find, was Lobel, but his new duties no longer allowed him as much time for the study of botany (*studio botanico*), an activity which he took very seriously indeed, for as he said 'nobody has ever been successful at medicine without a knowledge of simples' (Roels, 1592, no further date).[24]

Roels had certainly kept his eyes open while travelling: he mentions gathering white liliasphodelus in the Alps and seeing masses of yellow pulsatilla in 1590 at the foot of the St Bernhard on the side looking towards Switzerland. Once in Italy, Roels received a cyclamen with ivy-shaped leaves, and noticed consolida regalis with dark blue double flowers; in the environs of Padua he saw yet another consolida regalis with seeds of a special shape and leaves that also differed from the ordinary form. In Venice Roels visited the island Giudecca 'where many patricians have their gardens', and discussed rare plants with the German *hortulanus* of the richest garden on the island. And in Bologna he called on Ulisse Aldrovandi, who pointed out a botanical error of Clusius.[25] During the larger part of his stay in Italy Roels was based in Padua, where he familiarized himself with many of the plants in the botanical garden. From there he undertook various botanizing trips, some to the Tyrolean Alps, with a fellow student: Joachim Jungermann, the young naturalist who died on his way to go botanizing in Greece and the Levant with the same Casabona to whom Somer probably owed so many rare plants.[26]

It may have been the trip to Italy that opened Roels's eyes to the fascination of European wild plants. Roels is the only one of the Middelburg *naturalia* lovers in contact with Clusius to refer to the indigenous flora of Zealand. That interest does seem to have spread somewhat later, judging from a publication in 1610 by Casper Pelletier, a young town physician in Middelburg – and thus a successor (as of 1607) to both Roels and Lobel – who returned there in 1604 from his studies in Montpellier. In his work on plant names and their synonyms in various languages Pelletier listed some 1,600 plants, basically anything that grew on the island of Walcheren (where Middelburg is located), from the native flora to garden plants of foreign provenance and true exotics.[27] Roels himself, like Levenier, only wrote about indigenous plants to Clusius when he thought he had discovered something new, such as a 'small coastal plant' that was known neither to himself, Lobel or Hoghelande: 'it grows spontaneously and in abundance in clay soil which is washed every day by the sea' (Roels, 19 February 1591). He was interested in marine plants as well: a man in Flushing, who took a great delight in paintings and diverse shells, had told Roels about a plant (of which Roels sent a dried exemplar to Clusius) received from a sailor who said that it was widespread at the bottom of the northern seas (Roels, 12 June 1596). Besides being observant and curious Roels was also well read in botanical matters, and he had enough self-confidence to tell Clusius that a certain plant was newly discovered and unknown. He was not always right in this respect, but Clusius valued his knowledge and in several cases almost verbally quoted not only Roels's detailed plant descriptions – for instance of the Alpine pulsatilla – but also Roels's opinion that a plant had not been described before.[28] Especially after his return to

Middelburg exotica were Roels's main domain of expertise, however, in particular tropical fruits, beans and roots.[29]

Given the shipping contacts it is not surprising that there were direct connections between the circles at Middelburg and Bordeaux. Levenier and his rival garden owner Jacques du Casse made use of the services of Gerard Staels and Willem Parduyn to send Clusius in Leiden letters and packages with seeds and plants. Vice versa, Levenier's rose with the 300 petals has been a present from Willem Parduyn. But not only the *naturalia* travelled. Gerard Staels and 'several curiosi from Middelburg and elsewhere' actually must have visited Levenier's garden in Bordeaux, where they saw some special narcissi in flower which he had brought back from the mountains.[30]

All members of the Middelburg circle were part of the bourgeois elite and nearly all were rich. Some combined collecting *naturalia* and other curiosities with gardening. No women are mentioned. Physicians and apothecaries played a very prominent role. And aristocrats and their estates, castles or *maisons de plaisance* were almost completely absent from this setting.[31] In most of these respects the Middelburg circle resembles the cluster of plant lovers and collectors in the Poitiers-Bordeaux-La Rochelle area with whom Clusius was contemporaneously in touch. In terms of expertise both circles were far from homogeneous, and some of the rivalries – such as Levenier's statements of first discovery, and the irritation of Parduyn and Roels at Somer's claims to fame via Clusius – were connected with the considerable differences in expertise but not in ambitions. Some men had a great interest in *naturalia*, but a rather limited expertise: their ambition was mainly to possess and show off rather than to explore, investigate, discover, describe or identify. Others, such as Vertunien and Levenier in France, or Parduyn and Roels in Middelburg, were rather *curiosi*-researchers, which of course by no means obliterated a desire to display their rarities. Such differences are most clearly visible in their styles of describing, quality of observation, knowledge of plant names and botanical publications, the respect they showed each other, and their self-confidence in pronouncing identifications or disagreeing with Clusius.

But there were dissimilarities between the two circles as well, starting with the more important presence of the nobility in the French one, but extending also to the way in which their interest in nature manifested itself. Even though the exotic *naturalia* that Parduyn sent Clusius would have fitted immediately into Contant's collection, and Parduyn must have kept similar exotica in his own collection, there is a singular lack of further information about important cabinets or curiosities in Middelburg.[32] Furthermore, none of the *curiosi* in Middelburg seems to have actually created a private botanical collection-garden in Middelburg that could have stood comparison with those of, for instance, Coudenberghe in Antwerp or Levenier in Bordeaux. The gardens of the Parduyns and of Roels were undoubtedly of very good quality; there are general references to the many gardens in late sixteenth-century Middelburg; and for the first half of the seventeenth century several individual gardens are known.[33] Yet, there is no mention of special visitors nor of systematic plant observation or of garden catalogues. Although we should not neglect Pelletier's interest in both native and exotic plants growing in Zealand during the early years of the seventeenth century nor underestimate Roels's remarks about the local flora, the naturalists in Middelburg did not distinguish themselves by a great interest in local plants nor were they renowned for botanical field research. No expeditions of the kind that took Levenier to the Island of Oleron in exactly the same period seem to have been undertaken in the equally interesting delta of which the islands of Zealand formed part.

The Middelburg naturalists seem to have generally focused on rarities from afar – and they had the means, in terms of both wealth and access, to obtain them. This should, however, not be interpreted as an explicit difference in orientation between the French and Middelburg circles, let alone as a sign of a strong distinction between the categories native and exotic, but rather as the result of different practical circumstances. Rivalry in both circles concentrated around claims of first discovery or first ownership of rarities. For most Middelburg men rare exotic *naturalia* came within easy reach relatively suddenly in the course of the 1590s and were probably expected to bring more glory than indigenous ones. Roels's interest in special native plants while travelling in Europe and in exotic *naturalia* once based in Middelburg is the best example of how rarity and first discovery were far more important than a distinction between exotic and native.

All of this fits very well in a setting of rapidly increasing wealth combined with an influx of cultural traditions from the Southern Netherlands. But even while the prosperity of Middelburg continued during most of the first half of the seventeenth century and various new gardens were created there, events in Middelburg show how vulnerable such informal circles were when core members died or moved. The Middelburg circle flourished during the 1590s and disintegrated with the deaths – probably because of the plague – of both Willem Parduyn and the still young Roels in 1602.[34] We do not know how many of Clusius's other correspondents in Middelburg survived, but no more letters from Middelburg are known to have reached him after that year – with one exception. When Lobel returned to Middelburg from London in 1603, he reported to Clusius that he found the town much changed for the worse:

> I find everything here very much changed and the town dismal and depopulated, the practice of medicine so ruined by Roels's crowd and companions that, to nobody's great benefit, the charlatans and Pseudochemists or Badchemists have been elevated above all of the good physicians. (Lobel, 21 December 1605)

Lobel himself obviously was no supporter of new-fangled medical approaches, unlike Roels, whose early death – like that of his friend Jungermann a decade earlier – cut short what could have been a brilliant career as a naturalist.

Enkhuizen

Like Middelburg, the port of Enkhuizen in the province of Holland was affected in a major way by the voyages to the East Indies. Exotic *naturalia* were already arriving there in the early 1590s, and it too had its own chamber of the VOC as of 1602, while a small group of men could be found there during the late 1580s, the 1590s and the first few years of the seventeenth century who had an expert interest in voyages and other continents. Yet, in spite of these similarities, the character of that interest was rather different from that in the Middelburg circle. Key figures in the Enkhuizen group were the physician-collector Berent van den Broecke – better known as Bernardus Paludanus (1550–1633) – who moved to this town in 1586, and the explorer and author Jan Huygen van Linschoten (1563–1611). Paludanus is known as the creator of the earliest Dutch curiosity collection with a great international reputation. Linschoten spent five years (1583–8) in Portuguese service in Goa, published the famous *Itinerario* (1596) of which Paludanus was co-author in so far as botanical and medical topics were concerned, and took part in two journeys of 1594 and 1595 – promoted by Balthasar de Moucheron – to find a route north of Siberia to the Pacific. Other prominent men in the Enkhuizen group were the navigation expert and cartographer Lucas Janszoon Waghenaer (1533/4–1606) and Francis Maelson (1538–1601), an erudite town physician, diplomat, poet and advisor on navigation and exploration matters to Prince Maurits of Orange. Core interests of the Enkhuizen group were, therefore, navigation, voyages of exploration, cartography and the Indies; *naturalia* were, comparatively speaking, a side interest.[35] Clusius was personally acquainted with nearly all of these men, but he seems to have maintained friendly relations of exchange only with Paludanus, the single expert on *naturalia* among them.[36]

Comparing the expertise of Paludanus and some of the Middelburg collectors highlights some interesting differences even at the individual level. The four catalogues of Paludanus's collections (1592, 1603, 1617 and 1624) – made because he kept on selling items, groups of objects or even the whole of his collection, and continuing to expand or rebuild it – show that his interest in *naturalia* shifted in the course of time. From 1586 until 1592 the emphasis was on *naturalia* (in particular minerals, animals and marine plants) from countries that Paludanus had visited in 1578–9, such as Italy (where he studied medicine in Padua), Egypt and the Levant, as well as *artificialia*. After the start of his cooperation with Linschoten in 1592–3 exotic items from the African West Coast (such as ivory, pepper, tropical fruits, a nest of weaver birds and various *artificialia*), India and the Far East (e.g. birds of paradise from the Moluccas, Chinese paper and seeds of a tree from Goa that flowers only at night) began to occupy a more important place. So far, there were close parallels – chronologically and in terms of the geographical provenance of the objects – between the collecting activities of Parduyn, Roels and Paludanus, but the latter's geographical range was wider. Like Roels, moreover, he showed a particular interest in the uses – as medicine or food – of *naturalia*. By 1616–17 Paludanus's attention had further shifted, from the less durable plants and fruits towards the more robust *artificialia-ethnographica*, minerals, fossils and shells. In the meantime the share of East Asian objects increased.[37]

The differences between Paludanus and Roels in terms of interest and expertise concerning *naturalia* are even more interesting. In the course of the 1590s Paludanus changed from a physician-collector into a collector cum international broker of rare and exotic *naturalia* and *artificialia*, as has been convincingly argued by Van Gelder.[38] He seems to have developed his international network at least in part as a high-class circle of distribution, and he regularly sold off items and groups of objects, in particular to some German princes who were at that time beginning to expand their personal *Kunst- und Wunderkammern*. Paludanus's *Album amicorum*, which contains some 1,900 names of acquaintances and visitors from all over Europe (many of them from the highest social ranks) can perhaps also be understood as a marketing instrument, just as at least some of his catalogues were probably composed in order to facilitate the choice for potential buyers.[39] Secondly, expert as Paludanus undoubtedly was in some domains of nature, his letters to Clusius show no great passion for living plants or horticulture. There is little or no evidence of him ever undertaking botanizing trips, nor does his horticultural knowledge appear to have been special. In fact, almost nothing is known with certainty about his garden, which was possibly a 'herb garden' in Enkhuizen, or perhaps (according to a visiting German nobleman in 1593) a private botanical garden with rare flowers and exotica.[40]

Neither Paludanus's letters to Clusius, which only incidentally touch upon botanical issues (mainly about the naming of some plants and the identification of exotic fruits) in between matters of business, health, publications and affairs of the nobility, nor the twenty-six references to Paludanus in Clusius's printed works change the impression that living plants were not Paludanus's strong point.[41] Clusius mentions Paludanus rarely in connection with living plants (a few bulbous plants, a small lemon tree and seeds of the dragon tree), and never as an expert gardener. Most references concern beans or other fruits, and small branches, sometimes with fruits and leaves attached. A few of these had been brought back by Paludanus himself from his journeys, such as the strange habhel (juniper) fruit from the Middle East (1579), and the seeds of some kind of hyoscamus from Syria (1580), which Clusius managed to grow in his garden.[42] Others had originally been gifts to Paludanus. But in most of the cases in which *naturalia* had come from afar and not been collected by Paludanus himself, he knew next to nothing about them. When Clusius saw, for instance, a strange fruit as big as a fist at Paludanus's, the latter did not know its name and had no idea whether it came from India, America or Guinea; it was much the same with a number of exotic fruits that Paludanus gave or lent to Clusius.[43] Unlike Roels and Parduyn, Paludanus does not seem to have made an effort to find out more about his *naturalia*. Clusius's use of the term 'lavish and well-provided stocks' for Paludanus's collections may therefore be even more significant than Van Gelder has already suggested: these were indeed the stocks of a collector and *curiosa* seller who was also an erudite physician, but no researcher-naturalist.[44]

Exotic *Naturalia* and Commerce

The similarities and differences between Middelburg and Enkhuizen are suggestive. Aristocratic garden display on a grand scale was virtually absent in these smaller Dutch ports, as was the tradition of field research (although known to both Roels and Paludanus), while the investigation of the native Dutch flora only barely appears to have taken off in this period. There was a strong emphasis on *naturalia* brought in from overseas, first from the Mediterranean and the Levant, but increasingly also from Africa, the New World and Asia. Yet in spite of the many similarities and almost equally excellent access to *naturalia* from overseas, the cultural traditions of these two towns diverged, resulting in the growth of very different types of expertise. In Middelburg there was a widespread interest in living plants and horticulture: rivalry concentrated on precisely that aspect, and expertise concerned the growing of rare plants as well as their identification, besides the practical use of certain exotic *naturalia*. In Enkhuizen, on the contrary, *naturalia* played only a minor role: only Paludanus was interested in them, and they cannot have formed an important issue in local meetings or rivalry. Paludanus was, moreover, much more knowledgeable about minerals, shells, stones and useful vegetable products than about living plants or horticulture.

The interest in *naturalia* of Roels, Parduyn and Paludanus was bound up closely with the commercial boom of Middelburg and Enkhuizen and the Dutch voyages to the East Indies that brought incredible riches to the Dutch ports and themselves formed part of a major shift in world trade patterns. The link between the interest in *naturalia* and commerce is far less transparent, however, if we look beyond the rather mechanistic issues of 'means' (in the sense of wealth) and 'access' (in that of availability through shipping) to the character of this interest. Similar situations did not result in exactly the same types of interest or expertise in Middelburg and Enkhuizen. The interest of Roels and Parduyn in *naturalia*, whether exotic or indigenous, was that of the *curiosus*-researcher and their attention to the practical use of certain tropical plants as foodstuffs or medicines was not translated into any attempt to import or sell them or even prepare a market for them. They wanted to know rather than to sell. The case of Paludanus, who turned from physician-collector into broker of rarities, highlights borderlines and overlaps as well as important distinctions between a commercial and a non-commercial interest in nature. His interest in *naturalia* did not originate in economic motives nor was it an exclusively economic one. The items he collected and sold remained rarities, collector's items in the literal sense, much as antiquities or works of art are now and were then. Even though Paludanus converted rare *naturalia* into merchandise, he did not turn them into commodities of the kind that could be sold by the kilo. In all other cases that we have come across, commerce played a surprisingly small role, and curiosity a major one.

It is therefore impossible to generalize about exotic *naturalia* and the market economy beyond the fact that access to such *naturalia* was predicated on world trade and shipping.[45] That much has also become evident from our investigations of the Veneto and Naples, for instance. While some of these exotic *naturalia* – most famously the medicinal guayacum and the various spices from the East Indies – almost immediately became highly priced commodities, others remained rarities with no particular use apart from display or research. Some of the latter (we need think only of the bird of paradise or the horn of the unicorn and various kinds of shells) came on the market and therefore had a price, as rarities, however difficult it may have been to establish one. Many *naturalia* never came on the market, remained without a known monetary value and were of interest mainly to naturalists and collectors. The latter formed part of a gift economy in which money did not form the currency of exchange but which had its own rules and logic based on notions of obligation, gift and counter-gift, and honour. And many *naturalia* belonged to both economies: they could be sold, and they could be given away. Clusius's friends knew how to distinguish between such codes and how to operate in both of those idioms.

10 TOWN AND GOWN: LEIDEN AND THE CONVERGENCE OF EUROPEAN CULTURAL TRADITIONS

In Leiden too a remarkable cluster of plant experts, flower lovers and *naturalia* collectors emerged in the 1590s. Given the relative distance from the ports of Amsterdam and Middelburg, it is tempting to regard Leiden university as the principal motor of the interest in nature. But the university was only founded in 1575 – as a symbol of the independence of the new Dutch Republic – and the study of living nature there did not have a particularly quick start. In fact, no specific institutional interest seems to have been taken in natural history at all in Leiden university during at least the first decade of its existence. Medical botany was not taught there until 1587–8, and the botanical garden in Leiden – as a garden rather than as a project on paper – was created only in 1594, almost twenty years after the foundation of the university.[1] To get to grips with natural history in Leiden in the 1590s and early 1600s, we therefore cannot limit ourselves to the academic sphere and certainly not to that of book learning or teaching from books, but have to look at people and practices in a wider, local, regional and even international context.

Creating the *Hortus*

After a complicated history of first selecting a suitable plot and then finding the right prefect for the garden – during which negotiations in 1591–2 with Paludanus and Dirck Cluyt (1546–98), an apothecary from Delft, came to nothing – Clusius was eventually (1593) appointed as prefect of the *hortus*, with the same Cluyt as his second in command. The *hortus* was laid out and planted by these two men and some assistants in the spring and summer of 1594.[2] Not only its simple rectangular shape and four squares, but also its straight beds and the location of the long building which from 1601 onward served as both conservatory and gallery for the display of rare *naturalia* along one side of the rectangle, suggest that practical horticultural reasons strongly influenced the layout, as seems also to have been the case in the contemporary *hortus* of Montpellier.[3] Misfortune struck only a few years after its creation. The indispensable Cluyt died in 1598 at the age of fifty-two. Clusius's age and bad health no longer allowed him to take care of the practical management of the garden, and from the autumn of the same year the physician Petrus Paauw (1564–1617), who already taught medical botany, acted as prefect of the *hortus* and also took care of the on-the-spot teaching about medicinal plants. Clusius's only remaining duties were to honour the academic community with his presence and to work on his publications.[4] The joint reign of Clusius and Cluyt over the Leiden *hortus* was therefore a very short one, even if their ideas and planting determined its character for some time to come.

If Paludanus had become prefect of the *hortus*, part of his famous collections would have accompanied him to Leiden – since the invitation of 1591 explicitly referred to his collection of 'rarities, consisting of herbs, fruits, vegetation, animals, shells, minerals, types of soil, poisons, stones, marbles, corals et cetera'.[5] And that is precisely what happened in the cases of both Cluyt and Clusius. Cluyt owned a rich collection of dried plants and had a well-stocked private garden in Delft. Many plants from this garden eventually ended up in the Leiden *hortus*, as did plants and seeds from Clusius's personal garden and collection in Frankfurt, which comprised a considerable number of seeds or plants from Crete received from Casabona.[6] From the very start the Leiden *hortus* was not a collection of medicinal herbs, therefore, but a botanical garden which included rare plants from various parts of Europe and the Levant as well as some living exotica from other continents. It reflected the wide-ranging botanical interests of Clusius and Cluyt.[7] Similarly, during Paauw's directorship of the *hortus*, many items from his private collection ended up in the garden. After 1601 quite a few of them were on display in the *ambulacrum* – the covered gallery cum conservatory running the length of one side of the garden, where potted plants could be protected from inclement weather and many exotic *naturalia* and *artificialia* were on show.[8]

While it is not always clear in detail whether private collections were officially turned over to the university or simply integrated *de facto*, the process of incorporating private collections into the institutional setting has been noted before. De Jong has indeed argued that no clear division existed at this stage between the private collections that belonged to men who were working for the university and the collections of the university as an institution.[9] This process of incorporation was far more important and at the same time less self-evident, however, than it may have seemed up to now. In fact, during the first phase of their existence (1593/4–1602) both the plant collection of the Leiden *hortus* and the curiosity collections in the *ambulacrum* and the anatomical theatre depended completely on the incorporation of several different private collections which had originated outside a university setting and had been shaped by fashions, passions and fascinations that were not necessarily connected with academic purposes as they were defined then.[10] It also seems clear that Leiden university was interested in experts such as Clusius, Cluyt and Paludanus as much for their material collections and access to interesting *naturalia* as for their expertise. In this respect, therefore, the university operated very much like an aristocratic patron. Meanwhile, the distinction between private collection and university collection may have seemed unimportant from the university's perspective, but not necessarily from that of the private owners, as is evident from the fact that demands for financial compensation were made and that conflict and recrimination accompanied the incorporation of at least one such private collection. Financial problems arose between Cluyt and the Curatores (the governing body of the university) almost immediately after his appointment in connection with his valuable plant collection. Cluyt had been promised compensation for his plants, which according to his estimate were worth 1,500 guilders – a fortune. After months of discussion he eventually accepted the compensation of 400 guilders, but even after his death in 1598 there were still signs of resentment on the part of his family on account of his treatment by the university.[11]

The Medicine-Related Experts

The movement of living and dead *naturalia*, people and forms of expertise into the university sphere urges us to take a closer look at the background of some Leiden-based naturalists of the late sixteenth century. Three of them – the physician Petrus Paauw and the apothecaries Dirck Cluyt and Christiaen Porret – came from the same medicine-related background that played such an important role in Italy, the Southern Netherlands and France. The others were persons of independent means. What setting had shaped their expertise in the field of nature?

Paauw, who replaced Clusius as prefect of the *hortus* as of 1598, published its first official catalogue in 1601. By then the *hortus* comprised some 800 plants. Paauw and Clusius had had no contact before it became clear that Clusius was going to move to Leiden, and once there, Clusius did not get along with Paauw, whom he regarded as arrogant and jealous.[12] Yet, Paauw gave and lent exotic *naturalia* to the sometimes rather ill-tempered Clusius up to the last months of the latter's life, and Clusius's references to these gifts help us better to understand Paauw's expertise. As in Paludanus's case, very few of these *naturalia* were living plants, bulbs or seeds. Several of the fruits, nuts and (parts of) animals – such as an echinocactus, a lizard, the head of a crab, a bamboo on display in the *hortus*, the leg and claw of a dodo – that Clusius saw at Paauw's had been gifts to the latter by the governors of the VOC.[13] Paauw was especially interested in birds, perhaps on account of his name, which means peacock. Of course he had a bird of paradise, while some of the rarest birds in his collection had come from the far north and arrived as gifts from Henrik Høyer (*c.* 1565–1615/16), a German physician who had studied in Leiden and after moving to Bergen in Norway corresponded with both Clusius and Paauw about plants in Scandinavia and birds of the northern seas.[14]

Paauw himself had studied medicine in Paris, Orléans, Rostock and Padua. He used human and animal bodies for his dissections, studied the development of the foetus and was reputed to have a great knowledge of the eye and the human skeleton.[15] His professional expertise thus primarily concerned anatomy, animals and exotic curiosities rather than living plants, botanical identification or gardening. That is also what we can infer from various remarks by Clusius's friend Johan van Hoghelande in Leiden. Shortly before Clusius's arrival in Leiden, Hoghelande wrote to say that Paauw had started to work on a small part of what was to become the *hortus*, but in such a way 'that I hardly know whether he is doing it seriously or as a joke' (Hoghelande, 30 May 1593). Earlier, in August 1592, Hoghelande had already warned Clusius that Paauw's botanical knowledge was insufficient, and that he had confused an 'apium palustre Dodonaei' with an anemone (Hoghelande, 12 August 1592) – on the latter of which Hoghelande must have been something of an expert given the great variety of anemones that he had in his garden.[16] And yet earlier, in May 1592, he wrote that Paauw did

> not know any plant, however commonly known it is among herbarists. Some 12 days ago he was with Heurnius in my two gardens, so that I know with certainty how much he knows about it; nor did he deny that, because he and Heurnius together asked me if I was content to have them coming there often to practice. (Hoghelande, 20 May 1592)

Such remarks should be interpreted in the context of appointment rivalries, while Hoghelande was totally deaf, moreover, and had to communicate by notes, which could easily lead to misunderstandings and suspicions.[17] Nonetheless, the overall impression of Paauw remains that of an eminent and ambitious physician-anatomist and an excellent organizer rather than an expert on plants. But there is no doubt that he played a crucial part in the establishment – more particularly the institutionalization – of botany as a respectable field of study within the medical faculty of Leiden.

Clusius's collaborator, the apothecary Dirck Cluyt (also Theodorus Clutius), on the contrary, did most of the work in and for the *hortus*, taught there and donated many plants and seeds to it. But he died young and received little recognition for his work. Cluyt was born in Haarlem and practised there as an apothecary, but left in 1578 for Delft. There, he owned a large apothecary's shop (called the Pomegranate) as well as a walled garden in which he grew fruit trees, special roses, rare bulbs and medicinal herbs. In Delft he worked very closely with the physician Pieter van Foreest (1521–97), and he married the niece of

Foreest's wife.[18] It is to Foreest, in fact, that we should look if we want to know why Cluyt and his plant collection ended up in Leiden, and possibly even when and how Cluyt developed a passionate interest in plants that went far beyond their medicinal use. Foreest was already a physician of high repute in Holland around the middle of the sixteenth century. He acted as town physician of Delft from 1558 until 1597, and his star rose when he became the personal physician of Prince William of Orange in 1574. A year later Foreest was invited to take part in the inaugural procession of Leiden University, which had been founded on the personal initiative of the Prince. Foreest never accepted a position there himself, but it is likely that his recommendation helped to bring Cluyt to the attention of the Leiden Curatores.[19]

Foreest had first studied medicine in Louvain, where he seems to have been more fascinated by botanical excursions than dissections. During the early 1540s he went on a long journey to Italy, where he studied for three years in Bologna with the anatomist Berengario da Carpi and the botanist-physician Luca Ghini. Foreest also visited Venice, Ferrara and Padua, where he met Vesalius. He became friends with the already famous young German physician-botanist Valerius Cordus, with whom he undertook a botanical excursion in the area of Bologna. Foreest also joined Cordus and his companions on their long journey on foot to Rome in the summer of 1544, which proved fatal to Cordus.[20] Given the long-standing personal, professional and family connection between Cluyt and Foreest, it is probable that the older man with his wide international experience further stimulated Cluyt's interest in botany. We know that this interest went back to at least the early 1580s, because Cluyt visited the Antwerp garden of Peeter van Coudenberghe, while Lobel visited Cluyt in 1580 and saw pictures of plants drawn by Cluyt himself.[21] Six books of botanical watercolours as well as a huge collection of at least 4,000 dried plants were found among Cluyt's possessions shortly after his sudden death in 1598.[22] Well before meeting Clusius, moreover, Cluyt had already developed his own international contacts with Spain, Italy and elsewhere in order to obtain rare plants. He also established friendly exchanges of plants and gardening expertise with his fellow Leiden garden owner Johan van Hoghelande (D. Cluyt, 15 July 1593). Cluyt's input in the Leiden botanical cluster – which comprised botanical and gardening knowledge as well as living and dried plants and probably pictures of plants – thus originated in the private sphere of collectors-gardeners and physicians-apothecaries. If we are correct about Foreest's influence, its roots went back to an Italian-inspired tradition of botanical interest mixed with that of the Southern Netherlands.

In spite of a request by the students, the Leiden Curatores refused to appoint Cluyt's son Outgert (also Augerius Clutius; 1577–1636) as his father's successor. By 1598 Outgert Cluyt, just twenty-one years old, had already trained as an apothecary, matriculated (1594) in the faculty of philosophy and arts at Leiden, and had often assisted his father by teaching in the *hortus*. According to the students, he was the only person who understood the 'register' of the Leiden *hortus*; he had great ability and expertise in drying plants and the preparation of medicines, and his knowledge is said to have equalled or even surpassed his father's already in this period. The young Cluyt had, moreover, always shown himself willing to take the students on herborizing trips in the dunes, peat bogs and woods, and his Latin and Greek were good (unlike his father's). His travels also gave him excellent connections with botanical experts in Padua and Montpellier, and with the herborists of the Medici in Florence and those of the French king, father and son Robin.[23] The young Cluyt's career was by no means harmed by the Leiden decision, but it may have cost Leiden dearly. Outgert Cluyt could have become an eminent director of the Leiden *hortus*, who would have brought in crucial family continuity and international contacts – very much along the lines of father and son Robin in the French royal garden. That decision may also have prevented Cluyt Sr's magnificent private collection of dried plants and drawings from ending up in that of the university.[24]

The third of these medicine-related experts was the apothecary Christiaan Porret (1554–1627), a man who very closely resembles the apothecary-collectors of France and Italy.[25] He was in fact born in Paris, where his father Pierre was an apothecary and a representative of the Antwerp printer Christopher Plantin. Porret moved to Antwerp and subsequently to Leiden, where he was closely connected with the world of printers and university professors, in particular with those from the Southern Netherlands. Yet, Porret never seems to have been involved with the university, and his collection remained private property. When it was auctioned, upon Porret's death, the catalogue comprised more than 700 objects ranging from *naturalia* from the East Indies and Americas to weapons, minerals and albums with hundreds of drawings of fruits, flowers and herbs. There were some coins and quite a few paintings, but few antiquities. Porret appears to have been primarily interested in *ethnographica* and *naturalia*. Among the latter we find seven shells from Francis Drake's ship, a bird from Brazil with a very large beak, a herbarium printed in China, 60 small pen drawings of flowers and plants, 100 large pen drawings of trees and plants, 13 watercolours of capsicum which were perhaps connected with Fra Gregorio's gift to Clusius, 60 watercolours of fish and beasts, 50 watercolours of narcissi and hyacinths, drawings of both poisonous and edible mushrooms, and much else.[26]

Clusius visited Porret regularly and in his printed works drew a number of times on the exotic *naturalia* he received or borrowed from Porret or saw in his museum, such as branches, nuts and fruits brought back by Dutch ships from Africa, Bali, the Moluccas and other parts of the East Indies. There were also skins of exotic snakes and various lizards, the spiky orbis fish and a triangular fish just like the one owned by Willem Parduyn. Objects that Clusius particularly admired were the feathers and egg of an 'emeu' (in fact, a cassowary), which had been brought back from the Moluccas in 1603 by a Dutch fleet – the cassowaries themselves had died on board – some small stones found in the stomach of a dodo from Mauritius, and a large coral, which was, as Clusius wrote, literally a tree given its height of three feet and the circumference at the bottom of the 'trunk' of seven thumbs.[27] Perhaps Porret's collection was not as big or rich as that of Paludanus, but it was certainly a 'museum instructissimum'.[28]

As so often, we know less about his garden, which was located in town but not adjacent to his shop and house, where he also seems to have kept his collection.[29] Yet, Porret's collection of flower drawings and watercolours alone already indicates an interest that went beyond a simple desire for a few rarities or a merely professional interest in medicinal plants. His relative Raphelengius testifies that Porret was passionate about obtaining new and rare plants for his garden:

> that he, since this is the pleasure that we take in flowers – something not too far removed from the art of spices and drugs – never ceases to employ every type of effort and even expense to enrich and embellish his gardens with every possible rare flower.[30]

Porret himself made it clear that he had

> neither the leisure nor the means that are necessary to be well provided since I am very busy with my pharmacy and, not having abundant means, if I am as well furnished as others – perhaps not in quantity but in diversity – this is because I have gone to some expense to buy and thanks to the help of various friends and [plant] lovers who have liberally assisted me.[31]

Still according to Raphelengius, Porret 'had always been in the happy position to receive his share of the rarities which reached Clusius from all over the world', and he did indeed inherit a considerable number of Clusius's bulbs and tubers.[32] But he was by no means only a recipient. As an expert gardener he managed to bring certain bulbs into flower with which Clusius had no success, and he shared bulbs and information about rare plants with Clusius, just as the latter did with him. Porret gave him, for instance, a yellow lilium montanum, and (in 1605) the stem or flower of an unusual ornithogalum aethiopicum of which bulbs had been brought back by Dutch ships.[33] Porret also experimented – and he uses exactly that word – with various bulbs and seeds over the years, by changing the location, pots, positions, light and humidity, and he tested colour stability over the years.[34]

Clusius's references, which focus on bulbs and tubers, tell us that Porret had special anemones (several grown from seeds) and a very large variety of bulbs, such as rare narcissi, a summer-flowering hyacinth, colchicum vernum, various fritillaries, ranunculus and several pseudonarcissus of which some with double flowers. One of his irises had been collected by Dirck Cluyt's son on the Libyan coast. Among his rare items were a purely white fritillary, a white ranunculus and an auricula – all of them gifts from Levenier in Bordeaux. Other bulbs reached Porret from a Plantin relative in France.[35] Clusius observed intriguing colour variations in Porret's garden: nearly all of his 'hyacinthus stellatus baeticus' had blue flowers, for instance, but one produced white ones that verged on pink. Both Clusius and Porret bought the same bulbs from French plant sellers who had come from the Pyrenees and visited Leiden – at least one of them a fraudulent figure well known to Levenier – such as fritillaria lusitanica and various types of auricula.[36]

Although he never held any university function, Porret is said to have donated many plants to the Leiden *hortus*, and his value to the cluster of Leiden naturalists lay in his growing expertise as collector of exotic *naturalia* and gardener as well as access to his collection of living plants and items in his cabinet, plus excellent international family and professional connections.[37] His contacts included the Plantin circle in the Southern Netherlands and may have extended to the French royal apothecaries Jean Robin Sr and Jr. These contacts originated in the years before his emigration to the Northern Netherlands, and thus dated back to the 1570s or even earlier. It was probably thanks to Clusius, however, that he came to know his fellow apothecary Willem Parduyn in Middelburg, who called Porret his friend and sent him tropical roots via Clusius in 1596; the apothecaries and exotica-collectors Jean Mouton and Jacques Plateau in Tournai, who received gifts, such as 'emeu' feathers, from Porret; the apothecary Pieter Garet in Amsterdam; Levenier in Bordeaux, with whom he exchanged *naturalia* in 1598–9; and Matteo Caccini in Florence, with whom he corresponded and exchanged plants after Clusius's death.[38] Porret's influence on seventeenth-century Dutch gardens should not be underestimated either. In 1603–4 the young Petrus Hondius, who was studying in Leiden, visited his garden several times and befriended Porret. Later, as a minister at Terneuzen in Zealand south of the Scheldt river, Hondius went on long botanizing trips on foot and created one of the famous Dutch gardens of his time, which he celebrated in his poem *De Mouffe-Schans* (*c.* 1619), much as Contant had done with his collection and garden.[39] Porret, in short, was one of the key figures in Leiden natural history.

Noble and Mercantile Expertise

As a gardener and an expert on plants Johan van Hoghelande (1546/58–1614) was in a class of his own. Tobias Roels from Middelburg, one of his frequent visitors, described him as

> a man most diligent in making notes about plants, who has devoted himself completely to these matters, both because of his deafness which forces him to abstain

> from social intercourse and because he does not have any public or private business to occupy him. (Roels, 19 February 1591)

In so far as living plants were concerned, he of all persons in the Northern Netherlands of this period shared to the largest extent the range of interests and forms of expertise that characterized Levenier, Boisot and indeed Clusius himself. He was in personal contact with all three, and his relations with garden owners and plant lovers in the Southern Netherlands, Middelburg and London were particularly close.

Hoghelande was a wealthy landowner from Zealand and seems to have belonged to the lower nobility. He was born in Middelburg, where his grandfather had been town bailiff and owner of salt works. During his youth – before the partition of the Netherlands – the Hoghelande family was orientated towards the Southern Netherlands in terms of education, culture and careers. An uncle is said to have occupied a high court function in Brussels, and Hoghelande himself lived for some time in Malines during the mid-1560s.[40] Both Johan and his brother Theobald left Zealand around 1580, probably on account of the war. Whether religious issues played a part is hard to determine: Johan moved north, and lived until his death in the Protestant but religiously tolerant Leiden while remaining a Catholic.[41] For most of his life he seems to have lived on the proceeds of the family estates in Zealand. Theobald (*c.* 1560–1608), on the other hand, travelled in Central and Eastern Europe, and occasionally sent plants to his brother and Clusius. From 1593 onward he was mainly based in Cologne, where he published several works on alchemy and chemistry.[42]

Johan van Hoghelande's fascination with *naturalia* must certainly have been stimulated in the circles of the Malines elite, where he lived in the house of a member of the town council during the mid-1560s. It is hard to imagine that he would not have known of the famous garden which Brancion created in this same period, or at least have heard of Dodoens who also lived there. The young Hoghelande certainly knew of Clusius's return to Malines from Spain in 1565, but they never met during this period. As Hoghelande wrote in his first letter to Clusius of some twenty years later, he had admired Clusius's erudition already at the time (Hoghelande, 27 October 1583). This letter led to regular exchanges of plants and letters, especially from about 1588 onward, and in 1591–2 both Johan van Hoghelande and his good friend Princess Marie de Brimeu – who had a house and very well-stocked garden beside the Rapenburg canal in Leiden from 1590 to 1600 – were pulling ropes behind the scenes in Leiden to get Clusius appointed there.[43] Once Clusius had arrived in Leiden there was no need for correspondence, and the two men must have met frequently. Clusius communicated with the deaf Hoghelande by writing on the tablets which the latter always carried with him. In September 1607 Hoghelande was still inviting

Clusius, by then eighty-two years old, to come and see some special anemones in his garden.[44]

Hoghelande's letters are those of a sometimes irritable and suspicious man – probably on account of his deafness – but also portray him as a staunch friend, erudite, intelligent, sometimes stingy, occasionally generous and always passionately interested in flowers, 'because of the beauty of the flowers, which alone attracts many to this study, and makes the onerous work in the garden more pleasant and lighter, or because of their rarity' (Hoghelande, 27 October 1583). In most of his letters he alternates Latin and French from one sentence to the other, or even from one part of a sentence to another, because he felt not so much at home in Latin that he could use it throughout. The chatty style, his minuscule handwriting and the pages filled to bursting point convey that writing for Hoghelande was what conversation was for others. In his letters all of the different social groups can be found that were involved with *naturalia* in Holland at this time: Antwerp merchants who helped to transport *naturalia* and information; Leiden town and university authorities; scholars, collectors and naturalists in Leiden, Enkhuizen, Middelburg and Amsterdam (Roels, Lipsius, Porret, Cluyt, Paludanus, Paauw); publishers-printers (Raphelengius, Plantin); and aristocrats and noblewomen with an interest in plants and gardens (Marie de Brimeu, Madame de Brederode, Marnix de St Aldegonde). Even a Scottish student in Leiden is mentioned who had started a herbarium but did not manage to enlarge it since so many plants could not be found in his native country (Hoghelande, 20 May 1592).

In terms of botanical and gardening knowledge the exchanges between Hoghelande and Clusius covered a wide field. Clusius sent Hoghelande rare plants and a few exotica: from tulip seeds and bulbs (marked by labels and colour marks on their paper wrappings), tubers of cyclamen and 'leontopetalum', to seeds of Persian peonies, an opuntia cactus and American potatoes. By June 1591 Hoghelande had five potato plants growing in his garden. More advice about how to take care of plants that had come from a very different climate was welcome:

> I was glad that you gave me advice on how to take care of the *papas americanas*. I wish you had also done the same for the laurocerasus [Prunus laurocerasus, cherry laurel]. Because we often take so much care of our rare plants that the same thing happens to us that happens to a monkey with its young, for I fear that this is what happened to my *charam*. For it needed cool conditions instead of hot ones. (Hoghelande, 9 June 1591)

Hoghelande owned two gardens: one 'domestic garden' which probably was used as a kitchen garden; and one outside Leiden – far too exposed to the strong westerly winds to his taste – which can only be classified as an impressive, private botanical garden where he grew rare plants and experimented.[45] There are no catalogues or other lists which can help us imagine how big his collection was, but Levenier had heard from Hoghelande that he had more than 2,000 tulips (Levenier, 16 April 1600). The plants mentioned by Hoghelande in his letters perfectly fit the list of desirable, exotic or other fashionable flowering plants of the period: peonies, columbines, cyclamen, many anemones, a white rose centifolia, cyclamen aestivum, dens caninus, two types of arum of which one had leaves spotted with black and a pale yellow flower, numerous tulips, hyacinths and so forth. Like Porret he had a white ranunculus from the Pyrenees that had been a gift from Levenier. Roses and rare bulbs, anemones and cyclamen appear to have been his special favourites.[46] Hoghelande grouped plants not just for aesthetic effects, but by order and thus category (although we do not know by which criteria he defined such categories). And his whole garden seems to have functioned very much like a place for experiment and research as well as enjoyment. That much we can deduce both from his horticultural activities and from what Clusius briefly mentioned to his friend Caccini in Florence:

> There is a deaf man here with a most beautiful garden not far from town, in which each year the most beautiful anemone flowers grow spontaneously from seed spread in his garden by the wind ... which he had gathered and planted together in certain compartments, but taking them subsequently out of the earth in order to plant them by order, he discovered that some five hundred of them were spoilt.[47]

Both those 500 rotting anemones and Hoghelande's complaint that he had 'only' some 50 tulips flowering at a certain moment after many losses, suggest that his was a major collection in both size and quality. It was, in fact, one of the special sites in Leiden. Hoghelande mentions quite a few visitors who came to see his garden, some even from abroad. There were several women of high social position among them, and Hoghelande's remarks indicate that some of them not only admired flowers but were skilled gardeners themselves – which suggests that women may have been instrumental here, as in the Southern Netherlands, in spreading the culture of flowers.[48] Among them we find, of course, Marie de Brimeu, whose gardens in both Leiden and The Hague contained many rarities, but also the wife of Justus Lipsius who gave Hoghelande tulip seeds, the wife and daughter of town secretary Jan van Hout, and the wife of the nobleman and curator of Leiden university Janus Dousa. A very special occasion was the visit on 19 May 1592 of some important women at court in The Hague, the Princesses of Orange, Merode and Chimay (the latter was Marie de Brimeu), accompanied by all the children of the Prince of Orange.[49] The eminence of Hoghelande's plant collection is evident also from the fact that Jacques de Gheijn painted some of his flowers, in particular roses, in the years 1600–4; de Gheijn's picture of the centifolia rose may have been based on the exemplar in Hoghelande's garden.[50]

Hoghelande was a practical as well as an erudite gardener. He knew all of the major contemporary publications on plants, and was interested in identification of new plants as well as propagation, the best means to eliminate garden pests, and creating the right location and conditions for plants from different climates or soils. Hoghelande grew many bulbs from seeds, brought new soil into his garden in the hope that this would improve the conditions for tulip growing, experimented with drainage and irrigation systems, and transplanted some cyclamen from pots into the soil in order to inspect which condition was better for germination.[51] He also hoped to protect the new cyclamen aestivum plants which he was growing from seed, because 'those little beasts which are called piglets of St Anthony [i.e. woodlice] are making war on them and have munched some, which has forced me to put a couple of birds in the garden to protect the rest' (Hoghelande, 14 August 1591). Sustained observation of plants over the years was one of his strong points: 'I have seen a great variety in my double flowered delphiniums for I have had them with white, reddish-purple, scarlet, and variegated purple flowers, all of them from seeds scattered by the wind in my garden' (Hoghelande, 11 July 1592).

Hoghelande wrote to Clusius about his gardening experience and experimentation, and sent him many plants, bulbs and seeds: cyclamen aestivum, yellow ornithogalum, fritillaries, rosa muscata, crocus vernus with a dark red flower, hyacinthus stellatus, delphinium and 'bastard' columbines. The fifty or so references in Clusius's works to Hoghelande refer to many of these plants as well as to a double peony, iris damascena, moly, erica and a few plants (linaria cretica, hyoscyamus) grown from seeds from Crete, which had probably been gifts from Casabona or Belli to Clusius, passed on by the latter to Hoghelande.[52]

In some cases Hoghelande sent Clusius dried flowers (thalictrum), or extremely detailed descriptions and painted portraits.[53] Hoghelande had the plants portrayed at his home or in his garden, usually by a professional female painter ('painctresse'), who was, as Hoghelande complained, at a certain moment so busy with painting coats of arms for a big wedding party that she could hardly spare half an hour for the portrayal of a flower that had just opened. When she died, Hoghelande tried sending an 'orchis litoralis' to Cluyt in Delft in order to have it painted there by the same painter who had made the 'very nice drawing that Cluyt has sent to you' (Hoghelande, 12 August 1592). Hoghelande was by no means always content with the results, and sometimes added small descriptive details in his letters which the painter had missed or not represented exactly enough to his taste (Hoghelande, 12 August 1592). In several cases Hoghelande sent plants or their pictures to Clusius, just like Madame von Heusenstain from Vienna, because he wanted to know whether Clusius could identify them or confirm his own identification:

> the herbarists take [it] to be a kind of Ononis and call [it] Ononis levis. It is a dense ground creeper, which frequently occupies mounds of three or four feet, the flowers resemble those of the Ononis (as do also the leaves, except that they are both smaller) and they are of a beautiful red, with white lines crossing the lower lip. These add not inconsiderably to the charm of the flower for those who take the trouble to look closely. (Hoghelande, 14 August 1591)

Sometimes he thought that the plants he sent might be unknown to Clusius. Others were simply gifts which Clusius could share with his friends.

Almost none of the plants that Hoghelande sent to Clusius came from outside Europe and the Levant, although an occasional American seed slipped in which Hoghelande himself had received as a gift. His collection and expertise thus focused on rare plants from Europe and the Levant, and in particular on (preferably rare) native plants from the area in which he lived, for which he used the term 'herbes domestiques' (Hoghelande, 13 May 1592).[54] While Roels was the only one of the Middelburg naturalists of these years to show an interest in the native flora of Zealand, Hoghelande seems to have been a similar exception in the Leiden circle. His botanizing trips in the coastal dunes of Holland were probably solitary trips and by no means comparable to the full-scale expeditions undertaken by Levenier in the Pyrenees, but Hoghelande's observations were no less acute and he had as much of an eye for detail, colour and habitat as Levenier. He transplanted an orchid with small yellow flowers found in the dunes not far from Leiden to his garden – remarking to Clusius that it did not have two tubers, as depicted in Lobel's work, but only one. In its natural habitat this orchid multiplied by producing thin side-shoots.

> I have also been in the dunes to look for helleborines and I have brought home some plants, as well as I could, since it is very difficult to get a plant in its entirety, since they fix themselves deeply in the soil between the roots of that small kind of salix that is found in the dunes, which is a major obstacle to extracting them whole, and if you do manage to get them out whole, you will need a two-foot box to accommodate them completely, and if you were to carry them unprotected tied to a stick, they would be six times destroyed *en route*. (Hoghelande, 13 May 1592)

In Hoghelande we thus find a man from Middelburg who lived in Leiden and was closely connected with its university setting, but whose approach to gardening and interest in living plants had been influenced far more by the fashions of the aristocracy in the Southern Netherlands than by the opportunities available to the inhabitants of Dutch ports or even by the intellectual setting offered by Leiden university during the 1580s and 1590s.

Like Hoghelande, Princess Marie de Brimeu and Philips de Marnix, Lord of Saint Aldegonde, were representatives in Leiden of the elite gardening and botanical culture of the Southern Netherlands. And it is Marnix – aristocrat, theologian, Protestant, politician, diplomat and poet, but also garden owner, botanical *curiosus* and friend of Clusius – who introduces us to Daniël van der Meulen, the last of the men who should be considered with respect to the confluence of cultural traditions in the Leiden circle of naturalists. Marnix was based in the Northern Netherlands for a number of years after he (as burgomaster) had had to surrender Antwerp to the Spaniards in 1585. He lived in Leiden, but spent much time in Zealand, either in Middelburg or at the castle of West-Souburg which he had acquired in 1578 and where he must have done most of his gardening.[55] Marnix was known for his considerable expertise on rare plants and Roels even called him 'a first rank student of plants and (which is unusual among the nobility) a man of rare learning' (Roels, 10 April 1593). Marnix corresponded regularly with Clusius, but in so far as plants are concerned his exchanges with the merchant Daniël van der Meulen, who lived in Leiden from 1591 until his early death in 1600, are far more interesting.[56] In November 1593, for instance, Marnix sent Van der Meulen four quince trees, a purple lilac, broom, hyacinths, crocuses, irises, narcissi, muscari, peonies, yellow ranunculus, campanula, anemones, saxifraga and a daphne mezereum – all of which had come from Marnix's own garden in Zealand. In return Van der Meulen sent Marnix some small cauliflower plants grown from African seeds, among other things.[57]

Van der Meulen was born to an Antwerp family of merchants in 1554, left the Southern Netherlands for religious reasons, and married Hester della Faille in 1584 – a daughter of an even more important Antwerp merchant family.[58] His early trade was mainly overland with Venice, via Cologne and Frankfurt; textiles and raw silks formed the bulk, but spices were already prominent as well. After a period in Bremen, the couple moved to Leiden – perhaps partly because of its cultural and intellectual climate, although the good connections with the ports of Amsterdam and Middelburg may also have been relevant. There Van der Meulen added sea trade with Italy and other parts of the Mediterranean to overland trade. He furthermore owned shares in ships that travelled to the East and West Indies and the African West Coast, and – like Balthasar de Moucheron – took a personal interest in discovering the north-eastern route to the Far East. Van der Meulen became enormously wealthy during his Leiden years and has been called 'perhaps the greatest of all commercial émigrés from Antwerp to the North'.[59] He combined the roles of merchant, diplomat, patron, book and art collector, and banker.

His house was a typical example of his lifestyle. In 1593 he bought a large house on the central Rapenburg canal in Leiden and had it redesigned by the well-known Haarlem town architect Lieven de Key, another émigré. The result was a truly princely 'palazzo', set, somewhat incongruously, in a decidedly non-aristocratic town where some officials with Puritan leanings attempted to banish singing, dancing and other forms of display. Behind the 'palazzo' was a large garden, delimited at the end by smaller houses facing the next canal – all of which belonged to Van der Meulen as well. The part of the garden that was invisible from the main house served practical purposes: it comprised a chicken run, peat shed, bleaching ground and no doubt a kitchen garden. The section for special plants and ornamental flowers must have been located in direct view of the living room at the back of the house. Both the garden and the house were special enough to be visited (in December 1596) by members of the court in The Hague.[60]

Van der Meulen was an erudite merchant: he may have studied law, spoke several languages and certainly read Latin. His library was one of the largest private ones (more than 1,200 books) in Leiden, second only to that of Marnix. Van der Meulen's international connections were impeccable. He was involved in international diplomacy, sponsored foreign students to come to Leiden and was well enough versed in the arts to be commissioned by Emperor Rudolph II to buy paintings by Hieronymus Bosch and various other Dutch and Flemish painters.[61] Even more interesting for our purposes is the circle of erudite men and women with whom he corresponded or who were frequent guests at his house. Among the illustrious names we find Ortelius and the explorer Willem Barendsz (with whom he corresponded), as well as Clusius's friends Marnix, Bonaventura Vulcanius (professor of Greek), Scaliger and Lipsius. The mathematician Simon Stevin came to borrow books, and Van der Meulen sent Paludanus stones or minerals which he had obtained himself from a Parisian collector. A prominent musician was employed to teach the children. Raphelengius, the town secretary Jan van Hout, and the physicians Gerard de Bondt (Bontius) and Everhard Vorst (Vorstius) – both of them involved with the *hortus* – visited, and Marie de Brimeu too, frequently dropped in.[62] Van der Meulen is known to have taken the interests of the large Leiden community of émigrés from the Southern Netherlands very much to heart, and it is – given the chronology, his visitors, erudition, background, position, garden and the connection with a large number of Clusius's friends – unthinkable that Clusius himself would not have been among his guests. And, in fact, he was. During a large part of 1597, and throughout 1598, 1599 and 1600 (until Van der Meulen's death) nearly all letters from Giovanni Vincenzo Pinelli in Padua were sent to Clusius at Van der Meulen's home in Leiden, while both Pinelli and Clusius spoke of their great respect for Van der Meulen upon the latter's early death, at forty-five, from the plague.[63]

That evidence opens up a vista of Clusius and Marie de Brimeu participating in lively gatherings of intellectuals – predominantly émigrés and Protestants – at Van der Meulen's palazzo and discussing the best ways to grow rare plants or obtain them from Southern Europe, probably with the assistance of Van der Meulen's agent Noirot in Middelburg. The select circle to which Van der Meulen acted as host and patron is reminiscent of the Bruges humanists around Saint Omer and the Laurin brothers, and Pinelli's private academy at Padua. Van der

Meulen's circle comprised many men employed by the university, but its members must have been mainly chosen on the basis of personal affinity and interests as well as social standing. Erudition was important, but so were – almost certainly – a shared interest in art, voyages, explorations, nature, science, cartography and theology. A lavish display of style and wealth, finally, epitomized as much by his garden as his house or library, characterize the decade of Van der Meulen's life in Leiden as the high point of Southern Netherlandish cultural influence in those parts.

Converging Cultural Traditions

Contributions to botany by the persons discussed above were by no means always easily predictable in terms of their background or training. Paauw was more important in terms of organization, continuity and donations to the Leiden collections than as a botanical expert. The apothecary Porret and both Cluyts had great botanical and gardening expertise and shared these with Clusius; the plant donations of Cluyt were crucial in the early phase of the *hortus*, and Porret's important collection of exotic *naturalia* provided Clusius with numerous new items. The nobleman Hoghelande's vast knowledge, which he shared with Clusius, Porret, Marie de Brimeu and collectors and gardeners elsewhere, concerned living plants, gardening and rare wild plants in the coastal zone of Holland. Van der Meulen, Hoghelande and Brimeu stimulated the culture and fashion of gardening and collecting rare plants in a social setting where scholars, officials, merchants and aristocrats met. The Leiden circle of naturalists thus helped to fuse types of expertise that were heterogeneous in character, in terms of their geographical roots (Holland, Italy, the Southern Netherlands), social provenance (apothecaries, aristocrats, merchants, physicians, scholars) and the practices involved (collecting, gardening, field botany, book study). Above all, unravelling the various strands that determined the *couleur locale* of Leiden knowledge of living nature shows that persons whose expertise had been moulded to a major extent outside university spheres were crucially important – directly or indirectly – to the creation of an academic discourse on nature.

Three cultural traditions were particularly important. First, the aristocratic culture of the Southern Netherlands. The importance of scholarly traditions of the Southern Netherlands during the founding phase of Leiden university has been emphasized before, while the mere quantitative fact that one third of the population of Leiden in the last decade of the sixteenth century consisted of émigrés from the Southern Netherlands has its own weight.[64] But it has perhaps not been recognized earlier how crucially important the contribution of largely *non-academic* cultural traditions of the Southern Netherlands was to the shaping of the study of nature in Leiden. Without court fashion there might have been no academic discipline. A second important tradition, just as in Italy, France and the Southern Netherlands, was that of the urban, often port-based, circles of physicians and apothecaries where intellectual (university-fed) traditions met the practical experience of apothecaries. Less immediately visible, but nonetheless very important, was the example of the Italian universities and Montpellier, in particular with respect to the foundation of botanical gardens, the appointment of specialists to teach medical botany, and the organization of botanizing trips. Those few men discussed here who were not personally connected with the Southern Netherlands – Paauw and the Cluyts in Leiden, Paludanus in Enkhuizen and Roels in Middelburg – had all learnt from Italy.

The single tradition that seems to have been absent from this mixture is an indigenous Dutch tradition of gardening or collecting *naturalia*. In fact, no groups of *naturalia* lovers, collectors or expert gardeners are (as yet) known to have existed anywhere in the Northern Netherlands before the 1590s. Härting names only one garden before 1610 that may possibly be relevant in this respect – not coincidentally a castle garden.[65] Van der Veen's survey of curiosity collections in Amsterdam includes virtually none from before 1600, while nearly all owners of early seventeenth-century Dutch curiosity collections were émigrés from the Southern Netherlands.[66] The evidence of floral decorations on Dutch tiles has its own, parallel chronology which starts in the late sixteenth century: the upswing of the tile industry itself was connected with the influx of émigrés from the Southern Netherlands, while the floral patterns were mainly based on *florilegia*, which in this part of Europe started appearing after 1590.[67] Until further evidence is found, we can only conclude that there seems to have been no serious gardening or botanical tradition in Holland during most of the sixteenth century and that an interest in *naturalia* emerged there only at the very end of that century.

Even the apparent exceptions turn out to confirm this pattern. The oldest known collection of dried (medicinal) plants in the Low Countries – Petrus Cadé's herbarium of 1566 – was probably created by an apothecary connected with a hospital in the province of Brabant, which formed part of the Southern Netherlands until almost a century later.[68] And Adriaen Coenen (1514–87), the self-taught wholesaler in fish from Scheveningen near The Hague whose beautifully illustrated encyclopedic manuscripts of 1577–86 depict and describe marine life and many other natural and unnatural phenomena, was inspired by his noble patron, Cornelis Suys from Zealand – a relative of Jacob Suys, humanist friend of the Laurins and Charles de Saint Omer in Bruges. During the mid-1560s Coenen personally talked about fish and rare marine fauna with Saint Omer in The Hague and sent him dried sea birds for his famous collection.[69] In terms of cultural traditions, Cadé and Coenen thus were linked with the two main branches – the medical-pharmaceutical and the aristocratic – of the culture of natural history of the Southern Netherlands.

VI BEYOND PLACE

11 INTERLOCKING NETWORKS: LONDON AND AMSTERDAM

In 1592 the elderly apothecary and merchant in drugs and spices James Garet Sr (also Jacques Garret; 1519/20–*c.* 1594) wrote to Clusius from London that he had just returned from a trip of some nine weeks to the Northern Netherlands. He had paid a visit to their mutual friend Johan van Hoghelande in Leiden, and seen a double red ranunculus in his garden, but Hoghelande did not want to share it: 'he is so tight-fisted that I could not get it out of his hands, patience' (J. Garet Sr, 27 July 1592). Garet also referred to tamarind 'as it grows in its pods in some places in Calicut', exotic leaves and beans, and some very small melons – 'very green and with small white spots, like the coloquint' – which he had received from abroad; they had been so fresh that Garet had planted their seeds in his garden in London, where they had sprouted, so he hoped the two plants would bear fruit.[1]

Garet was the father of James Garet Jr (also Jacques Garret; *c.* 1552/5–1610) and Pieter Garet (*c.* 1552/5–1631), two men whom Clusius regarded during most of the 1580s, 1590s and early 1600s as among the most important persons in Europe where access to and information about exotic *naturalia* was concerned. The Garets are mentioned more often than anyone else in Clusius's *Exoticorum*: James Jr 40 times and Pieter 32 times, while James Jr also occurs 14 times in the *Rariorum.*[2] Seen against the backdrop of their respective cities London and Amsterdam, the Garet brothers are the key figures in this section. Previously, while exploring how local contexts influenced forms of knowledge and expertise concerning nature, we have found that many people who were seriously interested in *naturalia* participated in more than one cultural and geographical context and travelled internationally. They combined, transmitted and inserted forms of expertise which they had learned elsewhere into local settings, thereby helping to shape local cultures of natural history. Some individuals operated in far more social, cultural and geographical contexts than others; and some places – especially the fast-growing cities of London and Amsterdam in the late sixteenth century – had reached a critical mass in terms of interlinking networks. The Garets were such persons, and they found themselves in the right places at the right time.

Itineraries

By 1600 London and Amsterdam were very rapidly growing port cities for the trade with the East and West Indies. But as little as half a century earlier, in the 1540s or 1550s, neither had been particularly relevant in this respect, and Antwerp would have been the only possible choice for an investigation of how exotic *naturalia* ended up in Northern Europe. That had everything to do with the changing patterns and itineraries of long-distance trade. After the Venetian monopoly on Asiatic spices was severely undermined by the Portuguese arrival in India and the Far East in the early sixteenth century, Antwerp – linked as it was with Portugal, and part of the Habsburg empire – became the great spice stock market of Northern Europe, controlling drugs and spice quotations. But Iberian control of the spice trade in its turn weakened when the Dutch and British, who did not respect monopolies backed by the Pope, established their own direct contacts with the West and East Indies as part of economic warfare with Philip II. And the closure by the Dutch of Antwerp's port in 1585 removed an obstacle to the fast expansion of the North Sea ports.[3] The lives of the Garets were intrinsically connected with these shifts in the long-distance trade in drugs and spices, which during the early modern period predominantly concerned plants or plant-based ingredients. Enormous political and economic interests were involved in this trade, and knowledge of plants and their medicinal properties was itself of economic relevance.[4]

The Garet family arrived in London from Antwerp. Their history of the 1550s–70s makes clear how their international European network began to take shape. James Garet Sr was born in Louvain. In 1545 he became a citizen of Antwerp, where he worked as a druggist, spice merchant, apothecary and grocer precisely during its heyday as the spice stock market. Garet became a wealthy man, who eventually owned several houses, yards and gardens in Antwerp, trading among other things in refined sugar with Copenhagen and Germany.[5] Garet Sr's decision to leave the Southern Netherlands was taken long before the closure of the Antwerp port, and probably instigated by the political instability, religious persecution of Protestants and the general turmoil in the Low Countries during the 1560s. His choice of destination could hardly have been more fortuitous.

The Garets may have left for London already in 1569–70; by 1575 they were certainly established in Tower Ward, Lime (or Lyme) Street, near Leadenhall market, an area with a high percentage of Flemish and French Protestant émigrés. The Garets belonged to the Dutch Reformed congregation at Austin Friars and the family seems to have been well off.[6]

After their emigration, contacts with the Continent did not disappear. The young James probably visited Antwerp on a regular basis, and in 1593 wrote from there about business to the merchant Daniël van der Meulen in Leiden.[7] James Sr, who may well have become acquainted with Clusius during the 1560s in the Southern Netherlands, continued to work as an apothecary, druggist and merchant in London, and passed on both his professional interests and his fascination with *naturalia* to his sons. James Sr occasionally corresponded with Clusius and by 1581 at the latest, postal exchanges between Clusius and James Jr were well established; they would continue without major interruptions until at least 1601.[8] In 1576, James Jr was already married and working as an apothecary-merchant, living in his own house not far from his father. By early 1584 James Jr had his own apothecary's shop, owned a garden near London Wall at Aldgate, and worked in official partnership with his brother Pieter and his sister Lijntgen.[9] James Jr became well known in London as a professional apothecary, important intercontinental drugs merchant, perfumer and expert on exotic *naturalia* – whether medicinal or not – and it was said that he sold the best opium in town.[10] At least two of Garet's apprentices went on to study medicine on the Continent and eventually returned to England as physicians-apothecaries. And around 1591 his illegitimate son Ferdinando accompanied (perhaps as an apprentice) Jean de la Rivière, private physician to the famous Protestant Turenne (1589–94) and subsequently until his death in 1605 chief physician to Henry IV of France.[11] James Garet Jr was wealthy enough to reserve money for a dowry of some £300 for each of his daughters, and upon his death in April 1610 left his wife 'all and singuler my goodes, chattelles, silver, ready money, plate, jewelle ringes, householde stuffe, apparrell, moveables and other thinges whatsoever'.[12]

His brother Pieter's career was very similar and economically perhaps even more successful. In 1583 he lived in Candlewick Street, not far from his father, but he decided to emigrate again and establish himself in the most important port of the Dutch Republic; he became a citizen of Amsterdam in May 1593. There, Pieter was active as an apothecary, merchant in drugs and spices, sugar refiner and investor in property. Around 1605 he owned a sugar refinery in the heart of Amsterdam, as well as a large garden and warehouse or storage space just outside the St Anthony city gate. In 1614, two years after the Amsterdam city government had decided to widen the Herengracht and develop the first 'ring' of the city's expansion along this canal, he bought at least six plots of land beside the Herengracht, some of which his business partner, the wealthy burgomaster's son Hendrik Cromhout, was planning to develop.[13] Like his brother James, Pieter was a very well-respected apothecary with several apprentices, who received a notarial confirmation when they finished their training after three years. At times he accompanied a town physician during house visits and in 1618 he was among the apothecaries and physicians who signed a petition to the city government to be allowed to set up an organization of apothecaries, create their own *pharmacopoeia* and found a medicinal garden. Pieter died in December 1631 and was buried in the Nieuwe Kerk in Amsterdam. The value of his property at the time of his death was estimated at several thousands of guilders.[14]

London Naturalists

In London the Garets grew European and exotic plants in their garden, experimented, and exchanged information with fellow botanists, druggists, perfumers and spice traders. During the last decade of the sixteenth century and the early years of the seventeenth century James Jr in particular formed part of an important, informal London circle of naturalists, physicians and apothecaries based in or connected with Lime Street. It included the royal apothecaries Hugh Morgan and John Rich; the botanist-physician Lobel (almost Garet's neighbour in the parish of St Dionys Backchurch); the wealthy silk merchant and scholar James Cole (better known as Ortelianus, the nephew of the famous cartographer Abraham Ortelius) who in 1606 married one of Lobel's daughters; the physician and expert on insects Thomas Moffett (Mouffett); the physician, entomologist and botanist Thomas Penny; the alderman and mayor Sir James Harvey with his special garden; the surgeon-botanist John Gerard; and the senior clerk or president of Chancery Richard Garth, who took a great delight in Brazilian and other exotic plants.[15] Two more sons-in-law of Lobel, Simon Le Myre and Wolfgang Rumler – both of whom were later appointed to the post of royal apothecary – may have participated in it as well. Clusius was in personal touch via correspondence with many of these men or had met them in London during one of his visits.[16]

In her excellent discussion of the Lime Street naturalists, Harkness has rightly emphasized that most of these men have been overshadowed by John Gerard and his famous *Herball* and are largely forgotten, not because their knowledge was less relevant or expert than Gerard's (the contrary seems to have been the case), but because they did not publish. That point is particularly relevant to James Garet Jr, who warned the publisher of Gerard's *Herball* in 1597 that Gerard's manuscript was full of blunders and that many of the illustrations were incorrectly placed. Probably with the help of Garet, Lobel corrected more than a thousand mistakes in Gerard's manuscript, which itself was based on works by both Dodoens and Lobel, until further revision was stopped by Gerard and the whole case turned into a great scandal.[17] We may, however, conclude from these events that Garet regarded himself as an expert naturalist and not only as an expert on drugs, and was recognized as such by Lobel and the other members of the Lime Street circle.

Just as in Middelburg, Leiden and the Bordeaux-Poitiers area, different types of expertise came together in the Lime Street circle, while quite a few of its members had contacts all over Europe and drew on various cultures of natural history. Two examples – both concerning friends of James Garet with whom Clusius too was in personal contact – must suffice. Besides the Garets, Hugh Morgan (*c.* 1530–1613) was one of the greatest experts in England in so far as drugs and the art of growing exotic plants were concerned. He was appointed royal apothecary to Queen Elizabeth in 1583, and has been called 'perhaps the most outstanding of all the Tudor royal apothecaries'.[18] As a prominent member of the Grocers' Company of London, of which the apothecaries formed part, he was asked to act as inspector of the drugs on sale in their shops, while the medical garden at Westminster was stocked with plants under his supervision. Morgan had a house and apothecary's shop near Coleman Street, at an easy walking distance from Lime Street. He spent the last years of his life in his large house and garden in Battersea, at that time a country village on the Thames outside London.[19]

According to Lobel, who speaks of Morgan as a man with a keen interest in botanical research and the medicinal qualities of herbs, Morgan's garden, like that of Coudenberghe in Antwerp, was among the most beautiful, rich and famous in the whole of Western Europe. Lobel refers to Morgan at least twenty times, generally in connection with plants growing in the latter's garden, his explicit interest in West Indian plants and his contacts with sea-captains and merchants.[20] And Morgan wrote indeed about such contacts to Clusius, mentioning the explorers Raleigh (who liked tobacco so much), Cavendish (who brought back rhinoceros horns, some kind of white nutmeg, bezoar stones, Spanish civet and large quantities of costly oriental musk) and an anonymous Frenchman, who dealt in exotics and brought Morgan feathers of various colours, pulses, nuts, mother of pearl, horns and purple snails (Morgan, 22 July 1589). Morgan received new drugs from the Englishmen who returned from the first explorations of Virginia. And vice versa, the contents of the medicine chest which accompanied Martin Frobisher in 1576 on his voyage to discover the North-West Passage had come from Morgan.[21] Clusius was especially interested in the exotic fruits and spices which he received from Morgan or saw at his house in 1581. Some of these had been gifts to Morgan from men who had accompanied Francis Drake on his voyage around the world of 1577–80, such as a booklet of paper made of the paper tree, and some pieces of a bread made from flour of the 'flour tree' (it reminded Clusius of sago) – both from the East Indies. Morgan also gave Clusius various exotic beans and pieces of exotic plants or substances based on exotic spices (such as balsam, oils, sandalwood and a piece of the faufel or areca palm).[22]

Morgan's contacts were wide-flung and exotica were his main interest, but his background and training were English. The physician-botanist-entomologist Thomas Penny (*c.* 1530–88/9), on the contrary, spent important years of his life on the Continent, was inspired and taught by some of the most famous naturalists of his age, and concentrated on field research and rare European plants and insects, not on exotica.[23] He was a good friend of James Garet – Penny said about the cricket that Garet 'had produced their chirruping sound by rubbing together their torn-off wings' – and in his younger years of the great Swiss naturalist Conrad Gessner.[24] Penny studied theology at Cambridge and was ordained deacon in 1561. His interest in nature was already explicit in his youth, and when his Puritan leanings and outspoken preaching landed him in trouble, he left for the Continent in 1565 to study medicine. He was received hospitably in Zurich by Gessner, and botanized in the Jura and the Savoy, where he made notes on gentians and orchids. After Gessner's death from the plague in December 1565, Penny went to Montpellier, where he became close friends with Lobel. From Montpellier he undertook a highly original botanizing trip to Mallorca. Penny then proceeded to Orléans in order to study plant physiology together with the apothecary Noël Capperon, the expert on the fritillaria, and to Paris. By 1569 he was undertaking field trips in Germany and the Baltic, where he discovered the Cornus suecia which he gave to Clusius.[25]

Back in England, after 1570, Penny eventually settled in London and practised as a physician. He lived in Leadenhall Street – very close to Lime Street – while his garden was nearby, not far from Moorgate. During his sojourn on the Continent Penny had not only studied plants but also insects, and in the course of time he created a great collection of exotic and indigenous insects and drawings of insects with the help of many international friends. Penny undertook several field trips in England as well, and he was known as a great expert on the native English flora and the flora of the Jura and Mont Salève in Switzerland. He never published and died relatively young, whereupon the botanical part of his legacy ended up with Peter Turner, son of the most famous English botanist of the mid-sixteenth century, William Turner. All the insect material (a manuscript and some 500 drawings) was bought by Thomas Moffett, as James Garet wrote in 1589 to Clusius, 'and he takes it upon himself to finish the book on insects and have it printed ... he is my great friend' (J. Garet Jr, 9 September 1589).[26] Moffett cut out Penny's drawings of insects and pasted them in his own reworked version of Penny's text, the *Insectorum Theatrum*, of which a printed edition with rather crude woodcuts appeared only many years later (1634). Although Penny's work is thus only known via other authors, he is regarded as an outstanding observer of plants, with a great talent for description.[27] His description of the

Rubus chamaemorus (cloudberry), which he saw in northern England, is quoted at length by Clusius and conveys Penny's qualities as reporter. It also evinces his interest in plant habitat.

> It consists of stems twelve inches long on which alternately grow three, four, or rarely five leaves, rough in texture, not unlike those of mallow or rather a mulberry divided into five points and serrated on long pedicels, and springing out of two wings or processes embracing the stalk ... It spreads wonderfully and creeps very far, so that it quickly covers a wide area. It flowers in June and early July: the fruit is ripe in August. It loves snowy and open places and the tops of hills, and grows in great plenty among heather on mount Ingleborrow the highest in all England, twelve miles from Lancaster.[28]

The intensive contacts of the Lime Street naturalists with the Continent are equally evident from the visits that James Garet Jr received from colleagues and friends from the Low Countries and Germany. In 1589, for instance, Joseph Michaeli – by then the foremost physician of Antwerp and formerly private physician to the Prince of Orange and Archduke Mattias – stayed at his house and gave him a strange fruit with a hard shell and a nut inside which James planned to send to Clusius (J. Garet Jr, 9 September 1589). Many years earlier Michaeli and Clusius had both lived in the house of Rondelet at Montpellier.[29] Another friend, young Joachim Camerarius (1566–1642; the son of Clusius's friend of the same name) stayed in Garet's London house in the summer of 1592. Together they visited the country houses of some noblemen, when Paludanus showed up at Garet's as well. The three men made a plan to travel to Germany together and visit the estates of some noble families there, but nothing came of it. Garet's contacts with Continental botanists continued: Jean-Henri Cherler (*c.* 1570–1610), son-in-law of the Swiss physician-botanist Jean Bauhin (1541–1613), visited James Garet Jr in London in 1604, 1605 and 1606. And the Swiss botanist-anatomist Thomas Platter, who visited Lobel in London in 1599, probably also met Garet there.[30]

New types of interest in nature and new forms of research thus reached England via the arrival of émigrés from the Southern Netherlands, the visits to England of naturalists from the Continent, the journeys of Englishmen on the Continent, and, of course, via printed works. Natural history was as yet a relatively new tradition in England, which there, as on the Continent, was given a strong impetus by men with a medical and pharmaceutical background – such as William and Peter Turner, Penny, Morgan, Lobel and the Garets – on the one hand, and by some members of the high aristocracy, on the other hand. This new interest in natural history seems to have arisen not long before the middle of the sixteenth century and to have gained momentum especially during the 1580s and 1590s.

Aristocratic Gardeners in England

Few if any important English naturalists are known before William Turner (1508–68), whose main activities in England are dated between 1548 and 1568. Turner had been deeply influenced by his lengthy sojourn and training on the Continent and contacts there with men like Ghini in Bologna, Gessner in Switzerland, Fuchs in Germany and Coudenberghe in the Southern Netherlands.[31] No English works of botanical art or flower or animal painting are known until the *Florilegium* made by the painter-horticulturalist, entomologist and dealer in plants from the Indies Alexander Marshal around the mid-seventeenth century.[32] And few if any of the famous gardens belonging to British aristocrats from before 1560 are known as private botanical gardens or collections of rare plants. It must remain undecided here whether the latter is simply for lack of detailed information or because the typical knotted gardens of Renaissance England were stronger on decorative elements, topiary, banqueting houses, fountains, labyrinths and intricate patterns reminiscent of embroidery, than on rare and varied plants.[33]

One of the earliest private botanical gardens in England seems to have been that of Edward Seymour, Duke of Somerset, at Syon in Middlesex, which is mentioned by William Turner in his works of 1548 and 1568. Famous gardens of the later sixteenth century were those of William Cecil, Baron Burghley, at Theobalds in Hertfordshire, created after 1575 with the help of John Gerard; William Turner's own gardens at Kew and later Wells; John Gerard's own garden at Holborn with its 1,100 plants; and the physic garden of Lord Zouche at Hackney supervised by Lobel.[34] Even by 1605, however, Pieter Garet did not have a high opinion of the interest in rare plants of the English in general. He wrote to Clusius that Emanuel Sweert(s) (1552–1612) – a Dutch merchant in rare plants and nurseryman who later worked for Rudolph II in Vienna and is best known for his *Florilegium* – had left for England to sell his rare plants: 'I think that he will not find any customers there because the Englishman seeks no rarities but only good food and drink and being well dressed' (P. Garet, 29 August 1605).[35]

Members of the Lime Street community thus acted as personal consultants, and in Lobel's case also as supervisor, to two of the best-known aristocratic gardens of late sixteenth-century England. The link with the aristocrat and diplomat Edward la Zouche, eleventh Baron Zouche (1556–1625), is of particular relevance in connection with Clusius and the Lime Street naturalists. Zouche, who spent the main part of the years 1583–7 on the Continent and was there again regularly in later years during diplomatic missions, probably first met Clusius in the course of the 1580s. They were on terms of mutual friendship and respect, and corresponded about many different topics, in particular diplomacy, politics, gardens and plants.[36] Indeed, in November 1592 Zouche wanted Clusius to come with him to England and advise him on creating his garden:

> where I would have been well pleased to have your company, which I would have honoured like that of my father, being very desirous after my long labour to retire to a place where I can enjoy the conversation with an erudite man while creating a garden in which I can have as many things as I can find with the help of my friends. (Zouche, 11 November 1592)

He discovered, however, that Clusius had already decided to move to Leiden, but he still asked Clusius to help him obtain 'all the plants that according to you are rare in England', if possible both from Clusius's own garden and from Clusius's friends in Kassel, Heidelberg and Vienna (Zouche, 11 November 1592). It was Clusius who brought about the contact between Zouche and Lobel, which resulted in Lobel's appointment as Zouche's garden consultant-supervisor, and the friendly exchanges and gifts of seeds also continued afterwards. In 1597, Zouche thanked Clusius, for instance, for sending seeds to Lobel and tulip bulbs for his own garden in which

> I confess to take a great pleasure although I cannot find anyone who will organize it in the way that I wish, partly because there are so few honest people who know how to do this, and partly also because of the scant means that I have to remunerate them. Although it is true that I have become acquainted thanks to you with the good doctor Lobel whom I find very honest and well esteemed among us noblemen for his virtues. (Zouche, 22 October 1597)

And in the same year, Clusius received some further details about the garden that Zouche was starting to create in Hackney, a village within easy reach of London:

> where he has purposely bought some houses with several gardens, for his pleasure. He has given one house to Monsieur Lobel in order to come and stay there sometimes, with a garden to be used as kitchen garden. Monsieur Lobel is quite often there and is beginning to prepare a garden for Monsieur Zouche, because it is at present the greatest pleasure he has, and he desires that Monsieur de Lobel provide it with all sorts of simples. (L. le Myre, 15 July 1597)

Equally intriguing links existed between the Lime Street community and some aristocratic Englishwomen with a pronounced interest in *naturalia* and science. Louis le Myre (or Myriaens; d. 1635) was connected with two of them. He was the son of Flemish émigrés: his father was an apothecary in Frankfurt, and le Myre himself had been Clusius's own assistant in that town. Le Myre had trained as an apothecary and eventually left for England, where Clusius's recommendations helped him obtain a position with Zouche.[37] It was le Myre who occasionally informed Clusius about Zouche's garden and who also told him about the great liking of Zouche's wife (and cousin) Eleanor Zouche for the garden: 'she takes as much delight in simples and plants as my master himself' (L. le Myre, 20 February 1597). The connection with

Zouche and Lobel worked out well for le Myre. In 1599 he married Mary, the eldest of the beautiful Lobel daughters, and his career as an apothecary prospered thanks to the patronage of Lobel. Together with Wolfgang Rumler, another of Lobel's sons-in-law, le Myre was sworn in on 4 October 1607 as private apothecary to the new King of England, James I (Lobel, 18 June 1601). In 1600, the baptism of le Myre's first son at the French Church in Threadneedle Street in London had already shown clear signs of his social ascent. The principal witness besides Lobel himself was Margaret Russell Clifford, Countess of Cumberland, former bridesmaid to Elizabeth I, author of an alchemical recipe book, and one of the noble ladies to whom Thomas Moffett – the entomologist-naturalist from Lime Street – dedicated a publication.[38] Moffett was also one of the most important members of the Wilton Circle of which Mary Sidney Herbert, Countess of Pembroke, was the patroness.[39] The relevance of these English noblewomen for natural history – just like that of many male experts from Lime Street – seems to have been largely forgotten because they did not (or only rarely) publish or write about such matters. But on the side of the women too, we see two traditions joining and stimulating expertise concerning plants: the medical one, which shaded off into (al)chemical knowledge, and the one of aristocratic display and pleasure in gardens.

Harkness has emphasized the importance of London as a highly stimulating context in which new forms of rational, scientific thinking were developed – pointing to the Lime Street naturalists as one example – much as Cook has explored how the peculiar, commerce-dominated character of the early Dutch Republic generated a similar scientific mentality.[40] Here it should be emphasized that while the terms city and port immediately evoke a bourgeois cultural setting, neither London nor indeed Lime Street can be regarded as merely that. For many men based in Lime Street or closely connected with its circle, the court and various aristocratic houses and mini-courts in England were literally just a step away and moreover of great relevance to their careers.[41] In the Lime Street circle the interlocking of the various cultural influences and styles – of city, court and port – was crucial to the type of natural history which emerged there, just as the local culture of natural history in Leiden could not be understood without the influence of the elite fashions of the Southern Netherlands and Italy, and the new medical traditions of the Italian universities.

Amsterdam Entrepreneurs

Amsterdam was very different. It was fast transforming itself from a town into a city during the 1590s and early 1600s – in many respects thanks to the money, culture, connections and activities of émigrés from the Southern Netherlands. Almost within one person's lifetime the population grew from at most 30,000 in 1578 to at least 100,000 in 1622, and 160,000–175,000 in *c.* 1650. But there was no court and hardly any aristocratic tradition. This city's elite did indeed largely consist of wealthy bourgeois merchant and entrepreneurial families.[42]

Pieter Garet established himself in Amsterdam in 1593, the same year in which Clusius moved to the Dutch Republic. They were probably in fairly regular contact and Pieter may have visited Clusius in Leiden, since the latter was old and in rather poor health.[43] But Clusius was not too infirm to go botanizing once or twice in the dunes north of Amsterdam (1595) or to visit Amsterdam (1597, 1599), The Hague (1597) and Middelburg (1598).[44] In Amsterdam Clusius met physicians, merchants and several apothecaries who helped to provide him with exotica or were budding collectors themselves. A few examples help us to understand what was the nature of these exchanges, to what extent they indicate a growing interest in nature in Amsterdam, and whether there are signs in this city of a community of naturalists similar to the ones in Middelburg or London.

Clusius's exchanges concerning exotic *naturalia* with two important merchants were brief and strictly informative. Answering a request for information by Clusius concerning a bird of paradise *with* feet, the Amsterdam merchant in precious stones and jewels Jan van Weely sent him the following description in 1605.

> The bird of paradise was in every respect like the vulgar sort, somewhat flat, not of the round kind that they call papauw ... It had two legs or feet like those of a sparrow hawk or harrier, that looked unseemly and ugly, being pressed flat against the belly so that little more than the claws could be seen. The leg was dried and looked ugly too, so that the Indians very sensibly cut off the feet together with the leg, for it is the ugliest part of the bird, and in my opinion they all have similar feet. (Van Weely, 13 June 1605)

The lack of attention to every other aspect than its feet in this description is understandable given the widespread belief in Europe that these birds were footless, and the fact that Clusius very much wanted to know whether this was true. But description was all that Van Weely could offer. He had already sold the bird itself to Emperor Rudolf II, whose collection comprised a large number of birds of paradise both with and without feet (Van Weely, 13 June 1605).[45] The exchange between Van Weely and Clusius remained incidental: a request for information answered, no exchange of interest.

Similarly businesslike, and moreover explicit about his own lack of expertise in matters of *naturalia* was the entrepreneur Diederick (Dirck) van Os, a member of an elite family of merchants from Antwerp who had migrated north.[46] Van Os was one of the powerful initiators of the Compagnie van Verre, a forerunner of the Dutch East India Company, and after 1602 a governor of the Amsterdam chamber of the VOC. He had no expertise in the field of nature, but helped – via his brother-in-law in Antwerp and the latter's son in Seville – to forward letters between Clusius and the physician Castañeda in Seville. He also introduced Clusius to Dutchmen who had been to the Indies. According to Van Os, no one could tell Clusius more about tropical fruits, trees, branches and leaves than Lambertus Hortensius, a ship's surgeon who had been to the island of Banda and other parts of the East Indies. Hortensius, presumably, was supposed to succeed the ship's doctor and apothecary Nicolaas Coolmans (or Colius) in the role of Clusius's informant. Coolmans had died in August 1601 during the return journey of the so-called Fourth Voyage (1599–1601). The idea of asking the governors of the VOC to give one particular ship's doctor the task of gathering exotic – and possibly medicinal – *naturalia* had come from Clusius's Leiden colleague Paauw, who managed to get support from the board of Leiden university and the Estates of Holland.[47] In September 1602 it was again Van Os who made sure that a memorandum by Clusius was distributed to apothecaries and ship's surgeons bound for the East Indies in which they were asked to bring back small branches – between layers of paper – together with the leaves, fruits and flowers of many of the spices growing in the Indies, and of cotton, all trees that looked foreign, and the small trees that grow under water (coral). They were asked to gather information about the names of these plants, whether the plants concerned were evergreen or deciduous, what they could be used for, and if possible to supply a drawing. Small fishes were welcome too (Van Os, 17 September 1602).[48]

In spite of the official assistance from the side of the VOC and its predecessors, Clusius was dissatisfied with the material obtained via these channels.[49] This was in part because Coolmans's death and his incomplete notes obliged Clusius to deal with dried exotic *naturalia* about which he had virtually no information, generally not even an image or a description of the living plant. Many exotica were brought back 'unofficially' (and against VOC-regulations) from the Indies, moreover, and privately sold or given away. They probably came only to Clusius's attention if people were willing to share their material with him.[50] Although he was a merchant himself, Pieter Garet complained bitterly about the various deals that took place when a ship had returned from one of the voyages to Africa or the East Indies.

> Here there are more than 100 persons who take their small boats out to sea as soon as they hear that ships are arriving back from Guinea, the East or West Indies, and who buy everything so that I cannot put my hand on anything. And if I come across anything that looks rare they keep the price up high at such a level that it would be a shame to give so much for it ... and here there are some who buy at those high prices and take their things to Germany and to the court of the emperor in order to trade, so that it spoils my pleasure. (Pieter Garet, 9 February 1605)

After the publication of Clusius's *Exoticorum* in 1605, Peiresc showed himself unimpressed by what the Dutch ships had managed to bring Clusius from the East Indies. He compared it unfavourably with the tropical *naturalia* that he himself had managed to obtain via French mariners:

> But to tell you the truth, I was quite surprised to see that your Dutch mariners who travel so freely throughout the Indies have acquitted themselves so badly of their duty towards you, and that they have not supplied you with a larger number of curiosities, and especially that they have not been more careful to bring you not only the seeds or fruits but the branches themselves of the majority of the plants that they collect. I did not expect ours [i.e. our mariners] to come up with anything that you did not already know; but it is the opposite. (Peiresc, 15 February 1606)

In Amsterdam, the leading entrepreneurs directly involved in the companies that organized the voyages to the East Indies thus appear to have been only moderately interested in *naturalia*, except as merchandise. Only a very few of Clusius's other contacts in Amsterdam besides Pieter Garet mention a garden or an interest in living plants, moreover. Jan Popper, an official and merchant of the VOC in Amsterdam who promised to be personally on the lookout for things that might interest Clusius – 'No ship of our Company will come from the East Indies ... or I will keep an eye on everything and accommodate your desire in so far as possible' – was beginning to be affected by the interest in flowers and gardening (Popper, 30 October 1602). Clusius sent him bulbs for the two gardens behind his house which he had just started to create for himself and his family: a safe place where thieves could not easily enter, as he emphasized, where he hoped to grow many flowering plants and would attempt to raise plants from seeds that arrived from far away (Popper, 30 October 1602). David Mostaert (*c.* 1554–1615), yet another Protestant émigré from Antwerp, was explicitly interested in exotic plants and kept notes on them, which he promised to send to Clusius, as well as the plants themselves if they prospered. He was no merchant, but a notary and expert on church music and hymns, and the father of Daniël Mostart, the future town secretary of Amsterdam.[51] Mostaert owned a 'suburban garden' and also speaks of his 'horto pensili', probably a terrace or enclosed garden. He had received five kinds of exotic plants from a friend who had brought them from the 'further shores of Africa' (Mostaert, 7 September 1605). Among these were bulbs, some of which were already beginning to sprout. Others were dried flowers or stalks and leaves – such as a polyanthus with a white flower that looked like an asphodel, and a plant similar to a purple gladiolus – of which Mostaert sent various samples and brief descriptions to Clusius. He wanted the latter's advice about their identification and names, and concluded his letter with a short description of his dissection (he called it 'anatomisando') of a spiky American 'melocarduum' (Echinomelocactus). Mostaert had found 'in its wool or flower these seeds in follicles which I am sending to you, to see if they have perhaps not been observed by anybody before' (Mostaert, 7 September 1605). Both the fact that he took notes on his plants and his efforts at inspection and dissection show him to have been a *curiosus*, a budding naturalist with some interest in research.

Clusius may not have known every single person in Amsterdam with an interest in rare *naturalia* and exotica, but it is hard to imagine that he would not have been in contact with the principal ones. The rarity of remarks about gardening and growing rare *naturalia* in letters to him from inhabitants of Amsterdam – which matches an almost complete absence of references to Amsterdammers as sources of information about living plants in Clusius's works – suggests that gardening and the growing of living rare plants had not (yet) become a fashionable pursuit there by 1600–5, let alone a domain of special expertise. And that fits in perfectly with the evidence presented earlier on the late start of a Dutch gardening tradition.[52] Even the fashion of collecting curiosities may only have started in Amsterdam during the 1590s and early 1600s and manifested itself first and foremost in the circles of émigrés from the Southern Netherlands. None of the evidence available suggests that there was any circle of naturalists in Amsterdam during the period 1590–1610, but a letter to Clusius by the Dutch collector-merchant Jan Govaerts van der Aer gives us a glimpse of the relations between Amsterdam collectors of these years.

Jan Govaerts van der Aer (*c.* 1544/5–1612) was born in Leiden, but he moved to Amsterdam and later Haarlem. Govaerts was known as a wealthy merchant in textiles who owned shares in the VOC, but also as a collector of shells, a patron of the arts who commissioned several portraits (one by Hendrick Goltzius showing him holding a rare and beautiful turbo shell), and a member of the *Rederijkerskamer* (a literary circle) in Haarlem – in short as a figure of some weight in the cultural and economic life of these towns.[53] Govaerts's collection certainly comprised more than shells and may even have been a full-blown *Kunstkammer* with *naturalia*, art and other rarities. He was one of the private collectors, moreover, who showed or lent Clusius rare exotica, such as a scaly tropical fruit and a spiky seed pod.[54] When looking at his portraits, it is hard to avoid the impression that Govaerts was a rather pompous and self-satisfied man, and his letter to Clusius of 1597 does nothing to change that impression (Govaerts, 20 November 1597).

Govaerts met Clusius for the first time in person in Amsterdam in 1597, at the house of Dirck Clementsz Coornhert, a painter and engraver who sent some exotic *naturalia*, various bulbs and a special mussel to Clusius; Coornhert's son drew the sloth for Clusius.[55] As Govaerts explained, on that occasion he had not greeted Clusius as he should have done – given the fact that they had corresponded but never met before – since he had felt extremely annoyed that both

Coornhert and Pieter Garet had kept the information from him that Clusius had already been in town for four days:

> They eclipsed you from me. Yes, Corenhert [i.e. Coornhert] had been to my house and they [i.e. Garet and Coornhert] wanted me to give him [i.e. Coornhert] my curiosities. He said he would show them to a good friend. I said I do not let my things out of my office. I asked who this friend was. He did not want to mention your name to me, even though I knew well that it was you to whom he was taking these things, and thus he left me. I gave him some bits and bobs which he brought to you. See friend, thus they hide you from me. (Govaerts, 20 November 1597)

This letter would not have existed had Govaerts's vanity not been hurt, but it also reveals his real passion for rare *naturalia* and his attempts to be of service to Clusius: 'To sum up, I take pleasure in things that are rare and beautiful and am avid for colourful flowers, the more rare details they have, the more avid I become' (Govaerts, 20 November 1597). Govaerts offered to look for the rarities in which Clusius was interested, and planned to visit him in Leiden, bringing him presents and promising to tell him everything about the exotic *naturalia* that he owned. In the meantime he sent him a little box with two seeds and a root that had come from the East Indies, a 'papenschoen' (Calceolus Mariae), and some more seeds of which certain ones, as he had been told, were used in the Indies against headaches caused by the heat of the sun. He also included wood with a nice smell, a bit like camphor, that was very light because it grew during the night and was used as a medicine; black and white wood which had changed into stone; and some shells and artificialia such as weapons and exotic jewellery (Govaerts, 20 November 1597).

The squabbles between Govaerts, Coornhert and Pieter Garet are less petty than they may seem if interpreted as manifestations of rivalry over the honour that a citation in Clusius's work as donor of some rarity would confer. In that context the access itself to Clusius became valuable. These same squabbles are also the single piece of evidence of interaction between collectors of exotic *naturalia* in Amsterdam around the turn of the century. If this situation is anything to go by, rivalry appears to have set the tone, and there seems to have been no durable foundation for a liberal exchange or the sharing of knowledge and *naturalia* on the basis of exchange and friendship. Perhaps in Amsterdam around the turn of the century such exchanges were thwarted by the overwhelming economic competition coupled with the near absence of both an aristocratic-courtly culture and a setting where a learned tradition could establish itself.

12 SPANNING THE WORLD: DEALING WITH EXOTICS

Intercontinental Connections

To the Garets Amsterdam and London provided almost equally favourable settings for the development of connections which spanned the world. Since the very early 1580s at the latest and until at least the first years of the seventeenth century James Jr was very well informed about (and in several cases personally connected with) the voyages of British explorers. He managed to obtain many exotic *naturalia* that they brought back, gathered information about their names and uses, had access to manuscripts, printed works and early translations about the voyages and exotic *naturalia*, and was kept informed about the careers of the most important English commanders. He also tested and used new drugs in his professional preparations, and grew various newly imported plants in his garden.

In January 1584, for instance, James Garet Jr reported to Clusius that the voyage of John Drake (cousin of Francis) and Edward Fenton, which had set out for the East Indies and China in 1582, had not been a success. They had not gone further than Brazil, but nonetheless had 'brought back many beautiful things and I have made it my duty to get my hands on as many of them as I have been able to obtain for money' (J. Garet Jr, 26 January 1583, Old Style). James's contacts with Francis Drake too must already have been good, and early in 1584 Garet sent on presents of *naturalia* from Drake to Clusius, such as a small piece of gold from the mines in Valdivia (Chile).[1] As mentioned earlier, other presents from Drake or Drake's men reached Clusius via Garet's friend and colleague Hugh Morgan. Five years later, in July 1589, James had great hopes of obtaining many 'beautiful things' upon the return of a new fleet with a crew of some 400 men under the command of John Chidley and Paul Wheele; the fleet was leaving England for the Pacific and the presumed gold in the Chilean province of Arauco. Garet was actively involved this time. For three months he had hosted Chidley in his house in Lime Street, and he had sold drugs (to a total value of 260 angelots) for the medicine chest aboard the ship commanded by Wheele. In fact, Garet's own master servant, a Dutchman, joined the expedition and had been given the explicit instructions to 'cull all rare things', while 'the captain has promised me to make room on his gallery for a barrel or tub with soil in order to plant some plants in it so that, if the ship returns safely, we will have rare things' (J. Garet Jr, 28 July 1589). Garet himself had invested 80 angelots of his own capital in the expedition and hoped to share in its profits – unwisely as it turned out, because the expedition ended in disaster and shipwreck. Only six men survived.[2]

James Garet Jr's involvement in the Chidley voyage was probably triggered by the great success of Thomas Cavendish (1560–92), who had then only recently returned from his circumnavigation of the world (1586–8). Earlier, Cavendish had taken part in one of the first attempts to establish a colony in Virginia (1585). His voyage around the world and the capture of a Spanish ship laden with valuables off the west coast of Chile made him immensely popular in England; ballads were written about his voyage, and descriptions of it were soon published. Garet knew Cavendish personally and the latter visited Garet's house in Lime Street in 1589, together with three young boys (called 'Indians' by Garet) from Japan and the Philippines whom Cavendish had captured and brought back from the other side of the world. Also present at this meeting were Cavendish's ship's surgeon, and a certain Michael Sancius from Marseilles, who was known as one of the best coasters of the Pacific and had worked there for some thirty years, mainly for the Spanish. He too had been captured with his ship by Cavendish, and it was his advice that had enabled Cavendish to capture the Spanish ship which, besides treasures, also carried the Japanese and Philippine boys.[3]

Probably speaking a mixture of English, French and Spanish or Portuguese, these men sat in Garet's house in Lime Street discussing exotic plants, drugs and spices, their indigenous uses and names. One of the Japanese boys wrote down the name for star anise in his own language – 'they write starting from the top and then going down, as you will see from this handwriting' – and Cavendish's ship's surgeon gave some seeds of the star anise to Garet, who planted a few of them (without success) in his garden while sending others to Clusius: 'it has a strong aromatic flavour, as you will see when you taste it' (J. Garet Jr, 28 July 1589). The surgeon also gave Garet another kind of fruit that the boys called 'oregioella' and with which they used to 'prepare a drink with cacao' (J. Garet Jr, 28 July 1589).[4] The disasters of both the Chidley voyage and a new adventure by Cavendish – during which the latter died at sea in 1592, while only fifteen men returned alive in 1593 out of a crew of seventy-six – probably made Garet less inclined to take risks during the following years. No further mention is made of investments on his part, but his interest in explorations and exotic nature was not affected. In 1601 he again wrote to Clusius about some ships returning from the East and the aromatic leaves and other *naturalia* that he managed to obtain (J. Garet Jr, 7 September 1601).

Many of James Garet Jr's letters of the period 1589–91 briefly refer, moreover, to manuscripts, travel journals and publications in which both he, his friends in London and Clusius were interested. There is mention of a 'dispensatorium' by the Spanish physician Nicolás Monardes, books that might arrive from Spain, and of Latin manuscript translations of English treatises on Virginia and Florida commissioned by Garet's friend Richard Garth and made by a young London goldsmith called Harbles. The latter had based his translation on the copy of 'mister Haclet' – no one else than Richard Hakluyt, who had in 1588 returned to England after a long sojourn in France and published his *Principal Navigations* in 1589.[5] Garet sent Clusius the treatise about Virginia, and said that Garth himself would send Clusius a printed French treatise about Florida which was extremely hard to come by, as well as two books or brochures about Francis Drake's conquests of Santo Domingo and Cartagena, and his attack on Lisbon. Shortly afterwards he confirmed that there was as yet no printed treatise in Latin about Virginia, and sent two brochures to Clusius on behalf of Garth, one of which dealt with Virginia.[6] No further details are mentioned concerning these exchanges of publications and manuscripts, but even these fragments of information underpin Pieper's findings: manuscripts (and translations in manuscript form) played as important a part as printed works in the diffusion of news about the New World and the voyages of exploration.[7] They show, furthermore, that not only Clusius but also the Garets and the Lime Street community were among the hubs in the European networks of communication concerning exotic nature and the exploration of the New World.

From 1593 onward Pieter Garet in Amsterdam followed an almost exactly parallel course to that of his brother. Like James, he became a crucial source of exotic *naturalia* and information about exotica for Clusius because of his excellent contacts, expertise and curiosity. Pieter must have started to expand his contacts with Dutch explorers, merchants and captains immediately after his arrival in Holland. Already in 1596–7 he could send Clusius some oil made from Central American copal and a frond of a coconut palm.[8] Garet may also have acted as advising apothecary and provider of drugs for the medicine chests of Dutch ships bound for the East Indies. By 1600–2 he was in personal contact with the two admirals of the second Dutch voyage to the East Indies (1598–1600), Wybrant van Warwijck and Jacob van Neck, besides various captains, ship's surgeons and merchants who had been to the Indies. Because Van Neck's ship returned to Zealand, and not to Amsterdam, Pieter Garet had to tell Clusius that 'all rarities had remained there'; but Van Neck did give Garet three pickled mangoes from the stock which he had been eating himself during this voyage; they still tasted very good and Garet sent Clusius two of them (P. Garet, received October 1603).

Occasionally Pieter also acted as intermediary on Clusius's behalf. He suggested, for instance, that he could present a copy of Clusius's recently published *Exoticorum* (1605) to the governors of the VOC if the latter was unable to come to Amsterdam and do so himself (P. Garet, 29 August 1605).[9]

Gathering Evidence

Pieter's role as information broker was particularly important. He may have been actively involved in the distibution of the memorandum of 1602 with Clusius's instructions to apothecaries and ship's surgeons to bring back leaves, nuts, branches and so on from the East Indies. And he was instrumental in putting Clusius in touch with Jacob Cuelener, a merchant who had taken part in Van Neck's voyage. Because Cuelener had fallen ill in Amsterdam and was being treated by Garet, the latter had ample time to interrogate him – Garet even uses the term 'scherp ondervraagd' (closely questioned), which in a contemporary judicial context meant interrogation under torture. A particularly interesting episode involved hallucinatory experiences in the Far East after Cuelener and the crew of his ship had eaten a fruit that they called 'mad plums' and that Garet tentatively identified as belonging to the solanum family. The fruit looked somewhat like a plum and tasted deliciously like baked apples. Once the hungry sailors had seen that numerous monkeys were eating the 'plums', they too decided to try them. Some even ate hundreds. Later, on board, the effects made themselves felt:

> An Englishman on board climbed up on the poop deck and fell on his knees, crying 'don't you see the heavens have opened for me, and look at all the glories which are indescribable', while he raised his hands and prayed. Another saw the abyss of hell with all the devils and many strange sights; yet another saw straight through the ship down into the deep of the sea, and another heard a loud piece of music made by many instruments. Another trembled strongly, twitched his shoulders and hid his face from the horrible sights that he saw. And in all some 16 to 20 persons had seen different apparitions; others slept for three or four days without eating or drinking. When they were woken they begged to be left alone 'I am so happy and cheerful'. (P. Garet, 10 August 1603)

The merchant Cuelener provided not only Pieter Garet but also Clusius himself with a detailed written description of the fruits and trees.[10]

Pieter highly valued first-hand information and was avid for precise knowledge about exotic plants even when no commercial motives were involved. He tried his utmost to obtain corroboration of names and details by the men who returned from the East Indies to Holland:

> I have become acquainted with all ship's surgeons who travel to the East Indies, who have promised me to bring back curious things such as fruits, branches of trees, roots,

> and herbs with their names in the local language and their effects. (P. Garet, 30 January 1602)

Either on his own initiative or upon Clusius's request Pieter Garet interviewed the commanders and merchants of these ships about the plants, fruits, nuts and spices they brought back from overseas. Reporting back to Clusius, for instance, about the colour and shape of the flowers of the clove, he wrote that he had asked everyone he knew who had been to the East Indies, but had found that no one could tell him but Hans Vos and some other ship's captains who had actually seen them grow (P. Garet, received 4 May 1601). But when Clusius wanted some details about the famous dodo, apparently no one in Amsterdam could help in 1603. Garet ended up by referring Clusius to the publication of 1600 on the second Dutch voyage by Cornelis Claesz, which included a portrait of the dodo (P. Garet, received October 1603).[11]

Pieter was no exception in his quest for exact and detailed information concerning exotic nature, and his desire for eyewitness accounts or, more generally, for reports based on practical, personal experience. His methods were exactly the same as those of Roels and Parduyn in Middelburg. All of them went out of their way to check information about exotic plants and their uses with men who had visited those far-off regions or had lived there. All tried to compare various opinions, if possible. From London, James Garet Jr wrote to Clusius concerning the 'oregioella' that it was added while preparing a beverage by boiling 100 to 200 cacao beans in water, but he had not yet found out which tree produced this fruit: 'I will ask more thoroughly about this fruit ... the person who gave it to me does not have the kernel that was inside, because I have asked him; anyhow, I will ask those who have been there' (J. Garet Jr, 9 September 1589).

James's report to Clusius in which he describes his meeting with Cavendish, his ship's surgeon, the pilot Michael Sancius, and the three young Philippine and Japanese boys in Lime Street in 1589, shows that Garet used a range of different means and senses to gather and evaluate information. First of all, James looked carefully and listened to the various opinions proffered by these informants concerning the star anise (Figure 9):

> I have had the Indians that Candits [i.e. Cavendish] had brought with him here in my house and have shown them [the star anise], asking what it was. One who came from the Philippines called it *damor*. I also asked, separately, the other Indian who also came from those whereabouts, and he too called it *damor* in the presence of a Frenchman whom Candits has also brought back from the Indies, who has been there for thirty years and is a pilot of the South Sea. And they also told me that they boil the *damor* with water in order to drink it, as we do here with anise and water. (J. Garet Jr, 28 July 1589)

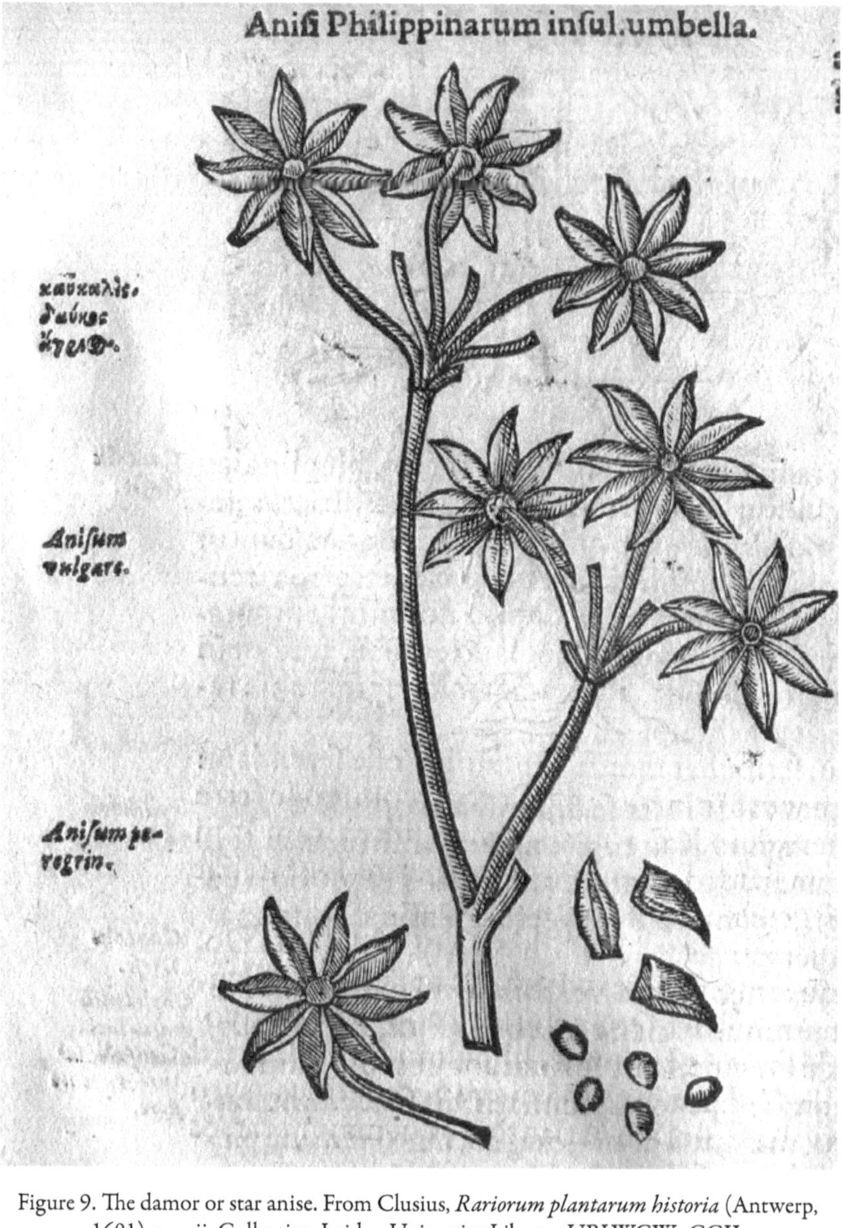

Figure 9. The damor or star anise. From Clusius, *Rariorum plantarum historia* (Antwerp, 1601), p. ccii. Collection Leiden University Library, UBLWGW_CCII.

This contradicted what Cavendish's ship's surgeon said about the name of the star anise: according to him it was called *cinchi*. But Garet attached more value to the fact that both his native informants had called it *damor* – separately, and thus independently, and in one case in the presence of an expert witness who had lived for thirty years in those parts of the world. He told Clusius: 'but I rather believe the Indians' (J. Garet Jr, 28 July 1589).

The evidence of the senses comprised more than sight and hearing. The taste, smell and touch of an expert apothecary formed professional instruments of verification and identification.[12] James Garet Jr tasted the star anise and wrote concerning another tropical fruit to Clusius that its pulp tasted slightly sour or bitter and almost melted in the mouth (J. Garet Jr, 28 July 1589, 9 December 1591). In another case, he wrote about a piece of wood:

> These last few days I have met some mariners, one of whom had a piece of wood in his hands of some two feet long. When I saw it I thought that it was a root of sassafras, but when I had it in my hand, I found that it was very light indeed and they told me that it was wood that preserves the fire as one does with a wick made of linen or those mushrooms that are called *vonchout* [lit. sparkwood] in Flemish. They told me that it grows like this naturally and is not at all rotten. (J. Garet Jr, 18 April 1601)

Pieter Garet, too, used to test and taste many of the exotic gums, resins and other substances that reached him:

> The one [gum] is different in smell from the other when one lights them with a candle; two kinds are well known to us, those are the gum anima which is very hard and clear, the other is the gum copal. When put on the fire these two gums produce a very sweet smoke. (P. Garet, 23 March 1601)

About a piece of wood that had been coloured red he speculated that the colorant might be lacquer, since it melted when singed, or otherwise 'their dragon's blood' (P. Garet, received October 1603). Among the most interesting gifts that he had received from a 'barbarian from the West Indies' was an intriguing piece of South American wood covered completely in a gum

> which smelt like mastic if held over the fire and scratched a little ... the Indians put upright 20 and 25 big branches as long as a man, vertically; underneath them they put stone pots in which they make a fire to extract the gum, which they use instead of tar to rub on their canoes; it [i.e. this wood] grows as tall as a man and a half with many wide-spreading branches. (P. Garet, 30 January 1602)

In their appreciation of first-hand information and their interest in eyewitness accounts the Garets also closely resembled Clusius himself. The latter's reasons for visiting Amsterdam precisely in 1597 and 1599 were connected with the return of the very first two major voyages undertaken by the Dutch to the East Indies. Ships from the so-called first voyage under Cornelis de Houtman, which had reached Banten and sailed along the north coast of Java, returned to Amsterdam in August 1597. And of the eight ships which sailed on the second voyage under the command of Jacob van Neck in 1598, the first four returned to Amsterdam in 1599. Strictly speaking there was no need for Clusius to be present upon their arrival in order to be among the first to get his hands on the exotic *naturalia*: friends could have assisted him in that respect. Clusius must also have gone to Amsterdam because he wanted to talk personally to men who had actually been in the Indies. Concerning an exotic branch with fruit which he had received from Porret in 1605, Clusius wrote, for instance, that he had been unable to check its provenance because he could visit neither the ship nor the person that had brought it back.[13]

Most of the Indians (here in the sense of inhabitants of India) whom the Dutch took with them on their ships to the East Indies or employed there upon their arrival seem to have had little to offer in this respect. Clusius calls them derogatively 'cheap slaves'. But Abdala from Gujarat (Guzaratensis) and Francisco Rodriguez from Bengal were indeed relevant to Clusius as informants.[14] The Dutch seem to have contracted Abdala's services in Java, where he had probably already been living for some years. He had picked up some Portuguese, and came back to Amsterdam with the Dutch ships in 1597. There, Clusius interrogated Abdala more than once about exotic fruits, and in particular about the names and uses of certain leaves, pieces of wood, nuts and nutmeg, while Abdala also identified some nuts as kemiri (or candlenut, Aleurites moluccana).[15] Clusius knew some Portuguese, but gestures were important as well. When asked about the use of a certain tropical wood, Abdala 'opened his knife and gave me a sign that it served to sharpen the blunt points of knives'.[16] The second Indian, Francisco Rodriguez, came originally from Bengal and was probably of mixed Portuguese descent. He may have been picked up in Madagascar – a stop on the route to the East Indies – and arrived in 1598 in Java with the Dutch ships under Van Neck. The following year he returned with them to Amsterdam. Rodriguez had seen the strange 'wool tree' ('lanifera arbor') growing near Banten and informed Clusius about betel, mangosteen and other tropical fruits.[17]

Both Pieter Garet and his brother James thus used not only methods but also terminology reminiscent of the contemporary judicial discourse, of which close questioning, the crucial value of eyewitness evidence, and the concept of the expert witness all formed part.[18] Their approach – which was shared by Roels, Parduyn and other friends of Clusius in different parts of Europe, as well as by Clusius himself – can be described as a comparative and critical evaluation of evidence, with a pronounced preference for direct observation and detailed description. The methodological similarities between these men cannot be explained by Clusius's influence, nor by a shared university tradition, given the fact that few of them (and almost none of the apothecaries) had gone to university or had Latin. The critical attitude of these men and their preference for first-hand evidence already formed part of their mental disposition before they were in frequent correspondence with Clusius. For all we know Clusius may, in fact, have been following *their* example, but it is much more likely that such epistemological modes were widespread in non-university-trained circles of European society long before anyone started speaking of a new science.[19] Judicial practice, which was part of a shared European tradition that was by no means limited to a learned discourse but extended far into everyday experience, may indeed have been very pertinent to these modes. And so were medical traditions – relevant to both physicians and apothecaries and again belonging to the domain of learning as well as that of everyday practice – with an emphasis on the critical comparison and weighing of observed evidence, and on practice-based criteria of observation and reliability.[20]

The Expertise of the Garets

Pieter Garet had a garden in Amsterdam for which he received plants (such as cyclamen and crocus vernus) and seeds from Clusius, but he hardly wrote about it in his letters, and he appears to have been less interested in gardening or experimenting with living exotic plants than in the composition, identification and effects of drugs, resins, gums, spices and various other exotic substances. Nonetheless, we find him in Sweert's *Florilegium* as the owner of two rare bulbs from Virginia both of which belonged to the amaryllis family: one with a white flower of which the outer edges were reddish and one with a totally white double flower.[21] In all likelihood these had been gifts from his brother James, who may have obtained them via Raleigh or others involved in the Virginia adventure.

The many references in Clusius's *Exoticorum* confirm this impression of Pieter's focused interest. His gifts included a piece of exotic wood from Punta El Rey in America, a box with various types of gums such as American copal and a gum from Guinea, a fruit from Bengal, cinnamon from Patana, yellow gum from China which was used as a purgative, mad plums and excellent green Chinese ginger. Clusius's references to him as a source of information make mention of various types of (aromatic) exotic woods, besides barks, resins and gums, both from South and Central America and from the East Indies, India and Guinea; several unidentifiable fruits, nuts and leaves from the East Indies; some exotic beans; and two special items: a beautiful piece of coral from the East Indies, and the tropical 'unicorn' fish. Garet had borrowed the latter two from a merchant and lent them to Clusius for them to be drawn.[22] Pieter Garet's sporadic descriptions of exotic plants in his letters show him to have been an observant man who could and did compare, also between exotic plants and indigenous European ones. He lent Clusius, for instance, a large leaf of an exotic water lily which he had borrowed from a friend:

> growing on the water, but carries a beautiful blue flower and they are in the Moluccas in every respect like those that are white here. It is strange that the leaves are so puckered, the stem is wound up and long. (P. Garet, 30 January 1602)

Concerning the West Indian guaiacum he remarked that 'the leaves were no bigger than those of the rubber tree ... the seeds you will [see] at the clusters of flowers, please note this, like the seed of shepherd's purse' (P. Garet, received October 1603).[23]

The exchanges between James Garet Jr and Clusius were more intensive, concerned a far wider range of *naturalia* and spanned a much longer period. James sent far more *naturalia* and information to Clusius than vice versa. But Clusius acquitted his debts to James in other ways: he performed small services for Garet, such as passing on greetings, delivering letters to a relative and helping to obtain a special steel mortar for Garet for the grinding of pearls, minerals and stones; he acknowledged James's information, help and expertise on numerous pages of his published works; and he gave Garet copies of his *Rariorum* and *Exoticorum*.[24] By 1583 the sending of packages with seeds, whole plants (such as ranunculus tripolitanus) and probably bulbs was already part of an established exchange, as was the role of James as the distributor of Clusius's gifts to other friends in London, such as Morgan and Garth.[25] Usually Garet thanked Clusius in general terms for his gifts so that we can identify only rarely what plants Clusius sent him. Garet mentioned tulips, for instance, and asked Clusius for information about the double white peony. He requested seeds of melons and kitchen herbs, the 'mirabillas de Peru' (probably mirabilis jalapa L.) and some tobacco, the golden yellow crocus which had died in his own garden, and a double colchicum about which he had read in Clusius's *Rariorum*.[26]

From what James Garet sent to Clusius we can infer that he was interested in and knowledgeable about exotic nuts, bark, fruits, roots, leaves and resins as well as living plants – rare as well as indigenous – which he tried to grow in his garden at Aldgate and with which he experimented. The first category was clearly connected with his professional interests as an apothecary. James Garet and Hugh Morgan together are said to have 'pioneered the importation of drugs or the cultivation of plants from the New World', which was at that moment still a very new and small business in England but grew exponentially in the years to come.[27] It is therefore hardly surprising that James sent both guaiacum and several types of sassafras to Clusius; both were regarded as anti-syphilitic. James is one of the rare men among Clusius's correspondents who often explained in his letters why he sent certain things, how he had obtained them and what he thought about them. In the case of the guaiacum he wrote:

> I have sent you with the present carrier a box with some exotics not knowing whether they will be of use to you. The guaiacum gum seems rare to me because it is with the seed or fruit because it is the first time I have seen it; there is a seed which I have put apart in a little box, still attached to its stalk as it has hung from the tree so that you can see how it grew. (J. Garet Jr, 18 April 1601)

And with respect to a tropical fruit he wrote in some detail to Clusius how he had obtained it, and why he thought it might be 'some kind of hyoscyamus' (yet another member of the solanum family)

> I found some small leaves among the aromatic fruits that make me think that the large leaves belong to the tree that bears this fruit. Later, at the shop of an apothecary, I came across a small bag with roots of a tree about which the apothecary said that they were the roots of a tree that bears the aromatic fruits – mariners had told him this. (J. Garet Jr, 7 September 1601)

Like Madame von Heusenstain and Hoghelande, Garet frequently made use of portraits of plants or their fruits in order to assist identification and further description by Clusius. He often suggested a tentative identification himself and asked Clusius for confirmation or alternatives. One of the only two extant plant portraits that were originally sent to Clusius by his correspondents – a drawing of 'piper caudatum' – was a gift from James Garet.[28]

> We have here a type of pepper that has come from Guinea which has a kind of tail, like the cubebe, it is called *piemento de Rama* or *Rana* and it is very hot in the mouth but has an unrefined flavour; if you want some of it I will send you a little. There is a person who has promised me a small branch of this pepper which I will have portrayed, God willing, because he will have to hand it over to me; there is also much *piper aethiopicum*. (J. Garet Jr, 19 January 1589)

Garet also had a drawing made of

> two types of plantain with double rosettes which differ from the other plantain and have not been described in any herbal as the single one has been. I will also send you the portrait of the piretrum silvestre with a double flower like the matricaria. We have here a plant that seems to be a kind of gnaphalium if one goes by the flower but it is ground covering and multiplies strongly. The flower and leaves are white like those of *Christusogen* or lychnis. (J. Garet Jr, 28 July 1589)

Elsewhere he refers to portraits of hyssop and of a rare plant that he had grown from seeds found in the waste material removed from spices. Garet did not make these portraits himself. As he explained to Clusius, it was James Cole, the nephew of Ortelius and one of the key figures in Lime Street, who drew all the portraits for him (J. Garet Jr, 28 August 1590, 20 July 1591).

James's investigations and efforts which went far beyond those *naturalia* that were professionally useful to him, distinguish him as a researcher-*curiosus*. They also extended to living plants and horticulture. There too Garet showed a desire for new, rare *naturalia*, as well as inventiveness and a willingness to experiment. He shared the current fascination for rare bulbs. A special lilium persicum (also called tusai) in his garden produced a record number of seventy-two flowers in 1583.[29] James was known in London for his superb tulips, but in his letters to Clusius there is far more mention of narcissi and especially of hyacinths, and he sent large numbers of hyacinth bulbs – many white ones, and a few red and blue ones as well as a hyacinthus hispanicus – to Clusius. Originally, each colour should have been wrapped in separate paper in order to distinguish them, but occasionally things went wrong. The markers which Garet had put next to the bulbs when they were in flower had been accidentally removed, for instance, or a servant had planted some bulbs so deep and marked them so indistinctly, that Garet's wife 'had the earth turned in my absence to sow vegetables and thus my 6 bulbs were lost' (J. Garet Jr, 28 August 1590). Garet had the habit of taking his hyacinths out of the soil in the summer, and of harvesting the new bulbs, remarking that 'they multiply better in gardens than they do in fields' (J. Garet Jr, 28 August 1590).[30]

Garet must have been one of the very first persons in Northern Europe to grow potatoes and eat them. Potatoes were known and to some extent grown in Southern Europe, but they were great rarities in Northern Europe until 1588–9; during the next ten to fifteen years they spread quickly in collectors' gardens.[31] Already in January 1589 James Garet Jr wrote that he was eating a lot of potatoes ('papos') and that they seemed very nutritious but had a rather earthy taste (J. Garet Jr, 19 January 1589).[32] He had grown these potatoes himself, so he must have already received some in 1588, perhaps directly from Cavendish and his men. In July 1589, he confirmed to Clusius that these were the real thing (now known as Solanum tuberosum) and not sweet potatoes: 'these are the real *papos* from Peru because I have asked the same Frenchman who has seen them growing last May' (J. Garet Jr, 28 July 1589).[33] This Frenchman, the pilot Michael Sancius, also told Garet that potatoes in Peru were often dried and used in soup or roasted in the ashes of a fire. Just as with the star anise, therefore, Garet received not only the plant or a part of it, but also relatively first-hand information about its use. The best way of growing this exotic tuber in a cold climate he had to find out for himself, however, by experimenting. He left some in the soil during the winter, for instance, but noticed that all those that had not been dug in deep had suffered from the frost. Meanwhile the others, which he had kept in his kitchen in an open box, began to sprout and developed their first leaves in April; when planted, these grew well. Some flowered in early June and although a few flowers had been spoilt by heavy rainfall, more promised to appear. Garet had one of these plants portrayed with its flowers; once the leaves started decaying in September he promised Clusius to dig up some plants and have the tubers depicted for Clusius 'just as they grew'. When Garet reported in September 1589 about the progress of his potatoes and sent Clusius a picture of a plant with its fruits, he remarked that these were already big and grew just like a small solanum (J. Garet Jr, 9 September 1589).[34]

His (correct) reference to the solanum in 1589 and his direct comparison in the same year between the potato and the equally exotic American tomato (also a member of the solanum family), give an indication of Garet's aptitude for comparative botany where new exotic plants were concerned:

> that these round ones had been sold to him [i.e. Garet] as *batates*, but are not that at all, and that they multiply quickly via the roots and have a great abundance of flowers with a flower like the violet anemone and fruit like that of the pommes d'amour [i.e. tomato], but not as big. (C. de Tassis, 25 January 1589)

Clusius did not follow his suggestion in this case. His description of the potato, which only appeared in print in 1601, does not connect it with the solanum family.[35] Pieter Garet's tentative identification of the 'mad plums' as belonging to the solanum family, the various other remarks by the Garets about plants that are now known to belong to the same family, and James's appropriate comparison of the potato and tomato seem too much of a coincidence. Undoubtedly they demonstrate that both Garets were well versed in plant comparison. But we may also speculate that they were particularly familiar with the solanum family on account of their work as apothecaries, and therefore better equipped than Clusius to recognize family resemblances. Not only the edible potato, tomato and aubergine belong to this family, but also many poisonous or narcotic plants, such as the belladonna (or deadly nightshade) and the datura.

In terms of quality the information provided by James Garet was high indeed, but the quantity of exotica he sent to Clusius was impressive as well. One package on its way via Antwerp to Frankfurt, for instance, contained some twenty to thirty kinds of recently arrived fruits and seeds from the Indies (J. Garet Jr, 9 December 1591). The variety was considerable, moreover. From the East Indies, the Pacific and Africa Garet sent Clusius, besides the star anise and 'oregioella' mentioned above: a picture and description of a branch of the lavender tree brought back in 1593 to England; agallochum or agalwood with its aromatic smell; a piece of East Indian bamboo; beans, resins and juices; an aromatic fruit somewhat like a clove; a tropical fruit like a fig; roots of the dracaena brought back by ships from the East; the picture of piper caudatum; a fruit from the Pacific; a baobab fruit which arrived with ships that had returned via Ethiopia; and a picture of a perfect *Wunderkammer* item, a Maldiva coconut, carved and decorated with silver, which had been found in a ship captured by the English in 1592.[36]

Naturalia from the New World were even more spectacular. Besides the potato Garet passed on roots of the gnaphalium americanum, the papaver spinosum (prickly poppy), the spike of panicum americanum one and a half feet long which he had received from an English sailor; the inner bark of the paper tree in which some Americans rolled tobacco in order to smoke it; a fruit of the macocqwer virginiensum (Crescentia cujete L.)[37] also known as the calabash or gourd tree, which had been brought back in 1591 from Virginia; various seed pods and fruits from Virginia as well as a root from Virginia with the label 'China species'; soap berries from Guyana; two fruits brought from Guyana by Walter Raleigh's men in 1598; a fruit from which a balsam could be made, received from the merchant Petrus de Frias, who claimed to have been given it by the viceroy of Peru; several Brazilian fruits, beans and seed pods; and a dried specimen of the 'herba mimosa' (Touch-me-not) from San Juan (Figure 10).[38] The latter had been brought back to England by George Clifford, Earl of Cumberland, after his short-lived capture of the citadel protecting San Juan, Puerto Rico, in 1598.[39] This 'sensitive plant' had been transported in a pot with its soil from Puerto Rico, but Garet did not manage to keep it alive in London and therefore sent Clusius the dried plant. Clusius used it as the basis for his description and illustration, and his description in its turn was used by Jean Bauhin and Jean-Henri Cherler, Bauhin's son-in-law and Garet's friend and visitor.[40] James Garet's *naturalia* thus tell the story of the late sixteenth-century English buccaneering feats, great voyages and attempts to found new settlements in Virginia from a slightly different perspective.[41] They also demonstrate that men like Raleigh, the two Drakes, Cavendish, Fenton and Clifford showed more attention to the living nature of foreign parts of the world than may have been realized until now, and were well aware who would be interested in these exotic *naturalia* back home in England.

Conclusion

To their displeasure and inconvenience men like Roels, Porret, Parduyn, the Garets and Clusius himself were very often faced with exotic natural objects about which hardly any information at all was available, not even concerning the area from which they had arrived, the name, the shape of the tree or bush from which a particular piece of bark, root or fruit had come, let alone about indigenous uses or ritual meanings. Roels's report about how manioc was grown, prepared and eaten, or Garet's information about the use of star anise or gum for tarring a canoe were relatively exceptional cases in which such information could indeed be obtained. This was certainly thanks to the presence of non-European informants in Europe, but also because Clusius's friends did their very best to obtain that information by questioning those who had witnessed these *naturalia* with their own eyes in their original surroundings, and piecing together frag-

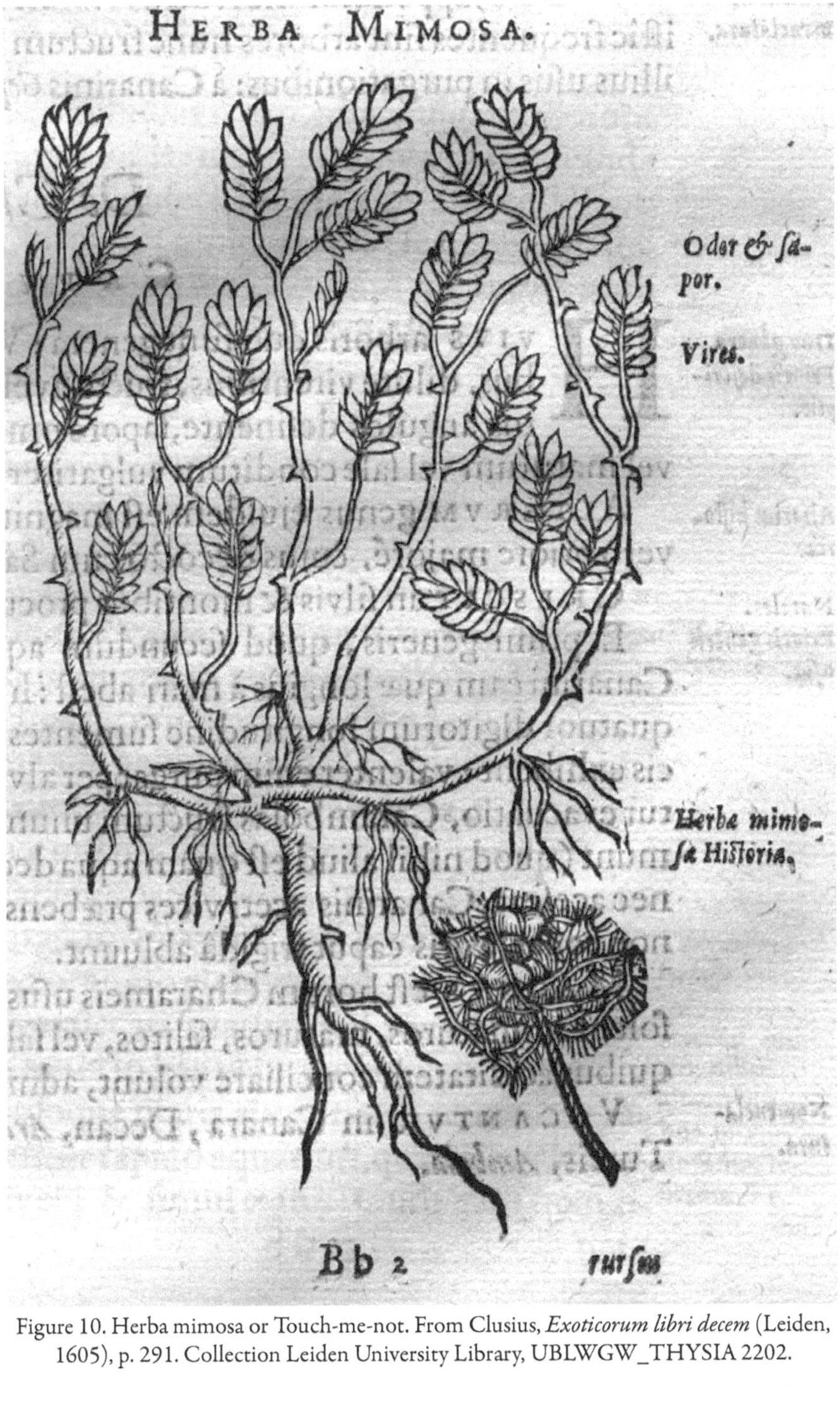

Figure 10. Herba mimosa or Touch-me-not. From Clusius, *Exoticorum libri decem* (Leiden, 1605), p. 291. Collection Leiden University Library, UBLWGW_THYSIA 2202.

ments of information about such *naturalia* from different sources. Neither the story of James and Pieter Garet nor that of, for instance, Roels tells us, therefore, a tale of how exotic *naturalia* were stripped in Europe of their local indigenous (and sometimes religious or ritual) meaning or use, wiped clean of their cultural complexities, and objectified or turned into products. Moreover, there is an important difference between an intentional 'stripping' of indigenous meaning and the mainly involuntary process with the results of which Clusius's friends were regularly confronted. In Clusius's world dealing with exotic *naturalia* did not so much involve objectification through the removal of meaning but rather description – which inevitably involved European categories – and only partially successful attempts at identification given the blank European ignorance in the face of unknown exotic nature.[42]

Unlike Paludanus, James and Pieter Garet did not become brokers in exotic curiosities. They operated either in the commercial sphere, importing and selling exotic and other drugs, or in the domain of gift exchange and curiosity, donating and exchanging information with friends who were as curious or as learned as they were. Neither the Garets nor the people with whom they were in contact had any problems in distinguishing between the role of apothecary-drugs merchant and that of apothecary-botanist, or between the idioms of gift exchange, honour and cooperation (which went with the latter role) and that of buying, paying and commercial competition, which went with the former.[43] The parallelism of those two idioms emerges clearly in one of Pieter Garet's letters:

> Sending you per courier a jar with very excellent green ginger from China and it is all that I could lay my hands on, please receive it as a gift. I have not seen any better in 20 years ... but I have not managed to obtain [more of this ginger] in order to send it to London ... here there are some who buy at those high prices and take their things to Germany and to the court of the emperor in order to trade, so that it spoils my pleasure. If there is anything here that I can do for you in friendship (P. Garet, 9 February 1605)

Garet thus presents a gift to Clusius, remarks on the potential market value of this kind of ginger, demonstrates his expertise – twenty years' experience – complains about others who drive up the prices, and concludes with a further offer of friendly gift exchange.

The interest of the Garets in *naturalia* must have started as a professional and commercial one: drugs and spices – from pepper to sugar, from guaiacum to cinnamon, cloves and opium – were their living. But it also comprised a collector's component of fascination with rare novelties and a scholarly element, which ran in the family just like the profession of apothecary and drugs merchant. The elder James Garet's interest in Hoghelande's double red ranunculus was nothing if not the desire of a *curiosus* for a rare plant. James Garet Jr in particular combined knowledge of published scholarly works and a practical interest in growing exotic plants with an awareness of the importance of (indigenous) nomenclature and of accurate description and depiction of the original habitat and use of plants. Comparison – including that between indigenous European and exotic plants – played a part in several of the descriptions by the younger Garets, and their involvement with the solanum family points to an explicit awareness of morphological resemblances between plants. They were attentive to taste, smell and touch as well as to visual experience, and they used portraits as a means of identification just as they valued first-hand eyewitness accounts of tropical plants above indirect reporting. They united a passionate interest in plants (especially exotica), unusual access to them via exceptionally wide-ranging networks, critical and investigative minds, a talent for observation and description, a close relationship with Clusius, and a clear awareness of the importance of their knowledge. It is hard, in short, to find better examples of expert knowledge which had been predominantly moulded by practice – and of the way in which such expertise contributed to natural history.

In all of this the networks of the Garets were much more than just infrastructure: they were actively developed, cared for and maintained, and as much shaped by the various commercial and non-commercial interests of the Garets as they helped to shape those interests. Interlocking, local knowledge met intercontinental access, non-commercial curiosity met trade, and practical expertise met erudition.

CONCLUSION

the art and discipline of herbs does not consist in words but in deeds and labour[1]

A Collective Enterprise

Natural history in sixteenth-century Europe was a 'collective enterprise' undertaken by a vast number of men and women of diverse social background with many types of expertise and a wide range of interests and purposes.[2] The term collective enterprise does not necessarily imply a concerted effort or the uniformity of individual aims, let alone an evenly spread awareness of that larger enterprise. Nor did the most acute awareness of that enterprise necessarily go with the greatest expertise or the widest ranging connections. Few members of Clusius's world – Clusius himself probably included – were explicitly bent on creating natural history as a discipline, although some probably realized that this was a consequence of their activities. A very large number of them, however, were aware that their joint efforts were contributing to an advance of knowledge concerning living nature and saw their own individual activities as part of this larger enterprise. And they realized that cooperation was indispensable to these efforts. Only a few limited themselves to simple exchanges of rare plants or the mere delivery of some interesting plant or animal to Clusius. The joint contributions of all of them, whatever their intentions, individual aims or awareness of this enterprise, were relevant to the formation of natural history as a scientific discipline, not because each and every one of these contributions met the criteria of 'science' – which cannot be applied anyway since they were only in the process of being established at the time – but because in a social, cultural, methodological and practical sense their joint efforts helped to create a respected and recognizable domain of expertise without which the formation of natural history and in particular botany as a discipline cannot be understood.

In Clusius's age one aim in particular was shared by a very large number of persons affected by the passion for nature: the discovery and identification of previously unknown plants. It fitted the enterprise of identifying and adding to classical *materia medica*; the desire of all apothecaries and many physicians to expand their knowledge of both indigenous and exotic plants and their potential medicinal uses; and the ubiquitous passion for rarity and collecting. This shared aim was therefore rooted in a confluence of developments in elite fashion, intellectual trends and new investigative practices. Together they resulted in many of the phenomena jointly known as the Botanical Renaissance. Socially, the passion for nature encompassed a very considerable and trend-setting segment of the (old and newly created) European aristocracy as well as many members of the professions connected with medicine and pharmacy, high-ranking clergy and a part of the rich mercantile elite. Culturally, it manifested itself in rivalry by means of display of rarity and the fashion of collecting *naturalia*. Its physical manifestations were the curiosity cabinets in which *naturalia* played an important part, gardens as living collections, herbaria with dried plants, and the so-called 'paper collections' comprising watercolours or drawings of plants, animals and minerals. In terms of practices, besides the varied activities which went with gardening and collecting, there was a growing interest in the exploration of living nature *in situ* – the countryside, ranging from the easily reached meadows near towns to the highest peaks, and from nearby parts of one's own country to far-off exotic places explored more or less thoroughly while travelling or living abroad. In terms of approach or methodology, finally, there was a strong emphasis on eyewitness or direct observation, experimentation, detailed description and reporting, an interest in contextualization and a pragmatic approach to living nature rather than an emblematic one.[3]

The ideology of friendship and of the mutual as well as free sharing of information and *naturalia* shaped the community of men and women who had a passion for living nature. To a certain extent this ideology created its own reality, not least because it was strongly underpinned by the explicitly recognized need for collaboration in the face of the great enterprise of coming to grips with an exponentially expanding mass of information about living nature and of gathering even more.[4] Free sharing and liberality were common, but they clashed both with the ambition of many of Clusius's friends to be recognized as the first person to have discovered or identified a particular rare plant – and to be publicly honoured by a reference in a published work – and with the attempts of garden owners-collectors to remain sole possessor of a certain rarity and thus keep rare things rare. The world of plant collecting and gardening differed structurally from that of the collecting of art or antiquities in the sense that plants could be propagated and shared, and by that process eventually became less rare. This process, incidentally, itself stimulated the continued search for rarities abroad and in the wild, as well as the creation of new rarities by experimentation with various methods of propagation in the garden. The fact that sharing was nonetheless a common practice among garden owners-collectors shows that considerations of honour often overrode those of economy: every collector, man or woman, had to balance in his or her mind the prestige that could be gained by liberality against that to be gained by being the sole owner of a very rare plant. Generally, it was better and more pleasant to share, not to get a reputation for stinginess, and to enjoy the participation in the continuous gift exchanges. But even the frequent fighting over claims of prime discovery of some rare plant – divisive as it may have been in personal terms – confirmed and underpinned the shared values of this community. The conflicts underlined that these were indeed issues worth fighting over.

The exchanges in this community concerned not only plants, seeds, bulbs, tubers, shoots, et cetera, but also information about their names, the best ways of growing them, propagation, and in some cases medicinal effects, smell or edibility. That information was continually checked and rechecked against practical experience (and some book learning) by all those who formed part of the community of exchange. Together those exchanges helped to create a body of increasingly documented, described, depicted and to some extent systematized knowledge about living nature, as well as the criteria of reliability and precision which helped its members to judge the validity of that knowledge. However, while circulation, transmission and adaptation of knowledge clearly were of crucial importance in this process, the Europe-wide exchanges between these men and women can be classified neither as dissemination or reception of knowledge – which have strong top-down overtones – nor as mere circulation. This was a collective enterprise: all of them were involved in the active production and exchange of knowledge. Individually they were developing new and varied types of expertise concerning living nature; together they were producing a new domain of knowledge. So, at the risk of simplification, one could say that this was not a case of science creating a network, but vice versa, of a world of overlapping and intersecting Europe-wide networks creating a science.[5]

Clusius's friends were very well aware of the considerable differences in expertise that existed in their world. In particular those men who were involved in serious rivalry over the discovery or identification of plants (such as Levenier or Belli) or those who found themselves in the emerging and as yet only half-defined role of expert consultant on botanical and gardening matters to members of the very highest aristocracy (for example, Plateau or Casabona), tended to define the knowledge of their rivals and those from whom they wanted to distinguish themselves as less valid, relevant or expert than their own. In terms of self-awareness and self-fashioning we may, therefore, speak of an emerging category of men – and in this case of only men – who saw their domain of knowledge as distinct, and recognized their own and each other's expertise as respectable, expert and professional. That awareness does not emerge from the letters by Clusius's women friends, in spite of the fact that neither their activities in the domain of natural history nor their contributions to Clusius differed drastically from those of many men. When analysing a very different setting in time and place – British geology and Darwin during the 1830s – Martin Rudwick described the social topography of science as graduated zones of ascribed competence, shading from the scientific elite, via accredited geologists and amateur geologists to the general public.[6] The distinction of those gradations is very useful as well in a late sixteenth-century setting in which the discipline of natural history (as more than just an offshoot of medicine) was beginning to establish itself. Clusius's world comprised all four types of competence, even if the terminology used was different and not homogeneous, but in his age great expertise on nature was more often found outside than within the academic world and not necessarily expressed in a professional identity or social position. Interestingly, in both Clusius's and Darwin's worlds, correspondence was the main instrument that enables us to 'place' individuals in such a graded topography.[7]

Places and People

Described in this way, and in spite of the recognition of different levels of expertise, Clusius's world is in some danger of sounding almost placeless and disembodied. The networks criss-crossing Europe, communication by means of letters between men and women who in many cases never met face to face, the fact that the shared passion for nature crossed social, religious, political and even gender barriers in a Europe scarred by violent rebellion and religious warfare, all too easily create the image of a deceptively virtual and relatively egalitarian community. We have seen that this is at least in part an illusion. All of these aspects were indeed important, but the whole of this book is also meant to show that such knowledge was fundamentally rooted in and shaped by geography, social position, local setting and gender.

Gender and social position sometimes turned out to be a hindrance rather than a help, albeit not so much for acquiring expertise but for its public recognition. The botanical knowledge of women may have been much more extensive and varied than was thought at first, but gender barriers nonetheless kept most of them from a public role as expert. Apothecaries were among the most impressive collectors of *naturalia* and greatest experts on these matters in Europe, but they have only half-heartedly been recognized as such and have in many cases been overshadowed by university-trained physicians. The importance of aristocratic patrons as prince-practitioners and experts in their own right has for many years hardly been taken seriously, partly because luxury and display were equated with frivolity and seen as incompatible with science. And private gardens – as distinct from university botanical ones – have as yet only rarely been studied as locations of knowledge since they are generally categorized under pleasure and leisure. The failure to publish appears to have been the main obstacle to recognition – especially in the long term, which tells us more about the focus of historians, however, than about history. Some of the highly expert and learned naturalists of their time, such as Roels, the Garets, Belli or Fra Gregorio da Reggio, have simply been forgotten because they did not publish or their works were lost. It should be emphasized that the historical 'invisibility' of such men has nothing to do with the intrinsic quality of their expertise. Nor does it always reflect a lack of reputation in their own time. Some of Clusius's friends (and most famously Peiresc) were known as experts on nature in a considerable part of Europe even without publishing, which emphasizes how a reputation as an expert was not merely established via printed works but also – and perhaps in a few cases as quickly or widely – via handwritten documents, including letters, and via word of mouth. We should therefore probably speak of invisibility to historians rather than of 'historical invisibility'. A change in historical search methods – less emphasis on printed works and more on manuscript sources – changes the historical picture, making it much easier to understand the interconnectedness between different types of knowledge and between the worlds of practice and of book learning. This study is only one attempt to do so; very considerable amounts of sources are still unexplored and hopefully will further change our image of early modern natural history.

Looking at the various parts of Europe investigated here in terms of what we might call a very basic geography of expertise, one type of setting – in the sense of a combination of geography and a socio-economic and cultural constellation – stands out as particularly favourable to the emergence of a strong culture of natural history and high-quality expertise in this field. Such an excellent setting existed where we find a combination of great ports or efficient transport routes over land (to help access to and distribution of exotics and other rarities) and a strong, aristocratic culture in which many wealthy noblemen and women rivalled with each other by means of display. Conditions were even better if a university or private academy existed nearby to propagate and stimulate the study of living nature and *materia medica*. The ports attracted the presence of a large number of physicians and apothecaries, many of them with a professional interest in *naturalia*. And even if some noblemen and aristocratic ladies became real experts on *naturalia* themselves, they still needed expert advice and assistance from highly qualified consultants for their collections, much as they employed expert art buyers and connoisseurs. The fact that plant knowledge was and remained so closely intertwined with medical science and pharmacy ensured its ties to the world of official learning and paved the road to its eventually becoming a discipline within academic science.

This is the situation we have found in the Southern Netherlands with the port of Antwerp, the court setting of Brussels and environs, and the rivalry of the high nobility – stimulated further by the competition between the old nobility and the *noblesse de robe* created by the Habsburg rulers. A similar situation existed in various parts of Italy, perhaps most clearly in the north-east with the port of Venice, strong political fragmentation which expressed itself in rivalry between princes, the university of Padua, and a high density of wealthy noblemen for whom high culture was directly linked with display. With some variations, comparable conditions obtained in London, France (linking the south-west and Paris) and Vienna: the bourgeois component was stronger in the former two, and the access routes varied from direct sea access to longer overland routes, but in each and every case there was a court or large group of noblemen and women interested in display, collecting and *naturalia*, and a considerable number of expert medicine-connected bourgeois professionals. Holland seems to have been the exception, at least in this period just before the Golden Age. Holland lacked an aristocratic culture, and no great gardens in the form of living collections were created there as far as we know until the seventeenth century. While its access to rarities from overseas had become excellent by the turn of the sixteenth century and several bourgeois collectors – physicians and apothecaries – created high-quality collections and developed eminent expertise in this field during the very last years of the sixteenth century, the culture of natural history manifested itself mainly in the smaller ports and in the setting of Leiden university. And it is likely that it would not have emerged even there if it had not been for the major cultural impetus of émigré aristocratic culture from the Southern Netherlands and the indirect influence of the Italian universities. Amsterdam concentrated on trade, at least for the time being, and on selling natural rarities to the major European collectors.

Without wishing to deny the importance of economic factors, all of this should make us think twice about trying to interpret collecting, the passion for rarity, and especially dealing with (and not *in*) exotic *naturalia* primarily in economic and market terms. Access in itself did not create a culture of collecting or a passion for nature, nor can rarity be studied mainly in terms of supply and demand. The cultures of collecting and studying *naturalia* were permeated by considerations of honour and the rules of gift exchange which were by no means the same as those of the money economy, even if both coexisted.

Clusius in Context

The exploration of Clusius's world – in the sense not only of a social phenomenon but of the context in which he operated – eventually leads us back to the question of what Clusius's place was in this world, and in particular of how his innovative work as a naturalist was related to that context. The key to this relation lay in shared practices, ranging from exchange to horticulture, from detailed observation to fieldwork, and from experimentation to particular forms of reporting. We can best come to grips with it by looking at those qualities for which Clusius has been most praised. Without wishing to go into the issue to what extent Clusius was a *modern* scientist, which is irrelevant here since it calls for an anachronistic application of later criteria, he has with reason been called innovative as seen from the perspective of natural history in his time.

First of all, Clusius's natural history – as represented not only by his printed works but also by his correspondence spanning half a century with more than 300 men and women all over Europe – was rigorously non-emblematic and non-metaphoric, and strikingly unimbued with references to natural philosophy. Nearly all of his exchanges are furthermore characterized by a high degree of specificity and a focused presentation of information concerning particular *naturalia*. Generalizing statements are extremely rare. Whereas the absence of an emblematic and metaphoric perspective on living nature has usually been regarded as a mark of Clusius's modernity as a scientist, that of an explicit natural philosophy has sometimes been seen as a lack – for no other reason, as far as I can tell, than that scientists of the seventeenth century are supposed to present themselves as having a natural philosophy. I suggest that all three of those 'absences', not only from his printed works but also from his correspondence, should be seen as the result of a conscious choice.[8] To Clusius, all three may well have been incompatible with the main purposes of natural history, i.e. accurate description and illustration based as far as possible on first-hand experience in the service of precise identification. Of course, that notion of natural history can be seen as a philosophy as well – and a very pragmatic and self-reliant one at that – but it is important that it is nowhere explicitly presented as such. This is hardly the place to go into these matters, but a further exploration of the similarities in this respect between Clusius and his world, and Pierre Belon and other members of the sixteenth-century French Enlightenment as analysed by Huppert would certainly be worthwhile.[9]

If these absences have been found striking in the case of Clusius, it is perhaps even more significant that hardly any of the hundreds of correspondents with whom he exchanged information on *naturalia* refer to either natural philosophy or emblematic interpretations of nature in their letters. The absence of references to natural philosophy in their letters may perhaps (also) be linked with the fact that most of the more than 300 men and women took part in the collective enterprise described above and thus had common goals even if they inevitably differed in social position, mentality and religion. Nonetheless, the possibility cannot be excluded that the importance of natural philosophy in sixteenth-century natural history has been overestimated on the basis of the published writings of a relatively small number of naturalists, and the influence of the later, seventeenth-century tradition of the 'new science' with its emphasis on such a philosophy. Given the fact that Clusius's correspondents formed a significant number of those in Europe who engaged in a serious way with natural history in his lifetime, it is impossible to regard them as an exceptional sample. Of course, this is not to suggest that natural philosophies, metaphors or emblematics played only a minor role in sixteenth-century society. But we should at least consider the possibility that many people at the time may have been well aware of when to use them and when not – which raises new questions about genres and styles. And the question should be reconsidered of whether some of the types of sources used thus far, in particular literary, painterly and religiously inspired ones, may have favoured an image that was much more 'metaphoric' than what we find in the exchanges of nature lovers and plant experts themselves.

If we leave aside Clusius's truly important role as translator and man who made knowledge about nature accessible throughout Europe, which has more to do with the distribution of knowledge than with its creation, the further characteristics for which his natural history is renowned can be summed up as follows: the effort of identifying as many new plants and animals as possible and thus expanding the known world of nature; a fascination with exotica and more generally, with all *naturalia* that were unknown before; an interest in *naturalia* for their own sake rather than for their potential use in medicine or food; a strong emphasis on eyewitness observation and basing reliability on direct experience; critical comparison and experimentation in the service of verification; a special interest in wild indigenous plants (and mushrooms), local names and knowledge; explicit attention to plant habitat; the use of field botany as a research method and the creation of regional botanical surveys; the use of gardens with living plants (and less often of herbariums with dried plants) as research instruments and locations; the conservation and propagation of rare plants in private and public botanical gardens in Europe; and a particular style of reporting, characterized by great attention to detail in both text and image, which was meant to convey reliability and credibility but also, in a practical sense, served identification.[10]

The present study should have made clear that literally all of these characteristics were shared by Clusius's friends. We need only think of the accuracy, critical comparisons and diligent attempts of Roels and the Garet brothers to verify information about exotic *naturalia*; the importation of rare plants from the Levant by Madame von Heusenstain, Louise Boisot, Jean Boisot, Jean de Brancion and Johan van Hoghelande, and their attempts to acclimatize and propagate them; the extended field research and attention to habitat of Levenier, Pona and Belli; and the detailed descriptive styles and analysis in the service of identification of Roels, Peiresc, Belli and Fra Gregorio in which there was often room for 'place' in the sense of habitat and occasionally even for aspects of plant behaviour.[11] These men and women were not imitating Clusius: all of these elements formed part and were the outcome of the contemporary culture of natural history which was fed by many different cultural traditions.

Looking at Clusius from the perspective of an art historian, Florence Hopper has pointed to two further facets of his work that deserve to be mentioned here: Clusius's aesthetic appreciation of beauty and colour; and the similarity of his descriptive imagery to dissection.[12] As we have seen, in these respects too Clusius did not differ from his friends. François Vertunien de la Vau dissected his crown imperial, Fra Gregorio did the same with his capsicums and David Mostaert with his woolly American cactus, while Peiresc opened up his stinking mushroom; each of them reported their findings in some detail. And the meticulous plant descriptions sent to Clusius by many others can only have been based on the close and internal inspection of plants, their flowers and fruits. The correspondence of a very large number of Clusius's friends – perhaps most explicitly the lyrical and precise descriptions of colour nuances and variation by Levenier and Boisot – furthermore speak of aesthetic sensibilities marshalled for the purpose of the investigation of living nature. For beauty and colour in particular were not 'merely' of aesthetic relevance. Colour was also a key element in plant identification, a recurring and often dominant factor in verbal description – given the fact that printed works had no colour illustrations – and a crucial element in rarity. Few people in Clusius's world were *not* intrigued by colour variation and the effects of propagation on colour instability. Many made great efforts or spent large sums to obtain that rarest colour variety of all: the white flower. And for at least some of Clusius's friends the hunt for rare colour varieties of wild plants was immediately connected with their aim to be known as the first discoverer of a new species.

There is, finally, yet another similarity between Clusius and the plant lovers and collectors among his correspondents, at the level of *modus operandi*. As Peter Mason has argued, Clusius was interested in scoops, scientific *primeurs*: he continued to add appendices with newly discovered plants or animals to his publications more or less as long as the printer-publisher allowed him, whether these had anything to do with the main topic of his book or not.[13] Clusius's interest in the honour of being the first person to publish a certain plant matched Levenier's in the prime discovery of new plants. But Clusius's custom of adding the latest items to his works is also strongly reminiscent of the practice of his friends who owned private botanical gardens. All were permanently striving to add new, rare items to their collections, and derive prestige from them. In that respect Clusius's works themselves somewhat resembled gardens with a special corner for the latest additions.

From Passion and Fashion to Natural History

Much of the history of the efforts of Clusius and his friends can thus be read as the curious tale of how a collective enterprise fed by cultural traditions as diverse as courtly fashions, the passion for collecting and rarity, and the interest in the *materia medica* of classical antiquity resulted in the creation of what we may call a shared and scientific methodology of precise observation (preferably in context), critical comparison, experimentation and verification, and particular styles of reporting. In that sense it tells us about the unplanned but important workings of how a scientific discipline came into being. The two principal results of the collective enterprise of natural history in Clusius's age, which had started more or less a generation earlier with the investigations of Bock, Fuchs, Gessner, Mattioli and others, were a vast increase of the known, and to some extent mapped and named world of plants and animals; and a deceptively modern-looking methodology, which we could have called reminiscent of that of the 'new science' of the seventeenth century if that 'new science' had not actually been its chronological successor. If this book had been about the history of science in the sense of long-term scientific development or progress, it would indeed have been tempting to describe men like Roels, Belli, Levenier, Hoghelande, Fra Gregorio and the Garets as forerunners of the scientific revolution.[14] But that would take us back to an approach in which the scientific revolution would form a point of reference, and even back to the very evolutionary perspective from which we have wanted to get away. And it would reduce the members of Clusius's world to something much less interesting than what they actually were: creators of new forms of expertise and the social roles to go with it.

One of the most intriguing aspects of the transformation of the types of knowledge that went with practical medicine, gardening and collecting into high-quality botanical expertise, lies in the incongruity between the (apparently) backward-looking character of at least part of the collective enterprise that motivated so many of Clusius's friends directly or indirectly, and its outcome.[15] As we have seen, identification was crucial to this enterprise. Without identification there was no way of determining rarity or novelty. And identification was only possible by setting off the newly found material against the plants known until then, i.e. primarily the body of classical *materia medica* since folk knowledge had not been systematically noted down or even collected beyond a regional level. That is why the enterprise of identifying plants and animals and assessing their newness as well as their possible family relations to known ones took Clusius's friends ever deeper into the observation, description and analysis of detail, form and colour. And precisely these practices led them to establish criteria of reliability and other aspects of a scientific methodology. It must remain undecided here whether we should regard this enterprise indeed as primarily backward-looking in character, since the first touchstone was classical *materia medica*, or as in fact forward-looking, given the open minds of the investigators, the flexibility of their ways of dealing with new evidence, and the fact that no classificatory principles were decided upon before looking at the evidence or indeed even after doing so.

By exploring the expertise of Clusius's correspondents we have been able to deduce which practice-based insights concerning both subject matter and methods were suggested to Clusius by his partners in exchange and helped to shape his own way of dealing with nature. The hundreds of men and women from all over Europe who formed Clusius's world and whom he scrupulously thanked in his printed works thus were no mere sources of 'raw', factual information that was interpreted, given new meaning and a new place in a theoretical approach by the great scientist Clusius himself. On the contrary, Clusius had no particular theory, and these men and women were his partners in exchange, representing the fine *fleur* of European collectors, plant experts and gardeners whose type of knowledge and interests were not structurally different from Clusius's own. In fact, it could be argued that Clusius was innovative as a naturalist not by *inventing* new, non-symbolic and non-emblematic forms of knowledge of nature, but by *propagating* the practical forms of knowledge of nature that he saw all around him and shared with his friends. Together, both the letters sent to Clusius and the references in his printed works therefore strongly suggest that the crux of Clusius's greatness as a naturalist lay in precisely the opposite of the 'isolated genius': he reached his position of eminence and helped to create a domain of expert knowledge that became the scientific discipline of botany because he gave prime importance to collaboration and knew how to create a community of exchange.

NOTES

Introduction

1. Emanuel, Prince of Portugal, 28 February 1602. Emanuel, who called himself Prince-Heir of Portugal (1568–1638), was a descendant of King Manuel I of Portugal. His father ruled Portugal for 33 days in 1580 as Antonio I. A Roman Catholic, Emanuel married (in 1597) a daughter of the Protestant Prince William of Orange, leader of the Dutch Revolt against Spain. The couple lived for many years in the Dutch Republic.
2. Brimeu, 18 September [after 1587 and before the end of 1593]; she is quoting Lipsius here. On Marie de Brimeu see Chapter 4.
3. Clusius to Caccini, 30 September 1606, in P. Ginori Conti, *Lettere inedite di Charles de l'Escluse (Carolus Clusius) a Matteo Caccini, Floricultore Fiorentino* (Florence: Olschki, 1939), p. 37. Obviously, this statement is also a show of modesty, but none the less significant for that. After Hunger's two volumes of 1927 and 1942 in Dutch and German respectively a new biography of Clusius in English is sorely needed: F. Hunger, *Charles de l'Escluse (Carolus Clusius) Nederlandsch Kruidkundige, 1526–1609*, 2 vols (The Hague: M. Nijhoff, 1927–42). The most recent publications devoting much attention to Clusius are B. Ogilvie, *The Science of Describing: Natural History in Renaissance Europe* (Chicago, IL: Chicago University Press, 2006); H. Cook, *Matters of Exchange: Commerce, Medicine, and Science in the Dutch Golden Age* (New Haven, CT: Yale University Press, 2007); F. Egmond, P. Hoftijzer and R. Visser (eds), *Carolus Clusius: Towards a Cultural History of a Renaissance Naturalist* (Amsterdam: Edita-KNAW, 2007); and J. de Koning, G. van Uffelen, A. Zemanek and B. Zemanek (eds), *Drawn After Nature: The Complete Botanical Watercolours of the 16th-Century Libri Picturati* (Zeist: KNNV-Publishing, 2008).
4. For full references to their works, see the works cited list in this volume. My central questions are very close to the ones dealt with in particular in P. Findlen, 'The Economy of Scientific Exchange in Early Modern Italy', in B. Moran (ed.), *Patronage and Institutions: Science, Technology, and Medicine at the European Court 1500–1700* (Rochester, NY: Boydell Press, 1991), pp. 5–24, although I have tried to move away from the concept of 'economy' and towards that of honour.
5. The term 'Botanical Renaissance' appears to have been coined in the twentieth century. Historians of science have assigned various dates to the different stages of the Botanical Renaissance; see especially A. Zemanek, 'Renaissance Botany and Modern Science', in Z. Mirek and A. Zemanek (eds), *Studies in Renaissance Botany* (Kraków: W. Szafer Institute, 1998), pp. 9–47, on pp. 11–13.
6. Ogilvie, *The Science of Describing*, esp. pp. x–xi, 14–15, 54–6; cf. 268.

7. P. Findlen, *Possessing Nature: Museums, Collecting and Scientific Culture in Early Modern Italy* (Berkeley, CA: University of California Press, 1994), p. 10.
8. For natural history in the sixteenth-century universities see the still excellent K. Reeds, *Botany in Medieval and Renaissance Universities* (New York: Garland, 1991).
9. Even though I am hardly dealing here with artists or artisans who 'make' things, in this respect my approach is close to that of Pamela Smith. In gardening and practical botany it is not so much making but doing that generates knowing. See esp. P. Smith, *The Body of the Artisan: Art and Experience in the Scientific Revolution* (Chicago, IL: Chicago University Press, 2004); and P. Smith, 'Making and Knowing in a Sixteenth-Century Goldsmith's Workshop', in L. Roberts, S. Schaffer and P. Dear (eds), *The Mindful Hand: Inquiry and Invention from the Late Renaissance to Early Industrialisation* (Amsterdam: Edita-KNAW, 2007), pp. 20–37.
10. Clusius to Caccini, 21 January 1609, in Ginori Conti, *Lettere inedite*, p. 95.
11. With respect to the contextualization of famous scientists, I found the following works in particular inspiring: M. Azzolini, 'Leonardo da Vinci's Anatomical Studies in Milan: A Re-Examination of Sites and Sources', in J. A. Givens, K. Reeds, and A. Touwaide (eds), *Visualizing Medieval Medicine and Natural History, 1200–1550* (Aldershot: Ashgate, 2006), pp. 147–76; M. Rudwick, 'Charles Darwin in London: the Integration of Public and Private Science', *Isis*, 73 (1982), pp. 186–206; J. Browne, *Charles Darwin: Voyaging* (London: Jonathan Cape, 1995); J. Browne, *Charles Darwin: The Power of Place* (London: Jonathan Cape, 2002); L. Koerner, *Linnaeus: Nature and Nation* (Cambridge MA: Harvard University Press, 1999); and M. Biagioli, *Galileo Courtier: The Practice of Science in the Age of Absolutism* (Chicago, IL: University of Chicago Press, 1993). A parallel development can be discerned in art history, with the increased attention to the workshops of artists, the art market, the role of the patron in deciding what a painter would paint and the role of the viewer/public.
12. D. Harkness, *The Jewel House: Elizabethan London and the Scientific Revolution* (New Haven, CT: Yale University Press, 2007) is an admirable example.
13. In total some 1,500 letters of the Clusius correspondence have been preserved, of which the bulk of more than 1,300 are kept in Leiden University Library, while about 200 are in the University Library at Erlangen in Germany. On Clusius's correspondence, see F. Egmond, 'Clusius and Friends: Cultures of Exchange in the Circles of European Naturalists', in Egmond et al. (eds), *Carolus Clusius*, pp. 9–48, with further references. On the formative aspects of correspondence among scientists, see A. Mosley, 'Astronomical Books and Courtly Communication', in M. Frasca-Spada and N. Jardine (eds), *Books and the Sciences in History* (Cambridge: Cambridge University Press, 2000), pp. 114–31. More generally on early modern correspondence, see F. Bethencourt and F. Egmond (eds), *Correspondence and Exchange in Europe, 1400–1700* (Cambridge: Cambridge University Press, 2007); H. Bots and F. Waquet, *La République des Lettres* (Paris: Belin / De Boeck, 1997); and C. Berkvens-Stevelinck, H. Bots, and J. Häseler (eds), *Les grands intermédiaires culturels de la République des Lettres: Etudes de réseaux de correspondance du XVIe au XVIII siècles* (Paris: Honoré Champion, 2005).
14. On network analysis and botanical networks, see E. Spary, 'Botanical Networks Revisited', in R. Dauser et al. (eds), *Wissen im Netz: Botanik und Pflanzentransfer des 18. Jahrhunderts* (Berlin: Akademie Verlag, 2008), pp. 1–18.
15. On connections between practical knowledge and science, see for instance, A. R. Hall, 'The Scholar and the Craftsman in the Scientific Revolution', in M. Clagett (ed.), *Critical Problems in the History of Science* (Madison, WI: University of Wisconsin Press, 1962),

pp. 3–23; W. Eamon, 'Unmannered Science: Natural Philosophy and Medical Practice in the Piazza', in L. Guzzetti (ed.), *Science and Power: The Historical Foundations of Research Policies in Europe* (Luxembourg: European Communities, 2000), pp. 63–8; J. Field and F. James (eds), *Renaissance and Revolution: Humanists, Scholars, Craftsmen and Natural Philosophers in Early Modern Europe* (Cambridge: Cambridge University Press, 1993); S. Pumfrey, P. Rossi and M. Slawinski (eds), *Science, Culture and Popular Belief in Renaissance Europe* (Manchester: Manchester University Press, 1991); R. Cooter and S. Pumfrey, 'Separate Spheres and Public Places: Reflections on the History of Science Popularization and Science in Popular Culture', *History of Science*, 32 (1994), pp. 237–67; (specifically on apothecaries) F. Egmond, 'Apothecaries as Experts and Brokers in the Sixteenth-Century Network of the Naturalist Carolus Clusius', *History of Universities*, 23 (2008), pp. 59–91; and Harkness, *The Jewel House*. The latter, in fact, argues that the 'modern' science propagated by Bacon was there in London all along. Cf. P. Findlen, 'Francis Bacon and the Reform of Natural History in the Seventeenth Century', in D. Kelley (ed.) *History and the Disciplines: The Reclassification of Knowledge in Early Modern Europe* (Rochester, NY: University of Rochester Press, 1997), pp. 239–60.

16. See further for a brief comparison of Darwin and Clusius, Egmond, 'Clusius and Friends'. On cooperation and exchange in the community of naturalists, see esp. Ogilvie, *The Science of Describing*; and G. Olmi, '"Molti amici in varii luoghi": Studio della natura e rapporti epistolari nel secolo XVI', *Nuncius*, 6 (1991), pp. 3–31, esp. p. 6. The latter has been an inspiration for much of my research on Clusius and his world.
17. One of the early and most stimulating volumes advocating this approach with respect to the life sciences was N. Jardine, J. Secord and E. Spary (eds), *Cultures of Natural History* (Cambridge: Cambridge University Press, 1996). For a survey of trends and debates, see J. Secord, 'Knowledge in Transit', *Isis*, 95 (2004), pp. 654–72. On embodied knowledge, see Smith, *The Body of the Artisan*; and Roberts et al. (eds), *The Mindful Hand*.
18. Thanks to a further grant from the Netherlands Organisation for Scientific Research (NWO), I plan to work specifically on visual aspects of early modern history during 2010–13.
19. Most explicitly in F. Egmond and P. Mason, *The Mammoth and the Mouse: Microhistory and Morphology* (Baltimore, MD: Johns Hopkins University Press, 1997).
20. R. Cobb, *A Sense of Place* (London: Duckworth, 1975).
21. Roberts et al. (eds), *The Mindful Hand*, p. 191.
22. See esp. J. L. Barona and X. Gómez i Font, *La correspondencia de Carolus Clusius con los científicos españoles* (Valencia: Universitat de Valencia, 1998), which contains nearly all of them, with comments; cf. J. L. Barona, 'Clusius' Exchange of Botanical Information with Spanish Scholars', in Egmond et al. (eds), *Carolus Clusius*, pp. 99–116; and L. Ramón-Laca, 'Las plantas americanas en la obra de Charles de l'Écluse: Primeras citas en las cartas de Juan de Castañeda', *Anales del Jardín Botánico de Madrid*, 57:1 (1999), pp. 97–107.
23. On Spanish natural science in this period and on its historiography, see J. Pimentel, 'The Iberian Vision: Science and Empire in the Framework of a Universal Monarchy, 1500–1800', *Osiris*, 15 (2001), pp. 17–30; J. Puerto, *La Leyende Verde: Naturaleza, sanidad e ciencia en la corte de Felipe II, 1527–1598* (Castilla y León: Junta de Castilla y León, 2003); J. Cañizares-Esguerra, 'Iberian Science in the Renaissance: Ignored How Much Longer', *Perspectives on Science*, 12 (2004), pp. 86–124; J. Cañizares-Esguerra, 'Iberian Colonial Science', *Isis*, 96 (2005), pp. 64–70; A. Barrera Osorio, *Experiencing Nature: The Spanish American Empire and the Early Scientific Revolution* (Austin, TX: Univer-

sity of Texas Press, 2006). Cf. for Portugal, P. Fontes Da Costa, 'Secrecy, Ostentation and the Illustration of Exotic Animals in Sixteenth-Century Portugal', *Annals of Science*, 66 (2009), pp. 59–82.

24. See esp. *Festschrift anlässlich der 400 jährigen Wiederkehr des wissenschaftlichen Tätigkeit von Carolus Clusius (Charles de l'Escluse) im pannonischen Raum* (Eisenstadt: Amt der Burgenländischen Landesregierung, 1973) on Clusius and his Austro-Hungarian contacts; and (for Clusius and his Hungarian patron Prince Batthiány) D. Bobory, *The Sword and the Crucible: Boldizsár Batthyány and Natural Philosophy in Sixteenth-Century Hungary* (Cambridge: Cambridge Scholars Publishing, 2009). A forthcoming dissertation in Dutch by Esther van Gelder (Clusius Project) specifically deals with Clusius and his role at the German and Austrian-Hungarian courts: E. van Gelder, 'Tussen hof en keizerskroon: Carolus Clusius en de ontwikkeling van de botanie aan Midden-Europese hoven (1573–1593)' (PhD dissertation, Leiden University, 2010).

1 The Garden of Europe

1. M. de Lobel, *Plantarum seu Stirpium historia* (Antwerp: Plantin, 1576); and M. de Lobel, *Kruydtboeck oft Beschryvinghe van allerleye ghewassen, kruyderen, hesteren ende gheboomten* (Antwerp: Plantin, 1581), introductions.
2. C. van Hulthem, *Discours sur l'état ancien et moderne de l'agriculture et de la botanie dans les Pays Bas*, 2nd edn (Ghent: Goesin-Verhaeghe, 1837), pp. 7–8.
3. See F. de Nave and D. Imhof (eds), *De Botanica in de Zuidelijke Nederlanden (einde 15e eeuw–ca. 1650)* (Antwerp: Snoeck-Ducaju, 1993).
4. Z. van Martels, 'On his Majesty's Service. Augerius Busbequius, Courtier and Diplomat of Maximilian II', in F. Edelmayer and A. Kohler (eds), *Kaiser Maximilian II: Kultur und Politik im 16. Jahrhundert* (Vienna: Oldenbourg, 1992), pp. 169–81; W. Blunt and W. Stearn, *The Art of Botanical Illustration* (Woodbridge: Antique Collectors Club, 1994), p. 73; J. Opsomer, 'Un botaniste trop peu connu, Willem Quackelbeen (1527–1561)', *Bulletin de la Société royale de botanique de Belgique*, 93 (1961), pp. 113–30; and J. Opsomer, 'Notes complémentaires sur les plantes envoyées de Turquie en 1557 par le botaniste Quackelbeen', *Bulletin de la Société royale de botanique de Belgique*, 103 (1970), pp. 5–10. On Busbecq and the introduction of the tulip, see A. Goldgar, *Tulipomania: Money, Honor, and Knowledge in the Dutch Golden Age* (Chicago, IL: University of Chicago Press, 2007), pp. 32–3.
5. O. Wijnands, 'Commercium Botanicum: The Diffusion of Plants in the 16th Century', in L. Tjon Sie Fat and E. de Jong (eds), *The Authentic Garden: A Symposium on Gardens* (Leiden: Clusius Foundation, 1991), pp. 74–84.
6. U. Härting (ed.), *Gärten und Höfe der Rubenszeit: Im Spiegel der Malerfamilie Brueghel und der Künstler um Peter Paul Rubens* (Munich: Hirmer Verlag, 2000), pp. 5–6; and C. de Maegd, 'Tuinbezit, tuinen en tuinlui: Een licht op de praktijk in de tijd van Vredeman de Vries', in P. Fuhring (ed.), *De wereld is een tuin: Hans Vredeman de Vries en de tuinkunst van de Renaissance* (Ghent: Ludion, 2002), pp. 69–87.
7. F. Garbari, 'The Prefetti of the Botanical Gardens from their Origins', in F. Garbari, L. Tongiorgi Tomasi and A. Tosi, *Giardino dei Semplici / Garden of Simples* (Pisa: Edizioni Plus, 2002), pp. 107–47, esp. pp. 113–14.
8. On expertise among gardeners in the Southern Netherlands, see De Maegd, 'Tuinbezit, tuinen en tuinlui', pp. 84–5. On collaboration between gardeners-designers and herbalists in the service of the Medici, see D. Filardi, *Orto de' Pitti: The Architects, Gardeners*

and Botanical Design of the Boboli Gardens (Florence: Centro Di, 2007), esp. pp. 113–18.

9. A. Postma, 'Een Zuidnederlandse hovenier in Spanje: Over Filips II, Jehan Holbecq en een nieuwe tuinkunst', *Tuinkunst: Nederlands Jaarboek voor de Geschiedenis van Tuin- en Landschapsarchitectuur*, 1 (1995), pp. 9–22; De Maegd, 'Tuinbezit, tuinen en tuinlui', pp. 82–4; and Puerto, *La Leyende Verde*, pp. 367–80; cf. *Felipe II. El Rey intimo: Jardín e Naturaleza en el siglo XVI* (Madrid: Sociedad Estatal Para la conmemoración de los centenarios de Felipe II y Carlos V, 1998).
10. See I. Dübber, *Zur Geschichte des Medizinal-und Apothekenwesens in Hessen-Kassel und Hessen-Marburg von den Anfängen bis zum Dreissigjährigen Krieg* (Marburg: Marburg University, 1969), pp. 131–45, 230–41.
11. Lobel, *Plantarum seu Stirpium historia*; and Lobel, *Kruydtboeck*, introductions.
12. The Flemish and Latin versons of his name are Karel van Sint Omaars and Carolus a Divo Odomaro.
13. *Exoticorum*, p. 171; *Curae*, p. 105; *Rariorum*, p. clx.
14. The first European cabinets of curiosities originated in Austria, Hungary and the Low Countries, from the 1520s onwards; Italy followed a few decades later. See O. Impey and A. Macgregor (eds), *The Origins of Museums: The Cabinets of Curiosities in Sixteenth- and Seventeenth-Century Europe* (Oxford: Clarendon Press, 1985); and A. MacGregor, *Curiosity and Enlightenment: Collectors and Collections from the Sixteenth to the Nineteenth Century* (New Haven, CT: Yale University Press, 2007).
15. For the original inventory, presented on 25 July 1569 by Saint Omer's widow Anne d'Oingnies, see Rijksarchief Bruges, Family Archive 367. With thanks to Jacques de Groote for making his transcription available to me.
16. *Exoticorum*, p. 171.
17. On the origins and history of the *Libri Picturati* A.16–30 with full references to the extensive literature, see F. Egmond, 'Clusius, Cluyt, Saint Omer: The Origins of the Sixteenth-Century Botanical and Zoological Watercolours in Libri Picturati A.16–30', *Nuncius*, 20 (2005), pp. 11–67; and F. Egmond, 'The Making of the Libri Picturati A.16–30', in De Koning et al. (eds), *Drawn After Nature*, pp. 13–21. The latter volume publishes for the first time all botanical watercolours in this collection. With respect to the new attribution of the collection the following publications in particular have been important: H. Wille, 'The Discovery of the Scientific Heritage of Karel van Sint Omaars (1533–1569): Libri Picturati A 16–31 in the Jagiellon Library in Krakow', *Scientiarum historia*, 22 (1996), pp. 67–80; H. Wille, 'The Albums of Karel van Sint Omaars (1533–1569) (Libri Picturati A 16–31, in the Jagiellon Library in Krakow)', *Archives of Natural History*, 24 (1997), pp. 423–37; and L. Ramón-Laca, 'Charles de l'Écluse and *Libri Picturati* A. 16–30', *Archives of Natural History*, 28 (2001) pp. 195–243, esp. pp. 197–203.
18. Such as *Libri Picturati* A20.052 (*Glycyrrhiza glabra*, a source of liquorice) and A23.024v (*Cassia fistula*, still applied against constipation). On medicinal plants in the *Libri Picturati*, see De Koning et al. (eds), *Drawn After Nature*, pp. 121–4.
19. De Koning et al. (eds), *Drawn After Nature*, p. 330 (A30.028). On ecological thought, see A. Ubrizsy Savoia, A. Zemanek and B. Zemanek, 'The Beginnings of Ecological Thought', in De Koning et al. (eds), *Drawn After Nature*, pp. 90–7; and A. Ubrizsy Savoia, 'Environmental Approach in the Botany of the 16th Century', in Mirek and Zemanek (eds), *Studies in Renaissance Botany*, pp. 73–86.
20. *Rariorum*, p. 198.

21. On Bruges Humanism, Goltzius and the Laurins, see N. Geirnaert and L. Vandamme, 'Cultuur en mentaliteit', in M. Martens (ed.), *Brugge en de Renaissance: Van Memling tot Pourbus* (Bruges: Ludion, 1998), pp. 33–42; W. Le Loup, 'Biografie van Hubertus Goltzius', in W. Le Loup (ed.), *Hubertus Goltzius en Brugge 1583–1983* (Bruges: Stad Brugge, 1983), pp. 9–12; H. de la Fontaine Verweij, 'De eerste "private press" in de Nederlanden: Marcus Laurinus en de "Officina Goltziana"', in H. de la Fontaine Verweij, *Uit de wereld van het boek*, 2 vols (Amsterdam: Nico Israel, 1976), vol. 1, pp. 69–83.
22. L. Guicciardini, *Descrittione di tutti i paesi bassi* (Antwerp: Plantin, 1581), p. 11.
23. L. Vandewiele, 'Wat groeide in de tuin van Pieter van Coudenberghe', in De Nave and Imhof (eds), *De Botanica in de Zuidelijke Nederlanden*, pp. 23–31.
24. *Rariorum*, p. clx.
25. Vandewiele, 'Wat groeide in de tuin'; G. de Munck and H. Wille (eds), *Peeter van Coudenberghe: Apotheker-botanicus (1517–1599) en tijdgenoten: De natuur als verzamelobject in de 16de eeuw* (Antwerp: KAVA, 1996); cf. M. de Cap, 'Un apothicaire belge au XVIe siècle. Pierre Coudenberg', *Journal de Pharmacie et de Chimie*, 3rd series, 39 (1861), pp. 448–57, esp. p. 454.
26. Letters addressed to Brancion were sent to 'Chez Mr de Brancion à la court du Roy' or to 'l'hotel de l'empereur a Malynes', the palace in the Keizersstraat in Malines that had served as Margaret of Austria's residence. Cf. Hunger, *Charles de l'Escluse*, vol. 1, pp. 100–4, 124. Brancion's death on 18 February 1575 is mentioned in Pontus Heuyterus, 26 February 1575, 9 May 1575.
27. R. Dodoens, *Florum, et coronariarum odoratarumque nonnullarum herbarum historia* (Antwerp: Plantin 1568), p. 295; A. Louis, *Mathieu de l'Obel (1538–1616): Épisode de l'histoire de la botanie* (Ghent: Story-Scientia, 1980), p. 84 (panicum indicum); *Exoticorum*, p. 322, cf. p. 334 (Peruvian plants).
28. All three mention him frequently in their works; Clusius mentions Brancion 22 times in his *Rariorum* and twice in his *Exoticorum*. Cf. Härting (ed.), *Gärten und Höfe*, pp. 8, 11, 191, 194.
29. Contacts between Aldrovandi and Brancion had begun before 1570, independently from Brancion's connection with Clusius. On gifts from Cortuso and Salviani, see *Rariorum*, pp. 179, ccxiv. See also Chapter 5.
30. *Rariorum*, pp. 186, 254–5, 265. On Brancion's Italian contacts, see also G. Masson, 'Italian Flower Collector's Gardens in Seventeenth-Century Italy', in D. Coffin (ed.), *The Italian Garden* (Washington, DC: Dumbarton Oaks, 1972), pp. 61–80, esp. pp. 66, 77. Cf. on tulip naming, see Goldgar, *Tulipomania*, pp. 107–12.
31. *Rariorum*, pp. 197, 263, 130.
32. Clusius to J. Camerarius, 4 August 1579, in Hunger, *Charles de l'Escluse*, vol. 2, p. 99.
33. Vandewiele, 'Wat groeide in de tuin'.
34. Hunger Archive, Section 19, University Library Leiden.
35. Hunger, *Charles de l'Escluse*, vol. 2, p. 99.
36. See Van Hulthem, *Discours*. On Van Veltwijck, see also J. Schnap and Y. Hersant (eds), *Nicandre de Corcyre, le voyage d'Occident* (Toulouse: Anacharsis Editions, 2002), p. 50.
37. See Postma, 'Een Zuidnederlandse hovenier', p. 20; De Maegd, 'Tuinbezit, tuinen en tuinlui', pp. 82–4. According to J. Morán and F. Checa, *El coleccionismo en España: De la cámara de maravillas a la galería de pinturas* (Madrid: Cátedra, 1985), p. 41, Yuste also had the first (or one of the first) *Wunderkammern* in Spain.
38. J. Balis (ed.), *Hortus belgicus* (Brussels: Bibliothèque Albert I, 1962), p. 28.

39. The Mariemont and Binche properties were raided and burnt down in 1554 by French troops. See Balis (ed.), *Hortus belgicus*, p. 28; B. van den Boogert and J. Kerkhoff, *Maria van Hongarije: Koningin tussen keizers en kunstenaars* (Zwolle: Waanders, 1993), esp. pp. 36, 162–5, 163, 269–301; K. de Jonge, 'Les jardins de J. du Broeucq et de J. Hollebecque à Binche, Mariemont et Boussu', in *Felipe II. El Rey intimo*, pp. 191–220; and J. Kerkhoff, *Maria van Hongarije en haar hof 1505–1558* (Hilversum: Verloren, 2008), pp. 127–89. On her collections, see J. von Schlosser, *Die Kunst- und Wunderkammern der Spätrenaissance: Ein Beitrag zur Geschichte des Sammelwesens*, 2nd edn (Braunschweig: Klinkhardt and Biermann, 1978), pp. 41–5; and P. Vandenbroeck, 'Amerindian Art and Ornamental Objects in Royal Collections, Brussels, Mechelen, Duurstede, 1520–30', in *America: Bride of the Sun: 500 years Latin America and the Low Countries* (Antwerp: Imschoot, 1991), pp. 99–119. Cf. A. Pérez de Tudela and A. Jordan Schwendt, 'Renaissance Menageries: Exotic Animals and Pets at the Habsburg Courts in Iberia and Central Europe', in K. Enenkel and P. Smith (eds), *Early Modern Zoology: The Construction of Animals in Science, Literature and the Visual Arts*, Intersections: Yearbook for Early Modern Studies, 7 (Leiden and Boston, MA: Brill, 2007), pp. 419–47.
40. D. Eichberger, *Leben mit Kunst: Wirken durch Kunst: Sammelwesen und Hofkunst unter Margarete von Österreich, Regentin der Niederlande* (Turnhout: Brepols, 2002), pp. 112, 114.
41. Ibid., p. 115, see esp. ch. VI.2.
42. See A. Lugli (ed.), *Albrecht Dürer: Viaggio nei Paesi Bassi*, 2nd edn (Reggio Emilia: Diabasis, 2005), p. 68; Van den Boogert and Kerkhoff, *Maria van Hongarije*, pp. 173–4.
43. Blunt and Stearn, *The Art of Botanical Illustration*, pp. 39–40.
44. O. Pächt, 'Early Italian Nature Studies and the Early Calendar Landscape', *Journal of the Warburg and Courtauld Institutes*, 13 (1950), pp. 13–47, on p. 32. The literature on the naturalistic representation of living nature is vast, but for some stimulating essays addressing connections between visual culture, art and natural science, see S. Kusukawa, 'Illustrating Nature', in Frasca-Spada and Jardine (eds), *Books and the Sciences*, pp. 90–113; L. Tongiorgi Tomasi, 'L'immagine naturalistica; tecnica e invenzione', in G. Olmi, L. Tongiorgi Tomasi and A. Zanca (eds), *Natura-Cultura: L'interpretazione del mondo fisico nei testi e nelle immagini* (Florence: Olschki, 2000), pp. 133–51; P. Smith, 'Art, Science and Visual Culture in Early Modern Europe', *Isis*, 97 (2006), pp. 83–100; and L. Zucchi, 'Brunfels e Fuchs: L'illustrazione botanica quale ritratto della singola pianta o immagine della specie', *Nuncius*, 18 (2004), pp. 411–66.

2 Expert Gardeners

1. In Lobel's *Plantarum seu Stirpium historia* the two most frequently cited men are Brancion (35x) and Mouton (30x); cf. Louis, *Mathieu de l'Obel*, p. 83. There are seven references to Mouton in the *Rariorum*, pp. 166, 182–3, 244, 265, 319.
2. *Rariorum*, pp. 164, 165, 172, 176, 189.
3. Cf. 21 January 1586.
4. Clusius's Spanish flora is still considered important for the study of Iberian botany; see L. Ramón-Laca and R. Morales Valverde (eds), *Charles de L'Écluse de Arras, Descripción de algunas plantas raras encontradas en España y Portugal* (Castilla y León: Junta de Castilla y León, 2005).
5. There is a gap in Houchin's letters to Clusius between 1586 and 1605, but the contents and letters by others show that the two men remained in friendly contact.

6. His death is mentioned in Plateau, 8 February 1602.
7. J. de Landtsheer, 'Justus Lipsius and Carolus Clusius, a Flowering Friendship', *Bulletin of the Belgian Institute*, 68 (1998), pp. 273–95, on pp. 283–4.
8. Boisot is mentioned 43 times, of which 31 concern bulbous or tuberous plants.
9. Clusius to Caccini, 21 January 1609, in Ginori Conti, *Lettere inedite*, p. 95.
10. *Rariorum*, pp. 153, 100, 202, 209, 161, 183, 213–15, 219, 221, 222, 223. See esp. J. Boisot, 7 August 1584, 4 August 1588; cf. Ginori Conti, *Lettere inedite*, pp. 83, 95.
11. De Landtsheer, 'Justus Lipsius and Carolus Clusius', p. 292.
12. See J. Boisot, 20 November 1584, 9 May 1586, 25 June 1592, 19 July 1591.
13. Goldgar (*Tulipomania*, esp. pp. 116–18) discusses the extent to which tulips were seen as art and as man-made objects (mainly in the seventeenth and eighteenth centuries). The earlier material fits in, but here I prefer to draw a direct parallel with the equally practical art of alchemy. Even though some naturalists and many apothecaries combined expertise in distilling and plant propagation, I have not come across any explicit references to 'plant alchemy'.
14. Cf. J. Boisot, 4 August 1588. On growing tulips from seed, see also Goldgar, *Tulipomania*, pp. 40–1, 116–17.
15. Several relatives figure in biographical dictionaries: his father had been advisor to Charles V. His brother Pierre occupied high administrative functions.
16. On gardeners and plants at Aranjuez, see Puerto, *La Leyende Verde*, esp. pp. 372–85.
17. See J. B. de Tassis, 28 January 1586, and the seven letters (1588–92) from C. de Tassis to Clusius.
18. Cf. C. de Tassis, 7 November 1592.
19. As reported in De Maes, 7 October 1602.
20. On the exchanges between Joachim Camerarius II and Francesco Calzolari, see G. Olmi, 'Per la storia dei rapporti scientifici fra Italia e Germania: Le lettere di Francesco Calzolari a Joachim Camerarius II', in G. Brizzi and G. Olmi (eds), *Dai cantieri della storia: Liber amicorum per Paolo Prodi* (Bologna: CLUEB, 2007), pp. 343–61, on pp. 343–4. On Felix Platter and the notion of science as a safe haven from political and religious troubles, see W. Jones, *William Turner: Tudor Naturalist, Physician, Divine* (London: Routledge, 1988), p. 16; and D. Hochlenert, *Das "Tagebuch" des Felix Platter: Die Autobiographie eines Arztes und Humanisten* (Tübingen: University of Tübingen, 1996), pp. 104–5. Cf. Reeds, *Botany*, pp. 13–14.
21. Clusius corresponded with other members of this group as well: Philips de Marnix, Lord of St Aldegonde, and Louis de Bergues d'Olhain, who married Houchin's sister. Further members were Jean d'Estourmel, who married Charles de Saint Omer's widow, Robert de la Marck and Georges de Bergues d'Olhain.
22. J. Roegiers (ed.), *Arenberg in de Lage Landen: Een hoogadellijk huis in Vlaanderen en Nederland* (Louvain: Universitaire Pers, 2002); L. Ramón-Laca, 'The Additions by the Count of Arenberg', in De Koning et al. (eds), *Drawn After Nature*, pp. 28–32; Masson, 'Italian Flower Collector's Gardens', p. 66.
23. J. Roegiers and H. Wille, 'Natuur en techniek', in Roegiers (ed.), *Arenberg in de Lage Landen*, pp. 371–7. On Arenberg's activities as *naturalia* collector, and correspondence with Caccini and other plant lovers, see esp. M. Zalum Cardon, *Passione e cultura dei fiori tra Firenze e Roma nel XVI e XVII secolo* (Florence: Olschki, 2008); and Masson, 'Italian Flower Collector's Gardens', pp. 66, 74–5.
24. Arenberg, 21 September 1595, British Library, London, Add. MS 24210.
25. Masson, 'Italian Flower Collector's Gardens', pp. 74–5.

26. Tovar, (received) 28 September 1596, published in Barona and Gómez i Font, *La correspondencia de Carolus Clusius*, pp. 72–6; cf. Barona, 'Clusius' Exchange of Botanical Information'.
27. Clusius wrote in May 1607 to Caccini that Arenberg had bought the town of Enghien from the French king, and had created a enormous garden there for which he had bought large numbers of plants from plant sellers; he would have asked Clusius to supervise these gardens if Clusius had not been so old (Ginori Conti, *Lettere inedite*, pp. 61–2).
28. Clusius to Caccini, 8 December 1608, in Ginori Conti, *Lettere inedite*, pp. 93–4.
29. With thanks to Paul Smith for his date of death and position as dignitary.
30. The list dated 'Anno 1584' belongs with Plateau's letters in the Leiden collection.
31. G. Gabrieli, *Contributi alla storia della Accademia dei Lincei*, 2 vols (Rome: Accademia Nazionale dei Lincei, 1989), vol. 2, p. 1020.
32. Plateau is quoted 52 times in the *Rariorum*. I have gratefully used the typescript MA thesis by E. Vasbinder-Ouwendijk, 'Carolus Clusius en Jacques Plateau' (University of Utrecht, 1979).
33. Croÿ was the brother of Arenbergh's wife Anne, and also the estranged husband of Clusius's best female friend Princess Marie de Brimeu (see Chapter 4). On Croÿ's estate near Brussels, see C. de Maegd, '"En ung sien jardin de plaisance au faubourgs de ceste ville": Het hof van plaisantie van Karel van Croÿ in Sint-Joost-ten-Node rond 1600', *Tijdschrift van Dexia bank*, 55:218 (2001), pp. 45–68; and De Maegd, 'Tuinbezit, tuinen en tuinlui', esp. pp. 71, 75, 78, 81.
34. See De Maegd, '"En ung sien jardin"', esp. pp. 56–62. For pictures and maps of estates and houses belonging to Croÿ, see Härting (ed.), *Gärten und Höfe*.
35. The two men had probably been in contact for years; Plateau used to sell ornithological curiosities to Croÿ.
36. I found some of the most inspiring works on patronage and prince-practitioners to be: B. Moran, 'Wilhelm IV of Hesse-Kassel: Informal Communication and the Aristocratic Context of Discovery', in T. Nickles (ed.), *Scientific Discovery: Case Studies* (Dordrecht: Reidel, 1980), pp. 67–96; B. Moran, 'Prince-Practitioning and the Direction of Medical Roles at the German Court: Maurice of Hesse-Kassel and his Physicians', in V. Nutton (ed.), *Medicine at the Courts of Europe, 1500–1837* (London: Routledge, 1990), pp. 95–116; B. Moran, 'Patronage and Institutions: Courts, Universities, and Academies in Germany; an Overview: 1550–1750', in Moran (ed.), *Patronage and Institutions*, pp. 169–83; M. Biagioli, 'Scientific Revolution, Social Bricolage, and Etiquette', in R. Porter and M. Teich (eds), *The Scientific Revolution in National Context* (Cambridge: Cambridge University Press, 1992), pp. 11–54; M. Biagioli, 'Etiquette, Interdependence, and Sociability in Seventeenth-Century Science', *Critical Inquiry*, 22 (1996), pp. 193–238; L. Sarasohn, 'Nicolas-Claude Fabri de Peiresc and the Patronage of the New Science in the Seventeenth Century', *Isis*, 84 (1993), pp. 70–90; and G. Olmi, 'Science and the Court: Some Comments on "Patronage" in Italy', in Guzzetti (ed.), *Science and Power*, pp. 25–45.
37. J. Boisot, 4 August 1588, 29 January 1597; Arenberg, 29 January 1597; cf. *Rariorum*, p. 280.
38. *Rariorum*, pp. 157–8.
39. The word in Nahuatl means 'ant flower': with thanks to José Pardo Tomás.
40. Tovar, 1 June 1596; cf. Barona and Gómez i Font, *La correspondencia de Carolus Clusius*, pp. 72–6.

41. Arenberg, 29 January 1597; J. Boisot, 10 July 1597. In June 1603 the Spanish cosmographer Zamorano in Seville wrote to Clusius about yet another Aztec lily, which he had sent to Clusius more than two years earlier but had been lost (or stolen) during transport; he knew that such flowers had been seen in the garden of Arenberg (Rodrigo Zamorano, 3 June 1603; cf. Barona, 'Clusius' Exchange of Botanical Information', p. 110). On Arenberg and the narcissus jacobeus, see also briefly Zalum Cardon, *Passione e cultura dei fiori*, p. 56.
42. On writing with one's own hand and the time involved as a courtly gift, see F. Bouza Alvarez, 'Letters and Portraits: Economy of Time and Chivalrous Service in Courtly Culture', in Bethencourt and Egmond (eds), *Correspondence and Exchange*, pp. 145–62.
43. Cf. Clusius to Caccini, 2 May 1607, in Ginori Conti, *Lettere inedite*, p. 57. I take Son Altesse to refer to the governor ruling the country on behalf of Philip II, i.e. Albrecht of Austria. Philip himself was in Spain at the time.
44. For *Libri Picturati* A21.085 see De Koning et al. (eds), *Drawn After Nature*, p. 207; Ramón-Laca, 'The Additions by the Count of Arenberg', p. 30. Cf. *Rariorum*, pp. 157–8.
45. S. Pumfrey and F. Dawbarn, 'Science and Patronage in England, 1570–1625: A Preliminary Study', *History of Science*, 42 (2004), pp. 137–88, use the expression 'ostentatious natural knowledge'.
46. Some ten years later the narcissus jacobeus still belonged to the domain of princes. Probably shortly before 1609 Lobel in England received a drawing of the plant from a British royal surgeon who had had seen it in the garden of the royal herborist Jean Robin Sr in Paris (Clusius to Caccini, 27 February 1609, in Ginori Conti, *Lettere inedite*, p. 102). During the 1630s the plant was still rare in Italy; see E. MacDougall, *Fountains, Statues, and Flowers: Studies in Italian Gardens of the Sixteenth and Seventeenth Centuries* (Washington, DC: Dumbarton Oaks, 1994), pp. 248–9, 275.
47. Postma, 'Een Zuidnederlandse hovenier', pp. 12–13.
48. Louis, *Mathieu de l'Obel*, p. 151; cf. p. 145.

3 Women in the Garden

1. See Dübber, *Zur Geschichte*; M. Hannay, '"How I These Studies Prize": The Countess of Pembroke and Elizabethan Science', in L. Hunter and S. Hutton (eds), *Women, Science, and Medicine 1500–1700: Mothers and Sisters of the Royal Society* (Phoenix Mill: Sutton, 1997), pp. 108–21; E. Tebeaux, 'Women and Technical Writing, 1475–1700: Technology, Literacy and Development of a Genre', in Hunter and Hutton (eds), *Women, Science, and Medicine*, pp. 29–62; L. Hunter, 'Women and Domestic Medicine: Lady Experimenters, 1570–1620', in Hunter and Hutton (eds), *Women, Science, and Medicine*, pp. 89–107; K. Keller, 'Kommunikationsraum altes Reich: Zur Funktionalität des Korrespondenznetze von Fürstinnen im 16. Jahrhundert', *Zeitschrift für Historische Forschung*, 31 (2004), pp. 205–30; and A. Rankin, 'Becoming an Expert Practitioner: Court Experimentalism and the Medical Skills of Anna of Saxony (1532–1585)', *Isis*, 98 (2007), pp. 23–53.
2. Useful in this field is L. Schiebinger, 'Women of Natural Knowledge', in K. Park and L. Daston (eds), *The Cambridge History of Science, vol. III. Early Modern Science* (Cambridge: Cambridge University Press, 2006), pp. 192–205. L. Schiebinger, *The Mind Has No Sex? Women in the Origins of Modern Science* (Cambridge, MA: Harvard University Press, 1989) does not discuss natural history for the relevant period. On specific

women and gardens, see also, for instance, A. Backer, 'Tuinkunst tijdens de Nederlandse Opstand: De prinses van Chimay: Marie de Brimeu en de "humanisering" van de bloem 1550–1605', *Cascade Bulletin voor Tuinhistorie*, 15 (2006), pp. 14–39; M. Sikkens-De Zwaan, 'Magdalena Poulle (1632–99): A Dutch Lady in a Circle of Botanical Collectors', *Garden History*, 30 (2002), pp. 206–20; cf. J. D. Hunt and E. de Jong (eds), *The Anglo-Dutch Garden in the Age of William and Mary* (London: Taylor & Francis, 1988).

3. These works were published by English elite women after about 1640; see Tebeaux, 'Women and Technical Writing'.
4. The only ones published are ten letters by Anna Aicholtz and excerpts of several letters by Marie de Brimeu in G. Istvánffi, *Études et commentaires sur le code de l'Escluse* (Budapest: private publication, 1900). For Brimeu, cf. also J. van der Gouw, 'Marie de Brimeu: Een Nederlandse prinses uit de eerste helft van de tachtigjarige oorlog', *De Nederlandsche Leeuw*, 64 (1947), pp. 5–49; and F. Hopper, 'Clusius' World: The Meeting of Science and Art', in Tjon Sie Fat and De Jong (eds), *The Authentic Garden*, pp. 13–36.
5. The only long-standing correspondences with men of which more than 20 letters remain are: Joannes Crato (36), Justus Lipsius (22) and Giovanni Vincenzo Pinelli (about 80). In the exceptional case of Joachim Camerarius II, 195 letters written by Clusius to him are extant and he seems to have written as frequently to Clusius.
6. J. van der Laen, 6 August 1603, 23 October 1603, 7 November 1605.
7. See Hunter, 'Women and Domestic Medicine'; C. Webster, 'Alchemical and Paracelsian Medicine', in C. Webster (ed.), *Health, Medicine and Mortality in the Sixteenth Century* (Cambridge: Cambridge University Press, 1979), pp. 301–33, on p. 306; M. Hannay, *Philip's Phoenix: Mary Sidney, Countess of Pembroke* (New York: Oxford University Press, 1990), p. 42; and Hannay, '"How I These Studies Prize"'. Two letters from Clusius to Sir Philip Sidney from 1576 mention politics, travel, friends and gossip but no plants or gardens (British Library, Add. MS 15912).
8. The single exception is Marguérite le Roy, wife of the Protestant bookseller Didier Royer (or Rougier) in Frankfurt, who wrote to him only about family and religious matters.
9. On the Zouches, see also Chapter 11.
10. Brancion, 3 August 1571; cf. 26 July 1571, 6 September 1571.
11. On the Boisots see also Chapter 2. While two of her brothers died in 1575–6 fighting as commanders of the Sea Beggars on the side of the Dutch rebels, in the 1590s two of Louise's daughters married into the Spanish aristocracy.
12. Cf. J. Boisot, 7 May 1582.
13. Its first European illustration is a watercolour (Codex Fuchs, Oesterreichische National Bibliothek, Cod. 11125, p. 391) made for Fuchs's main work on plants which was never published. On Bertolf and Dodoens, see also De Nave and Imhof (eds), *De Botanica in de Zuidelijke Nederlanden*, pp. 104, 143–4; and U. Härting, 'Over doel, gebruik en vorm van Nederlandse tuinen in de zestiende en zeventiende eeuw', in Fuhring (ed.), *De wereld is een tuin*, pp. 89–105, on p. 191.
14. Johannes van Heede, a citizen of Bruges, figures in Charles de Saint Omer's *Libri Picturati* (A29.019v).
15. See Dodoens, *Florum*, p. 296. De Maegd ('Tuinbezit, tuinen en tuinlui', p. 71), too, thinks that Madame Hopperus's role has been underestimated.
16. Hopperus was the grandson of the sister of Dodoens's grandfather; the family originally came from Friesland. Christine Bertolf was the daughter of the president of the Council of Friesland. For Hopperus's short biography, see http://dutchrevolt.leidenuniv.nl/Nederlands/default.htm [accessed November 2009].

17. Nor is he mentioned in Louis, *Mathieu de l'Obel*. Cf. *Rariorum*, p. 216.
18. *Rariorum*, p. ccxxv.
19. Harkness (*The Jewel House*, esp. p. 220) points to the practical knowlegde of ordinary women in sixteenth-century London and the relative openness to their expertise; women would later be barred from the Royal Society, as they already had been from universities.
20. At the top in terms of the number of references in Clusius's printed works are Anna Maria von Heusenstain (17) and Eva Ungnadin (10) from Vienna.

4 Female Experts

1. *Exoticorum*, p. 363.
2. All biographical information on Marie de Brimeu is based on her letters and on Van der Gouw, 'Marie de Brimeu'; cf. Backer, 'Tuinkunst'. She should not be confused with the contemporary Marie de Brimeu from Antwerp who died in 1604 and was the wife of C. Schetz; that Marie too was a plant lover and is mentioned by Clusius. With thanks to Hugo Soly.
3. These issues are recurring topics in the letters of both to Clusius.
4. See J. van Hoghelande, 12 August 1592; cf. Van der Gouw, 'Marie de Brimeu'. For Hoghelande, see also Chapter 10.
5. Brimeu's first extant letter to Clusius dates from 1571: in it she refers to both Brancion and her 'cousin' Johan van der Delft (d. 1579), who was also Brancion's cousin.
6. See Brimeu, 3 July 1603. She wanted extra watchdogs and guards to protect the precious plants and bulbs.
7. See Brimeu, 1 August 1596. Referring to this catalogue she asked Clusius's advice about which plants were the rarest.
8. On Brimeu's garden tapestries and the comparisons between tulip colours and cloth, see also Goldgar, *Tulipomania*, p. 118. MacDougall (*Fountains, Statues, and Flowers*, p. 233) discusses the practice of planting potted plants only when flowering, and thus creating colour patterns reminiscent of tapesties, in the Barberini garden in Rome during the 1630s.
9. Cf. Brimeu, 9 July 1592. She complains that the painter fails to do justice to the tulip's natural beauty.
10. Brimeu, 24 January 1592, 24 May 1593. The Brimeura amethystina (L.) Chouard, introduced in 1601 from south-west Europe, was named after her. See Tjon Sie Fat and De Jong (eds), *The Authentic Garden*, p. 84.
11. Blunt and Stearn, *The Art of Botanical Illustration*, pp. 88–9. On tulip 'portraits' and their naming, cf. Goldgar, *Tulipomania*, pp. 98–106.
12. For the Heusenstains and Clusius's role at the court in Vienna, see Van Gelder, 'Tussen hof en keizerskroon'.
13. Hunger, *Charles de l'Escluse*, vol. 1, p. 350.
14. For Aicholtz, see Van Gelder, 'Tussen hof en keizerskroon'. For his appointments in the Medical Faculty, see Acta Facultatis of Vienna on www.memoriamedicinae.meduniwien.ac.at [accessed November 2009].
15. Cf. H. Bloeme, 20 March 1600; he also informed Clusius that Anna Aicholtz died without issue, whereupon her second husband (Starzer) inherited her possessions.
16. See esp. Van Gelder, 'Tussen hof en keizerskroon'; Bobory, *The Sword and the Crucible*; and *Festschrift anlässlich der 400 jährigen Wiederkehr*.

17. See Heusenstain, 5 February 1605, 1 March 1606; cf. 10 June 1604.
18. A. Aicholtz, 8 January 1592; she refers to another female gardener in this letter as well, and elsewhere mentions a female apothecary or apothecary's wife involved in gardening.
19. On the introduction of the horse chestnut and the lilac, see H. Lack, 'Lilac and Horse-Chestnut: Discovery and Rediscovery', *Curtis's Botanical Magazine*, 17 (2000), pp. 109–37.
20. See Opsomer, 'Un botaniste trop peu connu'; Lack, 'Lilac and Horse-Chestnut'; and the unpaginated Appendix to Clusius's *Rariorum* in his *Exoticorum*.
21. Hunger, *Charles de l'Escluse*, vol. 1, p. 350.
22. Clusius's unpaginated Appendix to the *Rariorum* in the *Exoticorum* shows an illustration of this chestnut branch based on that drawing. On the horse chestnuts, see also Heusenstain, 17 June 1591, 5 February 1605. For the transport problems, see H. Bloeme, 28 April 1603; and P. Bloeme, 21 March 1604. The brothers Bloeme (émigrés from the Southern Netherlands) were not only transporters, but corresponded with garden lovers, such as Jean Boisoit, and had their own gardens, where they also cared for rare plants if these could not immediately be forwarded.
23. *Rariorum*, pp. 200, 242–3, 250, 253, 255. On politics, diplomatic missions and the arrival of plants from Constantinople, see also the letters to Clusius by Christoph Freiherr von Eck from 1588 to 1603. Anna Aicholtz too refers to the arrival of a courier from Turkey (5 May 1589).
24. On the ornithogalum, see *Rariorum*, p. 188.
25. Ibid., p. 264.
26. Ibid., p. ccxlviii.
27. Ibid., pp. 145, 157, 188, 198, 200, 242–3, 250, 253, 255, 258, 261, ccxlviii; and three references in the unnumbered Appendix to the *Rariorum* in the *Exoticorum*.
28. Blunt and Stearn, *The Art of Botanical Illustration*, pp. 89–90.
29. Ibid.; the important role of the women from the Southern Netherlands has also been suggested by Van der Gouw, 'Marie de Brimeu', p. 31.

5 Growing Expertise

1. On early news about America in Spain and Italy, see G. Symcox (ed.), *Italian Reports on America 1493–1522, Letters, Dispatches, and Papal Bulls*, Repertorium Columbianum, 10 (Turnhout: Brepols, 2001); on Spanish works on natural American history in Italy, see J. Pardo Tomás, 'Obras españolas sobre historia natural y materia medica americana en la Italia del siglo XVI', *Asclepio: Revista de Historia de la medicina y de la Ciencia*, 43 (1991), pp. 561–94; and on American *naturalia*, see P. Mason, *Before Disenchantment: Images of Exotic Animals and Plants in the Early Modern World* (London: Reaktion Books, 2009); and P. Mason, '*Americana* in the *Exoticorum libri decem* of Charles de l'Écluse', in Egmond et al. (eds), *Carolus Clusius*, pp. 195–219. For Italians wanting news from northern Europe about exotic places and *naturalia*, see for instance Pinelli, 10 November 1597, 31 December 1598, 27 January 1599; and, Pona 15 July 1603.
2. D. Coffin, *Gardens and Gardening in Papal Rome* (Princeton, NJ: Princeton University Press, 1991), pp. 210–11.
3. C. Acidini Luchinat (ed.), *Giardini Medicei: Giardini di palazzo e di villa nella Firenze del Quattrocento* (Milan: Federico Motta, 2000), p. 48.

4. H. Lietzmann, *Irdische Paradiese: Beispiele höfische Gartenkunst der 1. Hälfte des 16. Jahrhunderts* (Munich: Deutscher Kunstverlag, 2007), esp. p. 29.
5. E. Woodhouse, 'Spirit of the Elizabethan Garden', *Garden History*, 27 (1999), pp. 10–31, on pp. 14, 27. A comparative European chronology of collecting and gardening would be useful.
6. Masson, 'Italian Flower Collector's Gardens', on p. 64; C. Lazzaro, *The Italian Renaissance Garden: From the Conventions of Planting, Design and Ornament to the Grand Gardens of Seventeenth-Century Central Italy* (New Haven, CT: Yale University Press, 1990), pp. 20–1, 27–8 (and pp. 323–5 for a list of the plants most commonly found in sixteenth-century Italian gardens); and Coffin, *Gardens and Gardening*, pp. 206–9. Cf. MacDougall, *Fountains, Statues, and Flowers*.
7. Lazzaro, *The Italian Renaissance Garden*, p. 27, cf. p. 39.
8. MacDougall, *Fountains, Statues, and Flowers*, pp. 239–41; cf. Lazzaro, *The Italian Renaissance Garden*, p. 28.
9. Clusius to Caccini, 6 March 1608, in Ginori Conti, *Lettere inedite*, p. 53.
10. Clusius met some Italian naturalists in the context of the Viennese court. See A. Ubrizsy Savoia, 'Some Aspects of Clusius' Hungarian and Italian Relations', in Egmond et al. (eds), *Carolus Clusius*, pp. 267–92, on pp. 285–91. The reasons given by Clusius (business, family obligations) why he never visited Italy sound unconvincing. His real motive may have lain in the religious and cultural politics of Counter-Reformation Italy.
11. A large part of Clusius's exchanges with Italians took place in Italian. Cortuso (14 December 1566) complimented Clusius on his Italian.
12. It is uncertain whether Clusius was directly involved in Brancion's exchanges with Salviani and Contarini. On the papal physicians, cf. R. Palmer, 'Medicine at the Papal Court in the Sixteenth Century', in Nutton (ed.), *Medicine at the Courts of Europe*, pp. 49–78.
13. Pinelli and Clusius had met earlier, possibly in Vienna, through the composer Filippo di Monte (*c.* 1521–1604); see Pinelli, 19 March 1575. It is intriguing that Pinelli wrote invariably in Italian to Clusius (who wrote to him in Latin), especially given the fact that Pinelli's Latin (and Greek) must have been excellent.
14. On Pinelli, see M. Grendler, 'A Greek Collection in Padua: The Library of Gian Vincenzo Pinelli', *Renaissance Quarterly*, 33 (1980), pp. 386–416; A. Raugei (ed.), *Pinelli, Gian Vincenzo – Dupuy, Claude: Une correspondance entre deux humanistes*, 2 vols (Florence: Olschki, 2001); cf. Findlen, *Possessing Nature*, esp. pp. 369–70.
15. Borromeo (1564–1631) founded the Biblioteca Ambrosiana in Milan, of which Pinelli's library forms the core collection.
16. For the former, see Malocchi (then prefect of the Pisan *hortus botanicus*), 20 June 1606; for the latter, see Quattrami, 15 January 1597.
17. This may have been connected to the chronology of Clusius's contacts with Italy and perhaps also with Mattioli's reputation (cf. P. Findlen, 'The Formation of a Scientific Community: Natural History in Sixteenth-Century Italy', in A. Grafton and N. Siraisi (eds), *Natural Particulars: Nature and the Disciplines in Renaissance Europe* (Cambridge, MA: MIT Press, 1999), pp. 369–400).
18. See esp. G. Olmi, *Scienza e natura del secondo cinquecento* (Trento: Libera Università degli Studi, 1976); G. Olmi, *L'Inventario del mondo: Catalogazione della natura e luoghi del sapere nella prima età moderna* (Bologna: Mulino, 1992); M. Tagliaferri, S. Tommasini and S. Tugnoli Pattaro, 'Ulisse Aldrovandi als Sammler: Das Sammeln als Gelehrsamkeit oder als Methode wissenschaftlichen Forschens?', in A. Grote (ed.), *Macrocosmos in Microcosmo – Die Welt in der Stube: Zur Geschichte des Sammelns, 1450 bis*

1800 (Opladen: Leske & Budrich, 1994), pp. 265–81; Findlen, *Possessing Nature*; and A. Alessandrini and A. Ceregato (eds), *Natura Picta: Ulisse Aldrovandi* (Bologna: Editrice Compositori, 2007); and http://www.filosofia.unibo.it/aldrovandi/ with visual material and correspondence from the Aldrovandi collection.

19. The seven known letters from Aldrovandi to Clusius (1569–96) have been published in G. B. de Toni, *Il carteggio degli Italiani col botanico Carlo Clusio nella biblioteca Leidense* (Modena: Società Tipografica Modenese, 1911); and Istvánffi, *Études et commentaires*; cf. Ubrizsy Savoia, 'Some Aspects', esp. pp. 287–8. Some evidence of exchanges of seeds between Aldrovandi and Clusius can also be found in the Aldrovandi collection of the University of Bologna.
20. Many such gardens are discussed in P. Bussadori, 'Gli orti botanici privati padovani', in R. Bussi (ed.), *Di Sana Pianta: Erbari e taccuini di sanità: Le radici storiche della nuova farmacologia* (Padua: Panini, 1988), pp. 47–54; A. Paganelli, 'L'Horto Medicinale dell'Università di Padova nella storia della medicina', in Bussi (ed.), *Di Sana Pianta*, pp. 41–5; and K. Pomian, *Collectioneurs, amateurs et curieux: Paris, Venise, xvie–xviiie siècle* (Paris: Gallimard, 1987), pp. 81–142. Cf. M. Azzi Visentini, 'The Gardens of Villas in the Veneto from the 15th to the 18th Centuries', in J. D. Hunt (ed.), *The Italian Garden: Art, Design, and Culture* (Cambridge: Cambridge University Press, 1996), pp. 93–126 (although no plants are named here); and M. Azzi Visentini, *L'orto botanico di Padova e il giardino del Rinascimento* (Milan: Il Polifilo, 1984).
21. Cf. *Rariorum*, p. 130. On the Contarini gardens and collections, see Pomian, *Collectioneurs, amateurs et curieux*, pp. 86, 119.
22. Nicolò Contarini is frequently mentioned in G. Pona, *Monte Baldo descritto da Giouanni Pona veronese* (Venice: R. Meietti, 1617), pp. 37–8, 61. Cf. G. Gabrieli, *Il Carteggio Linceo* (Rome: Accademia Nazionale dei Lincei, 1996), pp. 1137–8; and Pomian, *Collectioneurs, amateurs et curieux*, p. 119.
23. See Pona, *Monte Baldo*, p. 20; Pomian, *Collectioneurs, amateurs et curieux*, pp. 105, 108, 119.
24. On Michiel and his albums, see esp. J. Pardo Tomás, 'Tra "Oppinioni" e "Dispareri": La flora americana nell'erbario di Pier'Antonio Michiel (1510–1576)', in G. Olmi and G. Papagno (eds), *La natura e il corpo: Studi in memoria di Attilio Zanca* (Florence: Olschki, 2005), pp. 73–100; and the still essential G. B. de Toni, *Pietro Antonio Michiel: I cinque libri di piante: Codice Marciano, Trascrizione e Commento* (Venice: Reale Istituto Veneto di Scienze, Lettere ed Arti, 1940); cf. E. Cappelletti, 'Plants Cultivated at the time of Aguillara', in A. Minelli (ed.), *The Botanical Garden of Padua 1545–1995* (Venice: Marsilio, 1995), pp. 163–71.
25. L. Beschi, *Onorio Belli Accademico Olimpico: Scritti d'Antiquaria e Botanica 1586–1602* (Rome: Viella, 2000), pp. 106–18, who mentions the gardens of the families Mocenigo, Priuli, Cappello and Mussati in Padua. On other Paduan gardens, see V. Dal Piaz and M. Rippa Bonati, 'The Design and Form of the Padua *Horto medicinale*', in Minelli (ed.), *The Botanical Garden of Padua*, pp. 33–54, on pp. 44–5; and Bussadori, 'Gli orti botanici privati padovani', and associated map on p. 130. For the coffee, see P. Alpini (also Alpino), *De Plantis Aegypti Liber* (Venice: De Francisis, 1592), pp. 36–7.
26. Bussadori, 'Gli orti botanici privati padovani', p. 52.
27. Olmi, 'Per la storia', esp. pp. 350–3.
28. Cf. Alpini, *De Plantis Aegypti Liber*, p. 73.
29. Fra Gregorio, 20 June 1606, which also shows that there must have been direct correspondence between Haller and Clusius. Matteo Caccini corresponded with Haller after

1600; see Bibliothèque Royale Brussels, MS III 893, with 99 letters by Haller. See Zalum Cardon, *Passione e cultura dei fiori*, pp. 51–2; and (also for the painter) G. Olmi, 'Lettere di fra Gregorio da Reggio, cappuccino e botanico del tardo rinascimento', in M. Beretta, P. Galluzzi and C. Triarico (eds), *Musa Musaei, Studies on Scientific Instruments and Collections in Honour of Mara Miniati* (Florence: Olschki, 2003), pp. 117–39, on pp. 126–7.

30. Undated postscript, belonging probably to Fra Gregorio, 9 March 1606 (cf. De Toni, *Il carteggio*, pp. 35–6).
31. Ibid. On the status and nobility of gardening, even for the aristocracy, and on soiling one's hands, see Coffin, *Gardens and Gardening*, p. 215.
32. Masson, 'Italian Flower Collector's Gardens', pp. 74–5.
33. The only exception is Pietro Antonio Michiel, but he died in 1577, two years after the Pinelli–Clusius correspondence started.
34. Clusius (*Rariorum*, p. 89 and elsewhere) calls him 'Patavinus patricius'. His Latin, and the references to his long experience of the medicinal effects of plants, suggest that he was a physician (e.g. Cortuso, 27 March 1569). See Minelli (ed.), *The Botanical Garden of Padua*, pp. 62–4; and on the Doge ordering his appointment, see Cortuso, 26 February 1592.
35. Louis, *Mathieu de l'Obel*, pp. 76–8; C. Longeon (ed.), *Gesner, Conrad, vingt lettres á Jean Bauhin fils (1563–1565)* (Saint Etienne: Université de Saint Etienne, 1976); cf. Reeds, *Botany*, pp. 111–33.
36. *Rariorum*, pp. 172, 179, 182.
37. On their first introduction, see Cortuso, 14 December 1566. Cf. G. Masson, 'Italian Flower Connoisseurs', *Apollo: A Journal of the Arts*, 88 (1968), pp. 164–71, on p. 166.
38. Cf. Cortuso, 14 February 1568, offering to send Clusius as many of these fossils as he can.
39. In his letter of 24 July 1568 Cortuso promises to send Clusius 'Greek herbs, new and beautiful plants'. For the Cortusa Matthioli, see *Rariorum*, pp. 306–7.
40. Later that same year Cortuso sent another 34 plants to Clusius, some of them in the form of pictures (Cortuso, 1569, no further date). For the Mexican poppy, see *Rariorum*, p. xciii.
41. See Minelli (ed.), *The Botanical Garden of Padua*, pp. 62–4; and Cortuso, 23 July 1593.
42. There are 26 references to Cortuso in the *Rariorum*, and 3 in the *Exoticorum*.
43. *Rariorum*, p. ccl.
44. See Girolamo (also Hieronimo) Calzolari, 6 October 1596. For the Calzolaris, see C. Salzmann, 'Francesco Calzolari, Student of the Flora of Monte Baldo, and his Shipments of Plants Addressed to Conrad Gessner in Zurich', *Gesnerus*, 16 (1959), pp. 81–103; and esp. Olmi, 'Per la storia'; and G. Sandrini (ed.), *Francesco Calzolari, il viaggio di Monte Baldo, con la Testimonianza sul Museo Calzolari di Ulisse Aldrovandi* (Verona: Alba Pratalia, 2007). Pona's seven extant letters to Clusius (1595–1606) have been published in De Toni, *Il carteggio*, pp. 40–52; cf. the many references to Pona in Pinelli's letters to Clusius.
45. Pona, *Monte Baldo*. Lobel found him extremely helpful and a 'gallant homme' (Lobel, 20 June 1602).
46. Of the nine references to Pona in Clusius's printed works only one concerns a plant of the Italian flora (*Rariorum*, p. 233).
47. *Rariorum*, pp. cccxxii–cccxlviii.
48. On the printing history of Pona's text, see Hunger, *Charles de l'Escluse*, vol. 1, p. 262.

49. For foreigners botanizing on Monte Baldo, see Jungermann, 1 April 1590. Cf. Findlen *Possessing Nature*, pp. 180–4; Olmi, 'Per la storia'; and Sandrini (ed.), *Francesco Calzolari*.
50. Pona, *Monte Baldo*, p. 177. Trachelio is probably the alpine Campanula trachelium L.
51. O. Huber, 'Die "Geographie der Pflanzen"', in *Alexander von Humboldt: Netzwerke des Wissens* (Ostfildern-Ruit: Hantje Catze Verlag, 1999), pp. 100–3. Cf. Salzmann, 'Francesco Calzolari'. On the new interest in plant ecology in the sixteenth century, see Ubrizsy Savoia, 'Environmental Approach'; and Ubrizsy Savoia et al., 'The Beginnings of Ecological Thought'. Ogilvie (*The Science of Describing*, pp. 221, 270) regards ecology as part of the 'contextual' information that was rooted in folk-biology, but was eliminated by Renaissance naturalists when they increasingly concentrated on taxonomy. The latter seems clear, but about its roots there are evidently different opinions.
52. All biographical information on Belli is based on Beschi's fundamental work (*Onorio Belli*), which also publishes all of Belli's known letters (in Latin). In the case of Belli's letters to Clusius (1593–1602), I have used the Leiden originals.
53. Beschi, *Onorio Belli*, esp. pp. 166–9. In Belli's first letter to Clusius he included plant pictures made from life by his own hand, as well as plants and seeds (4 April 1593). On the Venetian rule of Crete, see D. Holton (ed.), *Literature and Society in Renaissance Crete* (Cambridge: Cambridge University Press, 1991).
54. Beschi, *Onorio Belli*, p. 99, cf. p. 123.
55. On Crete and its variety of species, see Garbari et al., *Giardino dei Semplici*, p. 113. A long series of Italian and other naturalists visited Crete; one of the earlier ones (in 1547) was Pierre Belon. Giuseppe Casabona (see Chapter 6) spent some time on Crete during Belli's sojourn there (1590–1), and was an important source of Cretan plants in Italy and for Clusius; see L. Tongiorgi Tomasi, 'L'isola dei semplici', *Kos*, 1:5 (1984), pp. 61–78; cf. R. Palmer, 'Medical Botany in Northern Italy in the Renaissance', *Journal of the Royal Society of Medicine*, 78 (1985), pp. 149–57, on pp. 149, 151, 156.
56. Beschi, *Onorio Belli*, pp. vii–viii. The similarities between the approaches of Belli and Pierre Belon, who spent a much shorter time on Crete, however, are remarkable. See G. Huppert, *The Style of Paris: Renaissance Origins of the French Enlightenment* (Bloomington, IN: Indiana University Press, 1999), pp. 1–20; cf. G. Huppert, 'Antiquity Observed: A French Naturalist in the Aegean Sea in 1547', *International Journal of the Classical Tradition*, 2 (1995), pp. 275–83.
57. Beschi, *Onorio Belli*, esp. pp. xvii–xxxii. On the cannabis-like plant, see Belli, 15 October 1599.
58. Beschi, *Onorio Belli*, esp. p. xxxi. Belli was so critical of his fellow citizen Prospero Alpino, an expert on Egyptian and Middle Eastern plants, that their exchanges about the identification of certain plants ended in recrimination and, eventually, in a cessation of contact.
59. Beschi, *Onorio Belli*, pp. 149–53 (undated letter from Belli to Prospero Alpino).
60. The plant is probably Atractylis gummifera L.
61. This is an early description of such plant movements judging from C. Webster, 'The Recognition of Plant Sensitivity by English Botanists in the Seventeenth Century', *Isis*, 57 (1966), pp. 5–23. Belli's statement is the clearest of various examples (cf. also Chapter 12 on the mimosa pudica or 'sensitive plant' and Chapter 7 on Peiresc's mushroom) contradicting Ogilvie's conclusion that sixteenth-century naturalists did not notice plant behaviour and thought 'they could not initiate motion themselves, especially not motion directed to a specific end' (*The Science of Describing*, pp. 265–8, on p. 266). Whether they

subsequently did or did not proceed to call into question Aristotle's division between plants and animals cannot be used to argue pro or contra the quality of their observations.

62. *Rariorum*, pp. ccxcix–cccxiv; Beschi, *Onorio Belli*, pp. 205–45. Cf. Belli, 15 February 1596, on the honour of his writings being published together with those of Clusius.
63. Beschi, *Onorio Belli*, pp. xxxiii, 200–1. Many plants received from Belli are mentioned in Pona (*Monte Baldo)*, as growing in the gardens outside Verona.

6 Nature in the Garden

1. The exceptions are his contacts with Aldrovandi in Bologna (see Chapter 5) and Alfonso Pancio (or Panza) in Ferrara, court physician and *semplicista* of Duke Alfonso II d'Este from about 1570 until his death (*c.* 1575/9), and professor of medicine at the university of Ferrara (1550–74). His rather technical letters (11 letters, 1568–71) to Clusius focus mainly on plants and fruits obtained from abroad and their medicinal or other uses. Cf. Findlen, *Possessing Nature*, esp. p. 366.
2. G. Olmi, 'I cappucini e la scienza nell'età moderna', in G. Pozzi and P. Prodi (eds), *I Cappucini in Emilia-Romagna: Storia di una presenza* (Bologna: EDB, 2002), pp. 288–328; and Olmi, 'Lettere di fra Gregorio'.
3. Coffin, *Gardens and Gardening*, pp. 211–12, esp. p. 219. On the Villa d'Este, see I. Barisi, M. Fagiolo and M. Madonna, *Villa d'Este* (Rome: De Luca Editori, 2003), esp. pp. 23–9, 55–71.
4. Olmi, 'Lettere di fra Gregorio', p. 131.
5. Quattrami, 3 April 1598. By May that year Quattrami had not returned to Gubbio, however, and spoke of trying to establish a new garden in Modena (24 May 1599); cf. Masson, 'Italian Flower Connoisseurs', p. 166.
6. Quattrami, 15 January 1597, 24 May 1599; cf. Coffin, *Gardens and Gardening*, p. 219.
7. G. Belloni Speciale, 'La ricerca botanica dei Lincei a Napoli: Corrispondenti e luoghi', in F. Lomonaco and M. Torrini (eds), *Galileo a Napoli* (Naples: Guida, 1987), pp. 59–79, 62–4; and Gabrieli, *Il Carteggio Linceo*, pp. 175, 361, 890–1. For the Anguillara–Quattrami trips in the 1560s, see Coffin, *Gardens and Gardening*, p. 219.
8. Olmi, 'Lettere di fra Gregorio', p. 134.
9. See esp. Quattrami, 1 June 1599.
10. For biographical details and all letters of Fra Gregorio to others than Clusius I am relying on Olmi, 'I cappucini e la scienza'; and Olmi, 'Lettere di fra Gregorio'. For field trips, see Fra Gregorio, 14 March 1602, 9 March 1606, 20 June 1606.
11. Olmi, 'Lettere di fra Gregorio', pp. 137–8.
12. See Olmi, 'I cappucini e la scienza', p. 302; and Olmi, 'Lettere di fra Gregorio', p. 121.
13. Olmi, 'Lettere di fra Gregorio', p. 133, quoting Fra Gregorio to Aldrovandi, 1 October 1596.
14. *Curae*, pp. 95–108; ten further references to Fra Gregorio can be found there, besides two in the Appendices to the *Rariorum* in the *Exoticorum*. Cf. Olmi, 'I cappucini e la scienza', p. 302.
15. Olmi, 'Lettere di fra Gregorio', esp. pp. 125–6.
16. One volume of it still survives; see Olmi, 'Lettere di fra Gregorio', pp. 119–23. For Beyerlinck, see Fra Gregorio, 14 March 1602.
17. For Pisa, see Garbari et al., *Giardino dei Semplici*, p. 107. On the Medici, gardens and art, see Filardi, *Orto de' Pitti*, esp. pp. 17–28, and on collaboration between head gar-

deners-designers and herbalists pp. 113–18; Acidini Luchinat (ed.), *Giardini Medicei*; L. Tongiorgi Tomasi and G. A. Hirschauer, *The Flowering of Florence* (Aldershot: Lund Humpries, 2002); and Zalum Cardon, *Passione e cultura dei fiori*.

18. There are 22 extant letters from Clusius to Caccini (all published in Ginori Conti, *Lettere inedite*), but no letters from Caccini to Clusius. That is one reason why I have not attempted to explore Caccini's botanical expertise here; such an investigation would, moreover, have to take into account the extensive (early seventeenth-century) correspondence of Caccini with many plant lovers and garden owners in Europe in Bibliothèque Royale Brussels, MS III 893. This has been done admirably by Zalum Cardon, *Passione e cultura dei fiori*, esp. ch. 2, pp. 39–97; cf. Masson, 'Italian Flower Collector's Gardens'.
19. For Malocchi, see Garbari et al., *Giardino dei Semplici*, pp. 114–15, 158–9; cf. Malocchi, 20 June 1606, 8 November 1606.
20. See Olmi, '"Molti amici in varii luoghi"', pp. 25–6; L. Tongiorgi Tomasi and F. Garbari, *Il giardiniero del Granduca: Storia e immagini del Codice Casabona* (Pisa: Edizioni ETS, 1995); Garbari et al., *Giardino dei Semplici*, pp. 113–14, 149–50, 156–7; Tongiorgi Tomasi, 'L'isola dei semplici'; and Ubrizsy Savoia, 'Some Aspects', pp. 286–7. Cf. Chapter 5 on Crete and Chapter 10 on the Leiden *hortus*.
21. On contacts with Camerarius, see Olmi, '"Molti amici in varii luoghi"'. Jean Boisot refers to a plant gift from Casabona (J. Boisot, 3 August 1590; cf. 29 August 1590). For the Wilhelm IV connection, see note 24 below.
22. Wilhelm IV, March 1591; Casabona, 6 April 1590, 25 August 1590. On Casabona's preoccupation with patronage, cf. Olmi, '"Molti amici in varii luoghi"', pp. 16–23.
23. M. Cima and E. Talamo, *Gli horti di Roma antica* (Rome: Musei Capitolini / Mondadori-Electa, 2008), pp. 38, 28–33, 36–49. Cf. G. Di Pasquale and F. Paolucci (eds), *Il giardino antico da Babilonia a Roma: Scienza, arte e natura* (Livorno: Sillabe, 2007), esp. pp. 86–177.
24. A. Campitelli, *Gli Horti dei Papi. I giardini Vaticani dal Medioevo al Novecento* (Milan: Jaca Book / Musei Vaticani, 2009), pp. 106–19. For early seventeenth-century gardens in Rome, see esp. Zalum Cardon, *Passione e cultura dei fiori*, esp. pp. 99–153.
25. They were followed in his list by the gardens of the Orsini, Sforza, Medici, Pope Julius III and Cesi; see M. de Montaigne, *Journal de Voyage en Italie: Oeuvres Complètes* (Paris: Gallimard, 1962), p. 1235.
26. For plants in the gardens of the papal villas, see Coffin, *Gardens and Gardening*, esp. pp. 196–214, and on exotica in the Farnese garden, p. 208; cf. Masson, 'Italian Flower Connoisseurs', p. 169.
27. I. Baldriga, 'The Influence of Clusius in Italy: Federico Cesi and the Accademia dei Lincei', in Egmond et al. (eds), *Carolus Clusius*, pp. 249–65. Lobel was not very positive about Colonna ('I find him somewhat overpresumptuous'), but told Clusius that in writing he nonetheless referred to Colonna with reverence (Lobel, 20 June 1602).
28. For his biographical information I rely on the first modern study on Imperato by Stendardo, which has brought to light and re-evaluated archival material and integrates the older literature: E. Stendardo, *Ferrante Imperato: Collezionismo e studio della natura a Napoli tra Cinque e Seicento* (Naples: Accademia Pontaniana, 2001). See also A. Neviani, *Ferrante Imperato: Speziale e naturalista Napolitano con documenti inediti* (Rome: Accademia di Storia dell'Arte Sanitaria, 1936); and Belloni Speciale, 'La ricerca botanica'; cf. Findlen, *Possessing Nature*.
29. Coffin, *Gardens and Gardening*, p. 207.

30. Cf. Pinelli, 23 August 1589. On the Imperato–Pinelli connection, see further Stendardo, *Ferrante Imperato*, pp. 25–6.
31. Imperato was the first to describe the marine convulvulus, now known as Ipomoeae stolonifera. Cf. *Rariorum*, p. ccl.
32. His collection was probably first arranged for showing in *c.* 1566–7, and certainly expanded in *c.* 1586. See Stendardo, *Ferrante Imperato*, pp. 84–97, and for the garden pp. 15, 85; cf. Neviani, *Ferrante Imperato*, pp. 141–5.
33. Stendardo, *Ferrante Imperato*, pp. 89–92; cf. Findlen, *Possessing Nature*.
34. See Stendardo, *Ferrante Imperato*, pp. 62–5, 93; cf. Findlen, *Possessing Nature*, esp. pp. 225–9.
35. F. Imperato, *Dell' Historia Naturale* (Naples: C. Vitale, 1599). For discussions of the contents, see Neviani, *Ferrante Imperato*; Stendardo, *Ferrante Imperato*, pp. 67–79; and Findlen, *Possessing Nature*, esp. pp. 225–32.
36. Stendardo, *Ferrante Imperato*, pp. 87–91.
37. Pinelli, 27 July 1593, with a section of a letter by Imperato on bulbs and tuberous plants.
38. For the plant lists, see Pinelli, 3 December 1598; and Imperato, 28 June 1600; some were published in Imperato's work and referred to by Clusius as well (e.g. *Rariorum*, pp. 167, 192). Many other letters from Pinelli to Clusius refer to exchanges with Imperato.
39. Stendardo (*Ferrante Imperato*, esp. pp. 56–8) links the increasingly collaborative nature of natural history research with the growing amount of data and need of verification.
40. F. Calzolari to Aldrovandi, 20 September 1554; for the full text of all Calzolari letters to Aldrovandi, see http://www.filosofia.unibo.it/aldrovandi/; original BUB (Bologna), MS 382, III, cc. 28r–29r. This is also the gist of Olmi, '"Molti amici in varii luoghi"', see esp. pp. 6–7.
41. W. Eamon and F. Paheau, 'The Accademia Segreta of Girolamo Ruscelli: A Sixteenth-Century Italian Scientific Society', *Isis*, 75 (1984), pp. 327–42; cf. Findlen, *Possessing Nature*, esp. pp. 226–30; Stendardo, *Ferrante Imperato*, esp. pp. 23–33, 51–65; and G. Olmi, 'La colonia Lincea di Napoli, in Lomonaco and Torrini (eds), *Galileo a Napoli*, pp. 23–59.
42. Belloni Speciale, 'La ricerca botanica', pp. 75–8.
43. See Stendardo, *Ferrante Imperato*, p. 28, who emphasizes the Paduan predominance of *medicina filosofica* over 'the art of the apothecaries', while in Naples there was much more emphasis on collaboration between physicians and *aromatari*.
44. Neviani, *Ferrante Imperato*, p. 261.
45. See Olmi, 'La colonia Lincea di Napoli'; Stendardo, *Ferrante Imperato*, pp. 17–20, 30–40; Belloni Speciale, 'La ricerca botanica'; and Findlen, *Possessing Nature*, p. 232.
46. As suggested by Baldriga, 'The Influence of Clusius in Italy'. Clearly, there may have been other reasons as well.
47. Unlike some of their colleagues in England (see Chapter 11), the apothecaries discussed here were not in the direct service of the courts as far as I know. Some court connections did exist: Francesco Calzolari was rewarded for his services to Vincenzo Gonzaga, Duke of Mantua (see Findlen, *Possessing Nature*, pp. 348–9).
48. Olmi, 'Per la storia', p. 344.
49. R. Palmer, 'La botanica medica nell'Italia del Nord durante il Rinascimento', in Bussi (ed.), *Di Sana Pianta*, pp. 55–60, on p. 59.

50. As is also clear from, for example, Lazzaro, *The Italian Renaissance Garden*, p. 41; and Zalum Cardon, *Passione e cultura dei fiori*, p. xiii and *passim*. Cf. Ogilvie, *The Science of Describing*, pp. 151–64, on esp. p. 156.
51. The Paduan *hortus* was slighty different in this respect, because it was created from the start as a public garden. But the fact that the Veneto was ruled by a group rather than a single prince makes no difference with respect to the garden's representational function or the fact that it was directly controlled by the rulers. See W. Eamon, 'Court, Academy, and Printing House: Patronage and Scientific Careers in Late-Renaissance Italy', in Moran (ed.), *Patronage and Institutions*, pp. 25–50, on p. 35; G. Olmi, 'Science – Honour – Metaphor; Italian Cabinets of Curiosity of the Sixteenth and Seventeenth Centuries', in Impey and MacGregor (eds), *The Origins of Museums*, pp. 3–16; Pomian, *Collectioneurs, amateurs et curieux*, p. 83; and Garbari et al., *Giardino dei Semplici*, esp. pp. 101–3.
52. Coffin, *Gardens and Gardening*, p. 218. Campitelli (*Gli Horti dei Papi*, pp. 106–7), who does not refer to this statement by Coffin, points out that already from the thirteenth century a *simpliciarius* formed part of the papal staff and combined scientific and supervisory duties concerning the Vatican garden. However, her examples for the early centuries appear to refer to scientific duties in the field of medicine and medicinal plants rather than to the domain of 'botany' in the stricter sense of the term. The role of M. Mercati during the 1560s–80s and his exchanges of rare plants perfectly fit in with the pattern and chronology described here.
53. See Pinelli, 14 February 1592, in which he also refers to the information sent to him earlier by Clusius about Casabona and to Clusius's opinion that this behaviour was dishonourable. It is said that to please his patrons Casabona also had rare plants stolen from the gardens of other collectors (see Garbari et al., *Giardino dei Semplici*, p. 149).
54. As first argued by Zalum Cardon, *Passione e cultura dei fiori*, pp. 59, 71–3; cf. Masson, 'Italian Flower Collector's Gardens', p. 80. Among Caccini's customers of the period 1600–15 were Canon Levenier (Chapter 7) and the Duke of Arenberg (Chapter 2). Cf. note 19 above.
55. Findlen, 'The Economy of Scientific Exchange', p. 14, quoting P. Galluzzi.

7 Fieldwork in France

1. M. Séronie-Vivien, P. Coudroy de Lille and M. Favreau, 'Le plus ancien botaniste bordelais connu: Joachim Levenier dit Venerius (ca 1565–1619), *Bulletin de la Société Linnéenne de Bordeaux*, 140:33 (2005), pp. 163–9, which together with Levenier's letters is the source for the biographical information here (the authors did not know of the Levenier-Clusius correspondence); M. Verweij, 'Une lettre retrouvée d'un botaniste bordelais mal connu (Bruxelles KBR, ms. III 893/160)', *Bulletin de la Société Linnéenne de Bordeaux*, 140:33 (2005), pp. 159–62; and Ogilvie, *The Science of Describing*, pp. 67–9 (who did not know of Séronie-Vivien's work).
2. There are 18 extant letters from Levenier to Clusius (1597–1606). Two letters from Levenier to Matteo Caccini date from 24 November 1611 and 4 September 1612; the latter is published in Ginori Conti, *Lettere inedite*, pp. 118–19; the former in Verweij, 'Une lettre retrouvée'. With thanks to Michiel Verweij.

3. All of Clusius's 23 references to Levenier (as Venerius) are in the unpaginated Appendix to the *Rariorum* in the *Exoticorum*. In 1600–2 Levenier wrote at least three to four times a year to Clusius.
4. For a summary of the inventory, see Séronie-Vivien et al., 'Le plus ancien botaniste bordelais connu'. As the authors rightly point out, some of the plant names are very hard to interpret because the inventory was made by someone who was no botanical expert.
5. See Levenier, 26 November 1599, 8 November 1600, 8 July 1601, 20 August 1602.
6. Cf. Séronie-Vivien et al., 'Le plus ancien botaniste bordelais connu', p. 168. There is some degree of uncertainty given the incomplete list and the always tenuous identifications.
7. Levenier, 16 April 1600; and he wrote 'there is nothing I desire more than to obtain some carnations and rare roses' (26 November 1599).
8. See esp. Levenier to Caccini, 4 September 1612, in Ginori Conti, *Lettere inedite*, pp. 118–19; see also note 3 above.
9. Levenier to Caccini, 4 September 1612, in Ginori Conti, *Lettere inedite*, pp. 118–19. According to Zalum Cardon (*Passione e cultura dei fiori*, pp. 106–7), anemones were a rare and special item in the gardens of the Prince of Arenberg during this same period.
10. The treatise is discussed in Lazzaro, *The Italian Renaissance Garden*, p. 37.
11. For more information concerning the conflicts with his principal rival, the garden owner Jacques du Casse, who was also in contact with Clusius, see Ogilvie, *The Science of Describing*, pp. 67–9.
12. Nesmond to Levenier, 9 July 1601; a copy of this letter is included in Levenier, 28 July 1601.
13. See Levenier, 18 February 1600, 16 April 1600; cf. 18 December 1599.
14. Although the courtly metaphor of hunting was important as well, here I refer to the practice of hunting.
15. See Reeds, *Botany*, pp. 39–92; and G. Lewis, 'Clusius in Montpellier, 1551–1554: A Humanist Education Completed?', in Egmond et al. (eds), *Carolus Clusius*, pp. 65–98; cf. B. Ogilvie, 'Travel and Natural History in the Sixteenth Century', in B. Ogilvie, A. te Heesen and M. Gierl, *Sammeln in der Fruhen Neuzeit* (Berlin: Max-Planck-Institut für Wissenschaftsgeschichte, 1996), pp. 3–28.
16. Not only Levenier's letters but the whole of the Clusius correspondence show that issues of priority and honour played an important part in botany long before scientific societies such as the Royal Society were created. On priority disputes and issues, in particular in the early Royal Society, see R. Iliffe, '"In the Warehouse": Privacy, Property and Priority in the Early Royal Socety', *History of Science*, 30 (1992), pp. 29–68.
17. On Clusius's scoops and stop-press approach to publishing about exotic *naturalia*, see Mason, *Before Disenchantment*.
18. A more comparative analysis of the use of indigenous (non-European *and* European) knowledge of nature by early modern investigators as well as the clashes and non-translatability between the various types of knowledge would be worthwile. For various approaches, see R. Grove, 'Indigenous Knowledge and the Significance of South-West India for Portuguese and Dutch Constructions of Tropical Nature, *Modern Asian Studies*, 30 (1996), pp. 121–43; W. Vandewalle and K. Kasaya (eds), *Dodonaeus in Japan: Translation and the Scientific Mind in the Togukawa Period* (Louvain and Kyoto: Leuven University Press and International Research Center for Japanese Studies, 2001); J. Heniger, *Hendrik Adriaan van Reede tot Drakenstein (1636–1691) and Hortus Malabaricus: A Contribution to the History of Dutch Colonial Botany* (Rotterdam: Balkema, 1986); and L. Schiebinger and C. Swan (eds), *Colonial Botany: Science,*

Commerce and Politics in the Early Modern World (Philadelphia, PA: University of Pennsylvania Press, 2004).

19. My findings on the fascination with colour contrast with Ogilvie's remarks on a lack of interest in it (*The Science of Describing*, pp. 205–6).
20. Appendix to Levenier, 8 July 1601.
21. See Levenier, 7 November 1600, 25 June 1602, 20 August 1602.
22. This does not really fit with Ogilvie, 'Travel and Natural History', who states that persons in this generation of naturalists who noted that certain wild plants could not be acclimatized in the garden were not interested in causal explanations. The evidence of practice (cf. Hoghelande in Chapter 10) seems to me as significant as that of words.
23. C. Raven, *English Naturalists from Neckam to Ray: A Study of the Making of the Modern World* (Cambridge: Cambridge University Press, 1947), pp. 162–3.
24. Ibid., pp. 162–3.
25. The literature on Peiresc is extensive, but no special study has been devoted, as far as I know, to Peiresc and natural history. Most useful were to me L. Legré, *La botanique en Provence au XVIe siècle* (Marseille: H. Aubertin & Cie, 1901), pp. 69–139; D. Jaffé, 'Peiresc – Wissenschaftlicher Betrieb in einem Raritäten-Kabinett', in Grote (ed.), *Macrocosmos in Microcosmo*, pp. 301–22; and P. Miller, 'Description Terminable and Interminable: Looking at the Past, Nature and Peoples in Peiresc's Archive', in G. Pomata and N. Siraisi (eds), *Historia: Empiricism and Erudition in Early Modern Europe* (Cambridge, MA: MIT Press, 2005), pp. 355–97.
26. There are twelve extant letters from Peiresc to Clusius (1602–8), eight in Leiden and four in Aix-en-Provence; two of those four are identical in contents with two of the letters in Leiden.
27. Peiresc told Clusius about Pinelli's death (Peiresc, 18 January 1602). On Peiresc and his Italian contacts, see Hunger, *Charles de l'Escluse*, vol. 1, p. 270; Miller, 'Description Terminable and Interminable', p. 368, cf. 359; Findlen, *Possessing Nature.*
28. See esp. Peiresc, 25 January 1603, 27 February 1604.
29. On these various trips and visits, see Miller, 'Description Terminable and Interminable'; Legré, *La botanique en Provence*, pp. 83–6; Hunger, *Charles de l'Escluse*, vol. 1, p. 300, cf. 280.
30. These items were sent to Clusius in February 1605. See Hunger, *Charles de l'Escluse*, vol. 1, p. 280.
31. A. Schnapp, *The Discovery of the Past: The Origins of Archaeology* (London: British Museum Press, 1996), p. 134. For a different opinion on the function of Peiresc's collection-cabinet, see Jaffé, 'Peiresc', esp. p. 306.
32. Seseli (Seseli tortuosum L. according to Legré) is said to have been used for the preparation of theriac; tragacantha is now known as Astragalus Massiliensis Lam. See Hunger, *Charles de l'Escluse*, vol. 1, p. 274; Legré, *La botanique en Provence*, pp. 91–9.
33. List sent with Peiresc, 15 February 1605.
34. See Legré, *La botanique en Provence*, p. 82; the term in the contemporary source is 'botanicus'.
35. For details on these locations, see Legré, *La botanique en Provence*, pp. 83–4.
36. Clusius had sent Peiresc his *Fungorum brevis historia* of 1601 (Peiresc, 15 February 1605). The illustration, reproduced here as Figure 6, appears on the last page of the unpaginated appendix to Clusius's *Rariorum* in the *Exoticorum*. Cf. Legré, *La botanique en Provence*, pp. 106–9.

37. On descriptive precision, see Legré, *La botanique en Provence*, p. 91–2; and Miller, 'Description Terminable and Interminable'.
38. Schnapp (*The Discovery of the Past*, p. 134) uses these words for Peiresc's approach to antiquities.

8 French Collectors between Port and Court

1. José Pardo Tomás (personal communication, 2008; with many thanks) suggests as possible identifications: (for anonas from Oaxaca) the fruit anona common in Central America; (for the little, 'chico', zapote) a common fruit in Mexico; (for eloxochitl) the flower of a tree of the magnolia species used in many Mexican ceremonies; (for cempaxochitl) a flower used to honour the dead in ancient Mexico.
2. For a modern edition and comments, see P. Contant, *Le Jardin, et cabinet poétique* (1609), ed. M. Marache-Gouraud and P. Martin (Rennes: Presses Universitaires, 2004).
3. See A. Schnapper, *Le géant, la licorne et la tulipe: Collections et collectionneurs dans la France du XVIIe siècle* (Paris: Flammarion, 1988), pp. 223–5, who contrasts the modest value of Contant's collection with the vastly greater one of the apothecary Calzolari; and Contant, *Le Jardin*, esp. pp. 13–20.
4. Contant, *Le Jardin*, pp. 21, 40–1; the literary presentation of the garden as introduction or frontispiece to the collection does not necessarily reflect its actual use or location.
5. Schnapper, *Le géant, la licorne et la tulipe*, p. 392.
6. Contant, *Le Jardin*, p. 20.
7. For references to Clusius and a list of donors, see ibid., pp. 124–9, 259; for biographical details, see Schnapper, *Le géant, la licorne et la tulipe*, pp. 56, 223–5. On Palissy, see L. Amico, *A la recherche du paradis terrestre: Bernard Palissy et ses continuateurs* (Paris: Flammarion, 1996), esp. p. 29.
8. Schnapper, *Le géant, la licorne et la tulipe*, p. 222; and Contant, *Le Jardin*, pp. 100, 109, 168. For further references on apothecaries and natural history, see Egmond, 'Apothecaries as Experts and Brokers'; and S. Halikowski Smith, '"The Physician's Hand": Trends in the Evolution of the Apothecary and his Art across Europe, 1500–1700', *Nuncius*, 24 (2008), pp. 33–60.
9. Contant, *Le Jardin*, pp. 127, 258.
10. Schnapper, *Le géant, la licorne et la tulipe*, pp. 223–4.
11. Vertunien was the family doctor of the aristocratic Chastaigners, long-term patrons of Scaliger, who may indeed have put Clusius in touch with the Poitiers-Bordeaux-La Rochelle naturalists; see Vertunien, 16 October 1602; and Le Coq, 14 December 1599. For the Scaliger–Vertunien correspondence and biographical details on Vertunien, see R. Hawkins, 'The Friendship of Joseph Scaliger and François Vertunien', *Romanic Review*, 8 (1917), pp. 117–44, 307–27.
12. See Schnapper, *Le géant, la licorne et la tulipe*, pp. 223–4; and Hawkins, 'The Friendship', pp. 324–5.
13. Levenier claimed to have discovered this fritillary, but the one he sent to Clusius on 7 November 1600 was lost in a shipwreck.
14. Schnapper, *Le géant, la licorne et la tulipe*, p. 50.
15. See P. Mason, *Infelicities: Representations of the Exotic* (Baltimore, MD: Johns Hopkins University Press, 1998); A. Lugli, *Naturalia et mirabilia: Les cabinets de curiosités en Europa* (Paris: Adam Biro, 1998); L. Daston and K. Park, *Wonders and the Order of Nature, 1150–1750* (New York: Zone, 1998).

16. On the native–foreign distinction, see also Ogilvie, 'Travel and Natural History'. A. Cooper, *Inventing the Indigenous: Local Knowledge and Natural History in Early Modern Europe* (Cambridge: Cambridge University Press, 2007), is not reliable on the sixteenth century.
17. On Nérac, see Amico, *A la recherche du paradis terrestre*, p. 15.
18. See Nesmond to Levenier, 9 July 1601, sent from Nérac, which is included in Levenier, 28 August 1601.
19. For du Casse, see note 11 to Chapter 7, above, p. 242.
20. See Levenier, 20 October 1598, 26 November 1599, 7 November 1600, 10 February 1601.
21. Cf. Marnix, 4 January 1591, from which Clusius's response to the earlier letter can be deduced. On Marnix, see Chapter 10.
22. On the Robins, see Schnapper, *Le géant, la licorne et la tulipe*, pp. 41–2; and esp. M. Warner, 'Jean and Vespasien Robin, "Royal Botanists," and North American Plants, 1601–1635', *National Horticultural Magazine*, 35 (1956), pp. 214–20, who has pointed out that there has been some confusion between Robin's private garden and what would later become the official Jardin des Plantes, all the more since the collections were partly joint (pp. 215–6). I follow her interpretation here.
23. P. Bassani Pacht et al., *Marie de Médici, un gouvernement par les arts* (Paris: Somogy editions, 2003), p. 210.
24. See Warner, 'Jean and Vespasien Robin'; and Schnapper, *Le géant, la licorne et la tulipe*, pp. 41–2, 182–3.
25. Warner, 'Jean and Vespasien Robin', pp. 216–17; cf. Schnapper, *Le géant, la licorne et la tulipe*, pp. 41–2.
26. On this narcissus jacobeus, see note 41 to Chapter 2, above, p. 230. On the passion flower, see Warner, 'Jean and Vespasien Robin', p. 217.
27. Schnapper, *Le géant, la licorne et la tulipe*, p. 42.
28. P. Leith-Ross, 'A Seventeenth-Century Paris Garden', *Garden History*, 21 (1993), pp. 150–7.
29. See M. Bath, *Emblems for a Queen: The Needlework of Mary Queen of Scots* (London: Archetype Publications, 2008), pp. 2, 6; Blunt and Stearn, *The Art of Botanical Illustration*, p. 110; Schnapper, *Le géant, la licorne et la tulipe*, pp. 41–2, 183–4; Bassani Pacht et al., *Marie de Médici*, p. 210; and, on the connection with embroidery and patterns for silversmiths, MacDougall, *Fountains, Statues, and Flowers*, p. 226.
30. Quoted in Schnapper, *Le géant, la licorne et la tulipe*, pp. 183–4.
31. Warner, 'Jean and Vespasien Robin', pp. 216–19.
32. Levenier, 12 March 1603, mentions a trip in October 1602; cf. Plateau, 15 October 1603; and Robin Sr, 10 April 1599.
33. This list is often included in editions of Vallet's *Jardin du roi*.
34. P. Leith-Ross and H. McBurney, *The Fiorilegium of Alexander Marshal* (London: Royal Collection Enterprises, 2000). Marshall's insect album can be consulted online: http://www.ansp.org/library/getty_findaid/marshal941.xml (for the swordfly no. 19b). For the interest of Pierre Morin in shells and insects, cf. Leith-Ross, 'A Seventeenth-Century Paris Garden'.
35. Plateau was speaking here of Robin Sr's opinions.

9 Dutch Ports

1. *Rariorum*, pp. cccxv–cccxx; Roels's original letter (8 May 1597) no longer exists. There are seven letters in Latin from Roels to Clusius (1590–1602).
2. *Rariorum*, pp. cccxv–cccxvi.
3. The report itself indicates that Clusius had asked Roels for information about tropical *naturalia*. Earlier descriptions of manioc and how to deal with its poisonous juices were published in 1535 in Gonzalo Fernández de Oviedo y Valdes, *Historia general y natural de las Indias* (Seville: J. Cromberger; reissued in 1547 and 1557, and translated into Italian and French). With thanks to José Pardo Tomás.
4. *Rariorum*, p. cccxx.
5. Like most members of the Middelburg elite Roels was to some extent involved in commerce. In 1602 he invested a very modest sum in the newly founded Dutch East India Company (VOC). See J. Bruijn, 'Scheepvaart en overheid omstreeks 1600', in C. Lesger and L. Noordegraaf (eds), *Ondernemers en Bestuurders: Economie en politiek in de Noordelijke Nederlanden in de Late Middeleeuwen en vroegmoderne tijd* (Amsterdam: NEHA, 1999), pp. 77–83, on p. 83. On Clusius and the exotic, see K. van Ommen (ed.), *The Exotic World of Carolus Clusius 1526–1609: Catalogue of an Exhibition on the Quatercentenary of Clusius' Death, 4 April 1609* (Leiden: Leiden University Library, 2009).
6. On socio-cultural aspects of this period of quick expansion, see K. Heyning, *Return to Zealand: Masterpieces from the 16th and 17th Centuries* (Middelburg: Zeeuws Museum, 2008), esp. pp. 4–5, 78–81; V. Enthoven, 'Een symbiose tussen koopman en regent: De tweetrapsraket van de opkomst van de Republiek en Zeeland', in Lesger and Noordegraaf (eds), *Ondernemers en Bestuurders*, pp. 203–36; and J. Parmentier (ed.), *Noord-Zuid in Oost-Indisch Perspectief* (Zwolle: Walburg Pers, 2005).
7. On Moucheron, see H. den Heijer, 'Zuid-Nederlanders en het ontstaan van de Noord-Nederlandse handel buiten Europa', in Parmentier (ed.), *Noord-Zuid in Oost-Indisch Perspectief*, pp. 59–80; and L. Noordegraaf and T. Wijsenbeek-Olthuis, 'De wereld ontsloten', in R. Kistemaker and E. Bergvelt (eds), *De wereld binnen handbereik: Nederlandse kunst- en rariteitenverzamelingen 1585–1735* (Zwolle: Waanders, 1992), pp. 39–51, on pp. 40–1.
8. P. Meertens, *Letterkundig leven in Zeeland in de zestiende en de eerste helft der zeventiende eeuw* (Amsterdam: Noord-Hollandsche Uitgeversmaatschappij, 1943), pp. 12, 20, 429–30, 440; and esp. F. Hunger, 'Acht brieven van Middelburgers aan Carolus Clusius', *Archief Zeeuws Genootschap der Wetenschappen* (1925), pp. 110–33, on p. 112; cf. Enthoven, 'Een symbiose tussen koopman en regent', pp. 219–22.
9. W. Parduyn, 27 November 1593. Parduyn and Clusius eventually met: Clusius visited Middelburg in 1598, and Parduyn went to see Clusius in Leiden in 1599.
10. W. Parduyn, 1 November 1596; cf. Noirot, 6 Feburary 1601; and Combaud, 17 May 1599, 13 July 1599, 9 August 1599.
11. For references to Parduyn, see *Rariorum*, pp. 183, 238, lxxviii, lxxix, cclvii; *Exoticorum*, pp. 4, 6, 25, 30, 37, 49, 62, 70–2, 80, 142, 184, 237, 299; and *Curae*, pp. 47, 84, 93.
12. For Zealand and the Wild Coast, see W. Klooster, *The Dutch in the Americas 1600–1800* (Providence, RI: John Carter Brown Library, 1997), p. 61.
13. For Dutch shell collections after 1600, see H. Coomans, 'Schelpenverzamelingen', in Kistemaker and Bergvelt (eds), *De wereld binnen handbereik*, pp. 192–203; and Goldgar, *Tulipomania*, pp. 80–8.
14. On Clusius's gifts, see W. Parduyn, 16 December 1593.

15. The main sources about this circle are Hunger, 'Acht brieven'; Meertens, *Letterkundig leven*; Clusius's correspondence with Parduyn, Roels, De Jonghe and Noirot; and the references to inhabitants of Middelburg in Clusius's and Lobel's publications. Goldgar (*Tulipomania*, esp. pp. 20–30) focuses on their role as *liefhebbers* on the eve of the tulip craze.
16. *Rariorum*, p. cccxx.
17. Ibid., p. cccxx.
18. See note 15 above. De la Fosse was a nephew of the apothecary Jean Mouton at Tournai (see Chapter 2). Hunger ('Acht brieven') also names the specialized artists Balthasar van der Ast (1593/4–1657) and Ambrosius Bosschaert Sr (1573–1621). Goldgar (*Tulipomania*, pp. 25–6) also mentions the physician Caspar Pelletier and the apothecary Reymer van de Putte as members of this circle. None of these men figures in the Clusius correspondence.
19. Louis, *Mathieu de l'Obel*, pp. 201–3; cf. Meertens, *Letterkundig leven*, p. 440.
20. Cf. *Rariorum*, p. 135.
21. J. Somer, *Zee en Landt reyse naer de Levante, als Italien, Candyen, Cypres, Egypten, Rhodes, Archipelago, Turckyen en wederom door Duytslant*, 2nd edn (Amsterdam: Joost Hartgens, 1649), p. 10 (no copies of the first edition have survived). Cf. Goldgar, *Tulipomania*, pp. 20–3, who takes a more positive view of his knowledge of nature.
22. See W. Parduyn, 16 December 1593; and Roels, 2 January 1594.
23. Meertens, *Letterkundig leven*, pp. 161, 440.
24. Cf. Roels, 4 September 1590. Cf. Hunger, *Charles de l'Escluse*, vol. 1, pp. 174, 180.
25. For all of these plants, see Roels, 4 September 1590, 19 February 1591; *Rariorum*, p. 245.
26. Roels, 19 February 1591, and 1592, no further date. For Jungermann, see Chapter 5.
27. C. Pelletier, *Plantarum tum patriarum tum exoticarum in Walachria, Zeelandiae insula, nascentium synonymia* (Middelburg: R. Schilders, 1610). He was in contact with the Robins in Paris. On Pelletier, see P. Eldering, 'Caspar Pelletier: Ein Botaniker in den Anfängen des 17. Jahrhunderts', *Janus* 64 (1977), pp. 263–78; Goldgar, *Tulipomania*, pp. 25–6; and Meertens, *Letterkundig leven*, p. 419.
28. For the pulsatilla, Roels and Clusius, see note 25 above.
29. The following references to Roels in *Rariorum* (pp. 133, 183, 245, lxx) and *Exoticorum* (pp. 65, 237, 238) concern exotic fruits and roots.
30. See Levenier, 25 June 1602; cf. 16 April 1600.
31. Hunger ('Acht brieven') connects the large number of apothecaries and physicians in Middelburg with its role as a port for the long-distance trade. The single nobleman is the émigré Marnix de Saint Aldegonde; see Chapter 10.
32. Clusius (*Exoticorum*, p. 65) quotes Roels as referring to 'his museum in Middelburg', but this could point to a collection, a report on a collection, or even to the report on rarities that Roels sent to Clusius.
33. Goldgar, *Tulipomania*, p. 26.
34. Roels reported the deaths of both his wife and Willem Parduyn to Clusius (Roels, 6 February 1602). Roels himself died in that year as well (Meertens, *Letterkundig leven*, p. 440).
35. See, more generally, Cook, *Matters of Exchange*, esp. pp. 115–17, 120–30. For Paludanus, see R. van Gelder, 'Paradijsvogels in Enkhuizen: De relatie tussen Van Linschoten en Bernardus Paludanus', in R. van Gelder, J. Parmentier and V. Roeper (eds), *Souffrir pour parvenir: De wereld van Jan Huygen van Linschoten* (Haarlem: Arcadia, 1998), pp. 30–

50; K. van Berkel, 'Citaten uit het Boek der Natuur: Zeventiende-eeuwse Nederlandse naturaliënkabinetten en de ontwikkeling van de natuurwetenschap', in Kistemaker and Bergvelt (eds), *De wereld binnen handbereik*, pp. 169–91; and E. Jorink, *Het Boeck der Natuere: Nederlandse geleerden en de wonderen van Gods Schepping 1575–1715* (Leiden: Primavera, 2006), esp. pp. 275–85; and the older F. Hunger, 'Bernardus Paludanus (Berent ten Broecke) 1550–1633', *Nederlands Tijdschrift voor Geneeskunde*, 72 (1928), pp. 2–11; F. Hunger, 'Bernardus Paludanus (Berent ten Broecke) 1550–1633: Zijn verzamelingen en zijn werk', in C. Burger and F. Hunger (eds), *Itinerario: Voyage ofte Schipvaert van Jan Huygen van Linschoten*, 5 vols (The Hague: Nijhoff, 1910–39), vol. 3, pp. 249–68; L. van Wijk, 'Het album amicorum van Bernardus Paludanus', *Het Boek*, 29 (1948), pp. 265–86; and J. Theunisz, 'B. Paludanus en de brieven uit zijn Zwolse tijd', *Verslagen en Mededeelingen Vereeniging tot beoefening van Overijselsch Regt en Geschiedenis* 52 (1936), pp. 108–24. For Linschoten, see Van Gelder et al. (eds), *Souffrir pour parvenir*; E. van den Boogaart, *Civil and Corrupt Asia: Image and Text in the 'Itinerario' and the 'Icones' of Jan Huygen van Linschoten* (Chicago, IL: University of Chicago Press, 2003). On the North Holland school of cartography, see K. Zandvliet, *Mapping for Money: Maps, Plans and Topographic Paintings and their Role in Dutch Overseas Expansion during the 16th and 17th Centuries* (Amsterdam: Batavian Lion International, 1998).

36. Clusius sent greetings to Maelson and Linschoten in his letters to Paludanus. For the correspondence between Paludanus and Clusius, see A. Berendts, 'Carolus Clusius (1526–1609) and Bernardus Paludanus (1550–1633). Their Contacts and Correspondence', *Lias*, 5 (1978), pp. 49–64.
37. Van Gelder, 'Paradijsvogels in Enkhuizen', pp. 35–9, 49–50; Noordegraaf and Wijsenbeek-Olthuis, 'De wereld ontsloten', p. 46; cf. Van Berkel, 'Citaten uit het Boek der Natuur', pp. 171–5.
38. Van Gelder, 'Paradijsvogels in Enkhuizen', esp. pp. 34–5.
39. On the *Album*, see Van Wijk, 'Het album amicorum'; cf. R. van Gelder, 'Liefhebbers en geleerde luiden: Nederlandse kabinetten en hun bezoekers', in Kistemaker and Bergvelt (eds), *De wereld binnen handbereik*, pp. 259–92, on pp. 264–5.
40. Van Gelder, 'Paradijsvogels in Enkhuizen', pp. 39–40, and personal communication, 20 July 2008.
41. Hunger ('Bernardus Paludanus') and Van Gelder ('Paradijsvogels in Enkhuizen', p. 42) have also remarked upon Paludanus's unimpressive botanical expertise.
42. *Rariorum*, pp. 37, lxxxiv.
43. *Exoticorum*, pp. 28, 46–7. Cf. Berendts, 'Carolus Clusius', esp. pp. 59–60.
44. *Rariorum*, p. 175 (referring to 1594). I differ here from Jorink, who regards Paludanus's collection (in the singular) as a closed and almost 'hallowed treasure of learning' (*Het Boeck der Natuere*, p. 358).
45. The issue of market value to collecting, *naturalia* and science in the relevant period is perhaps most explicitly dealt with in P. Findlen, 'Inventing Nature: Commerce, Art, and Science in the Early Modern Cabinet of Curiosities', in P. Smith and P. Findlen (eds), *Merchants and Marvels: Commerce, Science and Art in Early Modern Europe* (New York and London: Routledge, 2002), pp. 297–323 (but see also the rest of this volume). I find myself closer to Goldgar's position in this respect (*Tulipomania*, esp. pp. 56–61). Cook argues that commerce and science share some of the same presuppositions and mentalities (*Matters of Exchange*, esp. p. 57), which is a rather different point. See also Chapters 10–12 below on dealing with exotics.

10 Town and Gown

1. In 1582 Dodoens (then 65) was appointed in Leiden as professor of pathology and general medicine, but he does not seem to have taught about plants; he died in March 1585. See H. Witkam, *De Financiën van de Leidse Universiteit in de zestiende eeuw*, 5 parts (Leiden: University of Leiden, 1979–82), part 3, p. 211.
2. H. Veendorp and L. Baas Becking, *Hortus Academicus Lugduno Batavus 1587–1937* (Haarlem: Enschedaiana, 1938); L. Tjon Sie Fat, 'Clusius' Garden: A Reconstruction', in Tjon Sie Fat and De Jong (eds), *The Authentic Garden*, pp. 3–12, on pp. 3–5; and G. van Uffelen, 'The *Libri Picturati* and the Early History of the *hortus botanicus* Leiden', in De Koning et al. (eds), *Drawn After Nature*, pp. 54–8. The first plans for the *hortus* date from 1591.
3. On the Montpellier *hortus*, see Reeds, *Botany*, pp. 72–92; cf. Peiresc, 25 January 1603, 27 February 1604.
4. Hunger, *Charles de l'Escluse*, vol. 1, pp. 241–2, 254; and on Paauw, see also Veendorp and Baas Becking, *Hortus Academicus Lugduno Batavus*, p. 65.
5. Quoted in Van Gelder, 'Paradijsvogels in Enkhuizen', p. 35.
6. On Cluyt, see notes 11 and 18 below. For seeds and plants from Clusius's own collection planted in the *hortus*, see Hunger, *Charles de l'Escluse*, vol. 1, pp. 198–201 (with a plant list); and Van Uffelen, 'The *Libri Picturati*'.
7. Tjon Sie Fat ('Clusius' Garden', pp. 7–8) mentions 1,060 species for 1594, of which about one third were medicinal; cf. A. Gogelein et al., *Leidse Universiteit 400: Stichting en eerste bloei 1575–ca. 1650* (Amsterdam: Rijksmuseum, 1975), pp. 170–8; and Van Uffelen, 'The *Libri Picturati*'.
8. See Gogelein et al., *Leidse Universiteit 400*, p. 178; E. de Jong, 'Nature and Art: The Leiden Hortus as "musaeum"', in Tjon Sie Fat and De Jong (eds), *The Authentic Garden*, pp. 37–60, on pp. 46–7; and Veendorp and Baas Becking, *Hortus Academicus Lugduno Batavus*, p. 42.
9. De Jong, 'Nature and Art', pp. 46–7.
10. On the *ambulacrum* and the collection (after 1617), see De Jong, 'Nature and Art', pp. 40–8, 55–6. For the anatomical theatre, see T. Lunsingh Scheurleer, 'Un amphithéatre d'anatomie moralisée', in T. Lunsingh Scheurleer and G. Posthumus Meyjes (eds), *Leiden University in the Seventeenth Century: An Exchange of Learning* (Leiden: Brill, 1975), pp. 216–77.
11. H. Bosman-Jelgersma, 'Dirck Cluyt: De eerste Leidse hortulanus', *Leids Jaarboekje*, 83 (1991), pp. 74–88, on pp. 83–5; and H. Bosman-Jelgersma, 'Dirck Outgaertsz Cluyt', *Farmaceutisch Tijdschrift voor België*, 53 (1976), pp. 525–48, on pp. 540–3.
12. The first letter from Clusius to Paauw dates from 28 March 1593. See Hunger, *Charles de l'Escluse*, vol. 1, pp. 215–16, 254; and De Landtsheer, 'Justus Lipsius and Carolus Clusius', p. 290.
13. Paauw published only on medical matters. He is mentioned in the *Rariorum* (1x), *Exoticorum* (20x; pp. 15, 16, 18, 26, 51, 88, 116, 129) and *Curae* (3x).
14. There are seven letters in Latin from Høyer to Clusius (1597–1604), all sent from Bergen, Norway; see K. Lundquist, 'Lilies to Norway and Cloudberry Jam to the Netherlands – On the Relationship, Correspondence and Exchange of Naturalia between Carolus Clusius and Henrik Høyer, 1597–1604', in Egmond et al. (eds), *Carolus Clusius*, pp. 145–69. Paauw received coral and several seabirds from Høyer (*Exoticorum*, pp. 104, 120).

15. Gogelein et al., *Leidse Universiteit 400*, pp. 107–8; cf. T. Huisman, *The Finger of God: Anatomical Practice in 17th-Century Leiden* (Leiden: Primavera Pers, 2009).
16. On Hoghelande's anemones, see Ginori Conti, *Lettere inedite*, pp. 43, 84–5, 91.
17. See Hunger, *Charles de l'Escluse*, vol. 1, p. 206. Hoghelande too was considered for an appointment in the *hortus*, but his deafness posed too much of a problem (Hoghelande, 18 March 1592).
18. On Cluyt, see Bosman-Jelgersma, 'Dirck Outgaertsz Cluyt'; Bosman-Jelgersma, 'Dirck Cluyt'; Veendorp and Baas Becking, *Hortus Academicus Lugduno Batavus*; and Tjon Sie Fat, 'Clusius' Garden'. Cf. Jorink, *Het Boeck der Natuere*, pp. 202–7.
19. Bosman Jelgersma ('Dirck Cluyt') mentions patronage of Cluyt by two of the highest Dutch politicians, Johan van Oldenbarneveldt and Rombout Hogerbeets.
20. See H. Bosman-Jelgersma (ed.), *Pieter van Foreest; de Hollandse Hippocrates* (Heiloo: Vereniging Oud Heiloo, 1996). On Foreest's Italian experience, see H. Houtzager, 'Opleiding in Italië en verblijf in Frankrijk', in Bosman-Jelgersma (ed.), *Pieter van Foreest*, pp. 81–5.
21. De Munck and Wille (eds), *Peeter van Coudenberghe*; and on Lobel's visit, see Bosman-Jelgersma, 'Dirck Cluyt'.
22. P. Molhuysen, *Bronnen tot de geschiedenis der Leidsche Universiteit deel I (1574–1610)* (The Hague: Nijhoff, 1913), p. 380 (these were not the *Libri Picturati* watercolours, however; cf. Egmond, 'Clusius, Cluyt, Saint Omer'). Anselm de Boodt mentions the painted herbarium in his letters to Clusius from May and October 1602, quoted in M. Maselis, A. Balis and R. Marijnissen, *De Albums van Anselmus de Boodt (1550–1632): Geschilderde natuurobservatie aan het Hof van Rudolf II te Praag* (Tielt: Lannoo, 1989), p. 50 and esp. pp. 204–5.
23. Molhuysen, *Bronnen*, pp. 380–1; Veendorp and Baas Becking, *Hortus Academicus Lugduno Batavus*, pp. 40–1; and H. Bosman-Jelgersma, 'Augerius Clutius (1578–1636), apotheker, botanicus en geneeskundige', *Farmaceutisch Tijdschrift voor België*, 56 (1982), pp. 167–74.
24. Outgert Cluyt went on to study medicine, travel in Germany, France and Spain, and act as deputy of the director of the Montpellier *hortus botanicus*. From his trips to Spain and North Africa he sent back many plants to the Leiden *hortus*. See Bosman-Jelgersma, 'Dirck Outgaertsz Cluyt', p. 527; and note 24 above.
25. See E. Moes, 'De sonderling-heden oft rariteyten ende wtgelesen sinnelickheden van Christiaen Porret', *Leids Jaarboekje*, 2 (1905), pp. 93–100; Bosman-Jelgersma, 'Dirck Outgaertsz Cluyt'; L. Voet, *The Golden Compasses: A History and Evaluation of the Printing and Publishing Activities of the Officina Plantiniana at Antwerp*, 2 vols (Amsterdam: Van Gendt, 1969–72); and briefly Goldgar, *Tulipomania*, pp. 72–3.
26. The catalogue was printed in 1628 in Leiden: *Catalogus oft register vande sonderlingheden oft rariteijten ende wtgeleseen sinnelickheden (...) die Christiaen Porret wijlen apoteker in zijn cunstcamer vergadert had.*
27. Porret is one of Clusius's most cited sources on exotics: *Rariorum* (5x), *Exoticorum* (25x; pp. 26, 45, 68, 86–7, 97–9, 113–14, 117, 119, 139, 142, 358, 374) and *Curae* (24x).
28. *Exoticorum*, p. 124.
29. H. Bosman-Jelgersma, 'De lotgevallen van een apothekersleerling in het 17de-eeuwse Leiden', *Leids Jaarboekje*, 79 (1987), pp. 62–81, on p. 68. He may also have owned a garden outside the city walls; see Moes, 'De sonderling-heden', p. 96.
30. F. Raphelengius to M. Caccini, 12 January 1610, University Library Leiden, BPL 2724.
31. Porret to M. Caccini, 6 January 1610, University Library Leiden, BPL 2724.

32. F. Raphelengius to M. Caccini, 12 January 1610, University Library Leiden, BPL 2724.
33. *Curae*, pp. 15, 41.
34. Porret to M. Caccini, [5?] January 1611, University Library Leiden, BPL 2724. He writes: 'je l'aij experimente troys ou quatre ans enca'.
35. *Rariorum*, pp. 246, 251–4; *Exoticorum*, p. 9; various references in the unnumbered Appendix to the *Rariorum* in *Exoticorum*; *Curae*, pp. 26, 28, 47, 114–16.
36. *Curae*, pp. 20, 37, 57.
37. On Porret as donor to the *hortus*, see Meertens, *Letterkundig leven*, p. 340.
38. Six letters from Porret to Caccini (1610–11) are in University Library Leiden, BPL 2724; on plant gifts in 1610 from Caccini to Porret, cf. *Curae*, pp. 117–18.
39. P. Hondius, *Dapes inemptae, oft de Mouffe-Schans, dat is de soeticheyt des buyten-levens, vergeselschapt met de boecken* (n.p., *c.* 1619).
40. Considering his stay in Malines, 1546 may be the more likely year of his birth. See Hoghelande, 27 October 1583, 13 May 1592; Hunger, *Charles de l'Escluse*, vol. 1, pp. 174, 184; and Meertens, *Letterkundig leven*, pp. 21, 23, 307.
41. Hoghelande, 18 March 1592: in the postscript he remarks that Clusius and he did indeed belong to different creeds, since he (Hoghelande) had stayed with the 'old religion'.
42. Meertens, *Letterkundig leven*, p. 439. Three of Theobald's letters to Clusius (1591–2) are extant, which discuss postal problems and a few plants and gardens.
43. Hoghelande's ten letters to Clusius cover the years 1583 and 1591–3; on exchanges of plants in 1588–9, cf. *Rariorum*, pp. 114, 249.
44. Clusius to Caccini, 25 September 1607, 6 November 1608, in Ginori Conti, *Lettere inedite*, pp. 43, 87.
45. The gardens are mentioned in several of Hoghelande's letters to Clusius, and in Clusius's letters to Caccini (i.e. Ginori Conti, *Lettere inedite*, p. 43).
46. See Hoghelande, 11 March 1591, 18 March 1592, 20 May 1592, 30 May 1593. On the anemones, see note 16 above.
47. Clusius to Caccini, 25 September 1607, in Ginori Conti, *Lettere inedite*, p. 43.
48. As already suggested in Van der Gouw, 'Marie de Brimeu'.
49. For these visits, see Hoghelande, 9 June 1591, 14 August 1591, 13 May 1592, 30 May 1593, 20 May 1592.
50. See Hopper, 'Clusius' World', pp. 16, 19–20; she does not mention a source for de Gheijn's visit.
51. See Hoghelande, 14 August 1591, 9 June 1591, 18 March 1592.
52. Clusius refers to Hoghelande in *Rariorum* (30x), *Exoticorum* (12x) and *Curae* (10x). For the two Cretan plants see *Rariorum*, pp. 323, lxxxiv.
53. See Hoghelande, 14 August 1591, 11 July 1592.
54. For seeds of a Peruvian tree, see Hoghelande, 27 October 1583.
55. H. Duits and T. van Strien (eds), *Een intellectuele activist: Studies over leven en werk van Philips van Marnix van Sint Aldegonde* (Hilversum: Verloren, 2001), esp. pp. 9–22; on Marnix and plants, see W. de Vries, 'Marnix als intellectuele tuinliefhebber', in Duits and Van Strien (eds), *Een intellectuele activist*, pp. 87–95.
56. Ten letters in French by Marnix to Van der Meulen (1593–7, none of them published in its entirety), are in Leiden Regionaal Archief inventory 96, no. 306; the majority was sent from the castle West-Souburg. Marnix's ten extant letters to Clusius (1581–94; all in Latin) were also mostly sent from West-Souburg.
57. De Vries, 'Marnix', p. 87.

58. Mainly economic historians have been interested in Van der Meulen and not much attention has been paid to his cultural role and contacts with the university in Leiden. Exceptions are J. Bangs and G. Winius, 'Daniël van der Meulen (= 1600), First Midas of Holland's Golden Age', in *Atti del II° Congresso Internazionale di studi storici: Rapporti Genova – Mediterraneo – Atlantico – nell'età moderna* (Genoa: Istituto di Scienze Storiche, 1985), pp. 345–71 (I have consulted the typescript version); and G. Jongbloet-Van Houtte (ed.), *Brieven en andere bescheiden betreffende Daniël van der Meulen, 1584–1600* (The Hague: Nijhoff, 1986), pp. xv–lxxxiv. Cf. A. Versprille, 'De geschiedenis van het huis van Daniël van der Meulen (Rapenburg 19)', *Leids Jaarboekje*, 35 (1943), pp. 158–69; A. Versprille, 'Hester della Faille', *Leids Jaarboekje*, 67 (1975), pp. 77–97; and J. Verburgt, 'Uit het particuliere leven van Daniël van der Meulen, heer van Ranst', *Leids Jaarboekje*, 29 (1936–7), pp. 73–88.
59. Bangs and Winius, 'Daniël van der Meulen'.
60. Versprille, 'De geschiedenis', pp. 158–63; and Versprille, 'Hester della Faille', esp. p. 87.
61. Bangs and Winius, 'Daniël van der Meulen'.
62. Vorst (1565–1624), who succeeded De Bondt (1536–99) in 1599 as professor of medical botany, had a great interest in exotic animals. See *Exoticorum*, pp. 96, 127–8, 144, 322. See Hunger, *Charles de l'Escluse*, vol. 1, pp. 249, 268, 307–11; Versprille, 'Hester della Faille', pp. 85–6; and Jorink, *Het Boeck der Natuere*, pp. 274–5.
63. See Pinelli, 24 June 1598, 13 September 1600. Cf. Jongbloet-Van Houtte ed., *Brieven en andere bescheiden*, p. lxxx.
64. Bangs and Winius, 'Daniël van der Meulen'.
65. At the Muiderslot (1609); Härting (ed.), *Gärten und Höfe*, p. 19.
66. Out of 90 Amsterdam collectors after 1600, 21 had a garden; see J. van der Veen, 'Dit klain vertrek bevat een weereld vol gewoel: Negentig Amsterdammers en hun kabinetten', in Kistemaker and Bergvelt (eds), *De wereld binnen handbereik*, pp. 223–58, on pp. 236, 250. More generally on Dutch collections, see R. van Gelder, 'De wereld binnen handbereik: Nederlandse kunst- en rariteitenverzamelingen, 1585–1735', in Kistemaker and Bergvelt (eds), *De wereld binnen handbereik*, pp. 15–38. Goldgar (*Tulipomania*, esp. pp. 54–6) also points to the importance of émigrés for flower and gardening culture in the Northern Netherlands.
67. E. Schaap, *Dutch Floral Tiles in the Golden Age and their Botanical Prints / Bloemen op tegels in de Gouden Eeuw: Van prent tot tegel* (Haarlem: Becht, 1994), pp. 1–24.
68. The complete herbarium can be consulted on the website of the Dutch Nationaal Herbarium: http://www.bio.uu.nl/~herba/Cade/ [accessed November 2009].
69. Coenen's complete *Visboeck* can be consulted on the website of the Koninklijke Bibliotheek: http://www.kb.nl/galerie/visboek/visboek.html [accessed November 2009]. On Coenen, see F. Egmond and P. Mason, '"These are People who Eat Raw Fish". Contours of the Ethnographic Imagination in the Sixteenth Century', *Viator*, 31 (2000), pp. 311–60; F. Egmond and P. Mason (eds), *The Whale Book: Whales and other Marine Animals as Described by Adriaen Coenen in 1585* (London: Reaktion Books, 2003); and F. Egmond, *Het Visboek: De wereld volgens Adriaen Coenen* (Zutphen: Walburg Pers, 2005).

11 Interlocking Networks

1. See J. Garet Sr, 4 April 1592, 27 July 1592. Coloquint probably refers to the bitter apple or wild gourd.
2. Dated references to them in Clusius's works span the period 1581–1601. Until Harkness (*The Jewel House*, pp. 15–56), little more than a few paragraphs could be found about

them, mainly in works on the drugs trade and the history of pharmacy. See, on Pieter, D. Wittop-Koning, *De handel in geneesmiddelen te Amsterdam tot omstreeks 1637* (Purmerend: J. Muusses, 1942); and D. Wittop-Koning, 'De voorgeschiedenis van het collegium medicum (1638) te Amsterdam', *Jaarboek Amstelodamum*, 41 (1947), pp. 51–63. On James Jr, see R. Gunther, *Early British Botanists and their Gardens* (Oxford: Oxford University Press, 1922), pp. 57, 237; Raven, *English Naturalists*, pp. 136, 170, 172, 185, 192–3; R. Roberts, 'The London Apothecaries and Medical Practice in Tudor and Stuart England' (PhD dissertation, University of London, 1964), p. 271; L. Matthews, 'Herbals and Formularies', in F. Poynter (ed.), *The Evolution of Pharmacy in Britain* (London: Pitman Medical, 1965), pp. 187–213, on pp. 196–7; B. Henrey, *British Botanical and Horticultural Literature before 1800*, 3 vols (Oxford: Oxford University Press, 1975), vol. 1, pp. 43–4; M. Pelling and C. Webster, 'Medical Practitioners', in Webster (ed.), *Health, Medicine and Mortality*, pp. 165–235, on p. 178; O. Boersma, *Vluchtig voorbeeld: De Nederlandse, Franse en Italiaanse vluchtelingenkerken in Londen, 1568–1585* (Kampen: Theologische Universiteit, 1994), pp. 140, 144. The information concerning the Garets presented here is based on this literature, their letters to Clusius and archival sources in Amsterdam, London and Antwerp.

3. F. Guerra, 'Drugs from the Indies and the Political Economy of the Sixteenth Century', *Analecta Medico-Historia*, 1 (1966), pp. 29–54. Cf. K. Bartels, 'Exotica et Transmarina: Arzneidrogen-handel und Apotheke in Mittelalter und früher Neuzeit', in C. Friedrich and S. Bernschneider-Reif (eds), *Rosarium Litterarum: Beiträge zur Pharmazie- und Wissenschaftsgeschichte* (Eschborn: Govi-Verlag, 2003), pp. 33–48; and R. Roberts, 'The Early History of the Import of Drugs into England', in Poynter (ed.), *The Evolution of Pharmacy*, pp. 165–86.
4. The shift of the centre of the drugs markets from Venice to Antwerp also entailed a transfer in scientific printing. Works of Garcia da Orta, Cristobal Acosta and Nicolàs Monardes (the first two on Asiatic, the last on American plants and drugs) were 'feverishly translated and reprinted in both drugs centres, Venice and Antwerp', and the translator of all three (into Latin) was Clusius who, according to Guerra, played 'the dual role of scientist and disseminator of economic information' ('Drugs from the Indies', p. 51).
5. See Stadsarchief Antwerp, Poorterboek (1545); and Schepenregisters, Certificatieboeken nos 7–31 (1551–69), in which records of Garet Sr's transactions can be found in almost every single year during 1551–69.
6. For dates of residence, London addresses, marriages, professions, payments and church membership of the Garets, see A. Jelsma and O. Boersma (eds), *Acta van het consistorie van de Nederlandse gemeente te London 1569–1585* (The Hague: ING, 1993); R. Kirk and E. Kirk (eds), *Returns of Aliens, Dwelling in the City and Suburbs of London, from the Reign of Henry VIII to that of James I*, 4 vols (Aberdeen: University Press, 1900–8), vol. 2; W. Moens (ed.), *The Marriage, Baptismal, and Burial Registers 1571 to 1874 and Monumental Inscriptions of the Dutch Reformed Church, Austin Friars* (Lymington: King, 1884); Raven, *English Naturalists*; and J. Hessels (ed.), *Ecclesiae Londino-Batavae archivum*, 4 vols (Cambridge: Typis Academiae, 1887–98). On Protestant émigré communities, see A. Pettegree, *Foreign Protestant Communities in Sixteenth-Century London* (Oxford: Clarendon Press, 1986).
7. J. Garet Jr to Daniël van der Meulen, 10 September 1593, Leiden Regionaal Archief, 96, inv. no. 390.
8. There are nine letters from J. Garet Jr to Clusius (1583–91).

9. James Jr had no children from his first marriage. After the death of his first wife (*c.* 1582) he married Jacquemijne (also Jacamine) Waingon, the young widow of another apothecary from the Southern Netherlands in 1583. See J. Garet Jr, 26 January 1583; and Jelsma and Boersma, *Acta van het consistorie*. For the garden, see Raven, *English Naturalists*, p. 170, 192. Cf. P. Wallis, 'Apothecaries and the Consumption and Retailing of Medicines in Early Modern London', in L. H. Curth (ed.), *From Physick to Pharmacology: Five Hundred Years of British Drug Retailing* (Ashgate: Aldershot, 2006), pp. 13–27.
10. Harkness, *The Jewel House*, pp. 15–56; on opium, personal communication with Harkness.
11. For the apprentices, see Roberts, 'The London Apothecaries', p. 271. About his son and Rivière, see J. Garet Jr, 9 December 1591. On Rivière (a friend of Vertunien in Poitiers), see H. Trevor-Roper, 'The Sieur de la Rivière: Paracelsian Physician of Henry IV', in A. Debus (ed.), *Science, Medicine and Society in the Renaissance: Essays to Honor Walter Pagel*, 2 vols (New York: Heinemann, 1972), vol. 2, pp. 227–50.
12. Last will of James Garet Jr (6 January 1609; probate was granted on 19 April 1610), National Archives, Public Record Office (PRO), Kew, Surrey, PROB 11/115). James's wife Jacamine survived him for almost 30 years (for her will and probate of 8 March 1640, see PRO, PROB 11/182).
13. He married Abigail Bucer from Sandwich, probably a daughter of the Dutch Reformed minister at Sandwich. On Pieter's marriage and children born in London, see Kirk and Kirk (eds), *Returns of Aliens*, vol. 2; Moens (ed.), *The Marriage, Baptismal, and Burial Registers*. On his Amsterdam citizenship, family information and burial, see Gemeentearchief Amsterdam, Poorterboek B, p. 253; Registers of Births, Marriages and Death of the Dutch Reformed Church in Amsterdam. On his houses, plots, gardens and sugar refinery, see Gemeentearchief Amsterdam, Notarial Archives, no. 33, p. 222; no, 54, p. 640; no. 55, p. 10; no. 34, p. 458; Kwijtscheldingen, no. 24, p. 119v; no. 30, p. 370v; no. 32, p. 135; no. 40, p. 90; Schepenkamer, no. 14, p. 28v; Huisverkoop Weeskamer VII, pp. 126, 141.
14. On the apprentices, see Gemeentearchief Amsterdam, Notarial Archives, no. 376, pp. 233, 388, 348; and on visiting the ill, Notarial Archives, no. 358, p. 357; further see Wittop-Koning, *De handel* and 'De voorgeschiedenis'.
15. Harkness, *The Jewel House*, pp. 15–56. On London (court) apothecaries and physicians, see R. Roberts, 'The Personnel and Practice of Medicine in Tudor and Stuart England. Part I: The Provinces', *Medical History*, 6 (1962), pp. 362–82; R. Roberts, 'The Personnel and Practice of Medicine in Tudor and Stuart England. Part II: London', *Medical History*, 8 (1964), pp. 217–34; and Roberts, 'The London Apothecaries'; F. Dawbarn, 'Patronage and Power: The College of Physicians and the Jacobean Court', *British Journal for the History of Science*, 31 (1998), pp. 1–19.
16. Clusius was in touch with Morgan, Lobel, Rich, Penny, Moffet, Le Myre and Garth. We know of no direct contacts with Cole, Gerard and Harvey.
17. See Harkness, *The Jewel House*, pp. 15–56, also with references to the older literature on the Gerard scandal.
18. L. Matthews, *The Royal Apothecaries* (London: Wellcome, 1967), p. 75.
19. Ibid., pp. 74–7; L. Matthews, 'Royal Apothecaries of the Tudor Period', *Medical History*, 8 (1964), pp. 170–80, on pp. 177–9.
20. Raven, *English Naturalists*, p. 116, discussing Lobel.
21. Matthews, *The Royal Apothecaries*, p. 77; for the medicine chest see J. Keevil, *Medicine and the Navy 1200–1900*, 4 vols (Edinburgh: Livingstone, 1957–63), vol. 1, p. 198.

22. There are 15 references to Morgan in Clusius's *Exoticorum*, and 2 in his *Rariorum*. The only plant in Morgan's garden mentioned by Clusius is a 'hieracium' (*Rariorum*, p. cxl).
23. Raven, *English Naturalists*, esp. pp. 153–71; cf. F. Hoeniger and J. Hoeniger, *The Development of Natural History in Tudor England* (Charlottesville, VA: University Press of Virginia, 1969), pp. 46–8; M. Welti, 'English-baslerische Beziehungen zur Zeit der Renaissance in der Medizin, den Naturwissenschaften und der Naturphilosophie', *Gesnerus*, 20 (1963), pp. 105–30; and Roberts, 'The London Apothecaries', pp. 55–6.
24. Raven, *English Naturalists*, p. 185.
25. Ibid., pp. 162–3. Clusius refers 14 times to Penny in his *Rariorum*, and 3 times in the *Exoticorum*, exactly the reverse of Morgan. For the cornus suecia (Chamaepericlymenum), see *Rariorum*, pp. 59–60.
26. On Moffett, see Harkness, *The Jewel House*, pp. 15–56; cf. Hannay, '"How I These Studies Prize"', pp. 112–14. See also Penny, 31 March 1583; and Moffet, 23 March 1590.
27. Raven, *English Naturalists*, pp. 154–5.
28. This English translation is quoted from Raven, *English Naturalists*, p. 154. On the Chamaemorus anglicana, cf. *Rariorum*, p. 118.
29. Michaeli may have originally come from Lucca. See R. Pollitt, 'The Wounding and Treatment of William of Orange in 1582', *Journal of the History of Medicine and Allied Sciences*, 37 (1982), pp. 135–58, on pp. 142, 155.
30. Hunger, *Charles de l'Escluse*, vol. 2, pp. 183–4, 438, letter from Clusius to Camerarius, 15 August 1592, which refers to a (no longer extant) letter from James Garet Jr to Clusius (30 July 1592). On Cherler, see Raven, *English Naturalists*, p. 192; for Platter (half-brother of the more famous Felix), see Welti, 'English-baslerische Beziehungen', p. 118.
31. Jones, *William Turner*; M. Rydén, 'William Turner the Botanist in a European Context', in G. Sorelius and M. Srigley (eds), *Cultural Exchange between European Nations during the Renaissance* (Uppsala: Uppsala University, 1994), pp. 173–83; and Raven, *English Naturalists*, esp. pp. 72–93.
32. See Leith-Ross and McBurney, *The Fiorilegium*. It contains pictures of flowers, insects and some birds. Marshal owned a copy of Moffett's book on insects to which he added his own drawings of insects.
33. See esp. R. Strong, *The Renaissance Garden in England* (London: Thames & Hudson, 1979); and R. Strong, 'The Renaissance Garden in England Reconsidered: A Survey of Two Decades of Research on the Period 1485–1642', *Garden History*, 27 (1999), pp. 2–9. He explicitly does not deal with 'practicalities of horticulture'.
34. Strong, *The Renaissance Garden*; Henrey, *British Botanical and Horticultural Literature*, vol. 1, pp. 38–9, 71–2; on Turner's garden, see Gunther, *Early British Botanists*, p. 247. On Theobalds, see M. Andrews, 'Theobalds Palace: The Gardens and Park', *Garden History*, 21 (1993), pp. 129–49. Gerard dedicated his *Herball* to William Cecil.
35. For the Dutch plant seller and nurseryman Emanuel Sweert(s) and his contacts with the aristocratic plant seller Matteo Caccini in Florence, see Zalum Cardon, *Passione e cultura dei fiori*, pp. 71–3.
36. In his letter of 22 October 1597 Zouche apologizes for the infrequency of contact. Zouche is reputed to have been obsessed by horticulture and to have spent the greater part of his fortune on plants and his garden; see Strong, *The Renaissance Garden*, p. 69.
37. See L. Le Myre, 20 February 1597. Cf. Zouche, 22 October 1597.
38. Hannay, '"How I These Studies Prize"', pp. 110–14.
39. See note 7 to Chapter 3, above, p. 231.

40. Harkness, *The Jewel House*; Cook, *Matters of Exchange*; cf. H. Cook, 'The Cutting Edge of a Revolution? Medicine and Natural History near the Shores of the North Sea', in Field and James (eds), *Renaissance and Revolution*, pp. 45–61.
41. See esp. Pumfrey and Dawbarn, 'Science and Patronage'.
42. See O. Gelderblom, 'From Antwerp to Amsterdam: The Contribution of Merchants from the Southern Netherlands to the Rise of the Amsterdam Market', *Review: A Journal of the Fernand Braudel Center for the Study of Economies, Historical Systems, and Civilizations*, 26 (2003), pp. 247–82; and M. Carasso-Kok (ed.), *Geschiedenis van Amsterdam. I: Een stad uit het niets: tot 1578* (Amsterdam: SUN, 2004), esp. p. 42.
43. Pieter's letters to Clusius cover the years 1601–5, but dated references to Pieter in the *Exoticorum* also include 1596 and 1597.
44. De Landtsheer, 'Justus Lipsius and Carolus Clusius', p. 290.
45. On birds of paradise in early European collections, see Mason, *Before Disenchantment*, pp. 134–6. Cf. Van Gelder, 'Paradijsvogels in Enkhuizen'.
46. See http://www.genealogieonline.nl/antwerpse-tak-van-os/ [accessed October 2009].
47. For Coolmans (also Colius), the 1599–1601 journey and the Paauw request, see J. Heniger, 'De Eerste Nederlandse wetenschappelijke reis naar Oost-Indië, 1599–1601', *Jaarboekje voor geschiedenis en oudheidkunde van Leiden en Omstreken*, 65 (1973), pp. 27–49; and Cook, *Matters of Exchange*, p. 129. Cf. *Exoticorum*, pp. 88, 89, 90, 91, and 92; and Van Os, 14 December 1601. On Hortensius see *Exoticorum*, pp. 14, 20, 78.
48. For the full text of Clusius's memorandum, see Heniger, 'De Eerste Nederlandse wetenschappelijke reis', pp. 44–5.
49. See esp. *Exoticorum*, Preface. Pinelli (24 October 1600) wrote that he was sorry to hear how badly Clusius had been served. On the role of the VOC with respect to scientific and botanical research in later centuries, see K. van Berkel 'Een onwillige Mecenas? De rol van de VOC bij het natuurwetenschappelijk onderzoek in de 17e eeuw', in J. Bethlehem and A. Meijer (eds), *VOC en Cultuur: Wetenschappelijk onderzoek en culturele relaties tussen Europa en Azië ten tijde van de VOC* (Amsterdam: Schiphouwer & Brinkman, 1993), pp. 39–58; and P. Baas, 'De VOC in Flora's lusthoven', in L. Blussé and I. Ooms (eds), *Kennis en Compagnie: De Verenigde Oost-Indische Compagnie en de moderne wetenschap* (Amsterdam: Balans, 2002), pp. 124–37.
50. See Heniger, 'De Eerste Nederlandse wetenschappelijke reis', pp. 37–9.
51. See E. Moes, 'David Mostart en zijn album amicorum', *Amsterdamsch Jaarboekje* (1904), pp. 29–41.
52. See Chapter 10. Carasso-Kok (*Geschiedenis van Amsterdam. I*, pp. 371–83) mentions no Amsterdam collections dating from before the 1620s.
53. An unusually large number of portraits of him have been preserved. See E. Reznicek, 'De achtste tronie van de schelpenverzamelaar', in A. Logan (ed.), *Essays in Northern European Art Presented to Egbert Haverkamp-Begemann* (Doornspijk: Davaco, 1983), pp. 209–12; L. Nichols, 'Jan Govertz van der Aar: On the Identification of Goltzius's Patron', *Nederlands Kunsthistorisch Jaarboek*, 38 (1987), pp. 241–55. With thanks to Roelof van Gelder for these references.
54. *Exoticorum*, pp. 25, 71.
55. For the sloth, see *Exoticorum*, pp. 110–11; and Mason, *Before Disenchantment*, pp. 131–4. For the exotica and bulbs sent by Coornhert, see *Exoticorum*, pp. 4, 129; and *Rariorum*, pp. 152, 227, cclviii.

12 Spanning the World

1. See J. Garet Jr, 26 January 1583 (Old Style). Clusius visited England several times; I have not found any evidence that he met Francis Drake in person.
2. See J. Garet Jr, 28 July 1589. The ships of this expedition failed to get through the Strait of Magellan, were forced to return to England, and finally shipwrecked off the coast of Normandy in 1590. An angelot was a French gold coin with the device of the archangel Michael piercing the dragon.
3. On Cavendish's feats, including the capture of the Spanish ship *Santa Ana* (laden with gold and precious cargo from Asia) off the west coast of South America, see 'The Admirable and Prosperous Voyage of the Worshipful Mr. Thomas Cavendish by Mr. Francis Pretty', in R. Hakluyt, *The Principall Navigations, Voiages, and Discoveries of the English Nation*, 2nd expanded edn (London: G. Bishop, 1598–1600).
4. On star anise and 'oregioella', cf. *Exoticorum*, pp. 42, ccii–cciii. Oregioella has sometimes been identified as cacao, but Garet's description suggests that it was a substance added to a cacao drink.
5. See Garth, 8 July 1589, 20 December 1589.
6. See J. Garet Jr, 19 January 1589, 9 September 1589, 28 August 1590, 9 December 1591.
7. R. Pieper, *Die Vermittlung einer neuen Welt: Amerika im Nachrichtennetz des habsburgischen Imperiums 1493–1598* (Mainz: Philip von Zabern, 2000); and Z. Barbarics and R. Pieper, 'Handwritten Newsletters as a Means of Communication in Early Modern Europe', in Bethencourt and Egmond (eds), *Correspondence and Exchange*, pp. 53–79.
8. Copal is a type of tree resin, often used as incense in Central America. See *Exoticorum*, pp. 192, 297.
9. Pieter himself had also received a copy from Clusius.
10. Clusius (*Exoticorum*, pp. 53–4) quotes what is probably a mix of Garet's and Cuelener's descriptions.
11. On this voyage, its publication history and the dodo, see P. Moree (ed.), *Dodo's en galjoenen: De reis van het schip Gelderland naar Oost-Indië, 1601–1603* (Zutphen: Walburg Pers, 2001).
12. This information once more underscores the relevance of the work by Pamela Smith, esp. *The Body of the Artisan*, and 'Making and Knowing'.
13. *Exoticorum*, p. 358. Cf. Pona, 15 July 1603; and Pinelli, 10 November 1597.
14. For the 'slaves', see *Exoticorum*, p. 39. For Abdala Guzaratensis, see *Exoticorum*, pp. 8, 11, 15, 25, 38, 39, 57, 73; for Francisco Rodriguez, see *Exoticorum*, pp. 12, 13, 25, 233, 282, 284. For further information, see J. Pardo Tomás, 'Two Glimpses of America from a Distance: Carolus Clusius and Nicolás Monardes', in Egmond et al. (eds), *Carolus Clusius*, pp. 173–93, on, pp. 180–2. We know extremely little about their persons and background, and Clusius's own dating is occasionally contradictory.
15. *Exoticorum*, pp. 8–9, 15, 28, 39.
16. Ibid., pp. 8–9.
17. Ibid., pp. 12, 233, 282, 284.
18. On issues of proof, see, for example, J. Chandler et al. (eds), *Questions of Evidence: Proof, Practice and Persuasion Across the Disciplines* (Chicago, IL: Chicago University Press, 1994).
19. These methods are more usually connected with the Baconian 'new science' of the seventeenth century. The evidence presented here, but also, for instance, by Harkness (*The Jewel House*), Cook (*Matters of Exchange*) and Huppert (*The Style of Paris*), suggests that

claims of the 'new scientists' to have invented a new science should be studied as a highly succesful self-promotion strategy rather than as factual statements. That topic goes far beyond the limits of this study, but see also Ogilvie, *The Science of Describing*, pp. 256–61; Cook, 'The Cutting Edge of a Revolution?'; and P. Dear, 'Narratives, Anecdotes, and Experiments: Turning Experience into Science in the Seventeenth Century', in P. Dear (ed.), *The Literary Structure of Scientific Argument* (Philadelphia, PA: University of Pennsylvania Press, 1991), pp. 135–63.

20. I am not referring here to legal or medical *theories*, however important, but to judicial and medical practices (which were not merely rooted in learned discourse and theory but also incorporated popular traditions) that formed part of everyday life in early modern Europe for a large part of the population.
21. E. Sweert, *Florilegium ... tractans de variis floribus et aliis indicis plantis ad vivum delineatum in duabus partis et quatuor linguis concinnatum* (Frankfurt: A. Kempner, 1612–14). The former bulb was probably Zephyrantes atamasco, and the latter possibly the double-flowered Zephyrantes candida. For the page of Sweert's *Florilegium* with these identifications, see http://www.bulbnrose.org/Amaryllis/Sweerts/sweerts.htm [accessed November 2009].
22. For the latter two, see *Exoticorum*, pp. 120, 143. For further references to Pieter's gifts, see all of his letters to Clusius, and *Exoticorum*, pp. 7, 8, 9, 13, 15, 18, 30, 33, 34, 35, 68, 71, 74, 78, 79, 80, 81, 82, 85, 87, 120, 143, 184, 192, 297, 314.
23. I assume that '*garetse palm*' refers to the rubber tree, since the term *karet* or *garet* is still used in Indonesia for rubber.
24. See J. Garet Jr, 20 July 1591, 7 September 1601.
25. See J. Garet Jr, 26 January 1583 (Old Style). Garth gave some new South American (probably Brazilian) specimens to Clusius, who in turn gave him a root of Solomon's Seal.
26. See J. Garet Jr, 9 September 1589, 28 August 1590, 9 December 1591, 18 April 1601.
27. Pelling and Webster ('Medical Practitioners', p. 178) state that the annual value of imported drugs had not reached £1,000 by 1600, but that twenty years later it had increased to £8,000.
28. It is now in the Plantin Museum, Antwerp; the other one is a watercolour of the potato in the same museum. See De Nave and Imhof (eds), *De Botanica in de Zuidelijke Nederlanden*, p. 127, 145.
29. Clusius's own tusai in his Viennese garden had some 50 flowers in 1581; Wilhelm IV of Kassel's tusai had 40 flowers in 1580. See *Rariorum*, p. 128.
30. See also J. Garet Jr, 28 July 1589, 9 September 1589.
31. In 1588 Clusius was also corresponding about the potato with Jacques Plateau (see Plateau, 3 September 1588). The first printed botanical descriptions of the potato are by Gaspard Bauhin and date from 1596 and 1598.
32. Cf. *Rariorum*, pp. lxxix–lxxxi.
33. Cf. Hunger, *Charles de l'Escluse*, vol. 1, p. 167, vol. 2, p. 172. Clusius had received his first potatoes and some seeds of the potato plant in January 1588 from Philippe de Sivry, prefect of Mons in the Southern Netherlands, who in 1587 had received a potato plant ('taratoufli') via the legate of the Pope to the Southern Netherlands.
34. Cf. J. Garet Jr, 28 July 1589, 28 August 1590.
35. *Rariorum*, pp. lxxix–lxxxi.
36. For the specifically named items, see *Exoticorum*, esp. pp. 7, 173, 18, 16–17, 11, 84, 184, 29, 21–2, 192–3.

37. For this identification, see Ramón-Laca, 'Las plantas americanas', p. 99.
38. For the specifically named items from the New World, see *Rariorum* pp. 328, xciii, ccxvi; and *Exoticorum*, pp. 6, 23, 83, 43, 25, 57, 60–1, 291.
39. Clifford was the husband of Margaret Russell Clifford (See Chapter 11).
40. *Exoticorum*, p. 291. On mimosa, Clusius, Bauhin and Cherler, see Webster, 'The Recognition of Plant Sensitivity', p. 9. Ogilvie (*The Science of Describing*, p. 266) is not correct in stating that this plant (or its behaviour) was unknown in the Renaissance.
41. There is no evidence of contact between the Garets and John White. See K. Sloan (ed.), *European Visions: American Voices* (London: British Museum, 2009).
42. On stripping of meaning, European categories and objectification, see esp. D. Bleichmar, 'Books, Bodies, and Fields: Sixteenth-Century Transatlantic Encounters with New World Materia Medica', in Schiebinger and Swan (eds), *Colonial Botany*, pp. 83–99, speaking mainly of Monardes; cf. L. Schiebinger and C. Swan, 'Introduction', in Schiebinger and Swan (eds), *Colonial Botany*, pp. 1–16, on p. 7; and see H. Cook, 'Global Economies and Local Knowledge in the East Indies', in Schiebinger and Swan (eds), *Colonial Botany*, pp. 100–18 (for the 1630s–40s). For different approaches, see, for example, J. Pardo Tomás and J. Lopez Pinero, *La influencia de Francisco Hernández (1515–1587) en la constitución de la botánica y de la materia médica modernas* (Valencia: Universitat de València, CSIC, 1996); and Grove, 'Indigenous Knowledge'.
43. Here I differ from Ogilvie (*The Science of Describing*, p. 14) not only in my regarding various people whom he calls outsiders as relevant to or indeed as members of the community of naturalists, but also in stressing that many of them were capable of distinguishing between and operating in both idioms.

Conclusion

1. Lobel, *Kruydtboeck*, unpaginated preface.
2. The term is also used by Ogilvie (*The Science of Describing*, pp. 51–3), but I borrow 'collective enterprise' here from Rudwick, in particular because he regards such an enterprise as not necessarily held together by a central theory ('Charles Darwin in London', esp. p. 196). Reeds speaks of a 'community of botanists' (*Botany*, pp. 7–12). Harkness emphasizes that not only those who published took part in the Republic of Letters (*The Jewel House*, esp. pp. 45–7, 343–4).
3. Throughout this book I have tried to let the examples of what people did when they experimented in the garden speak for themselves in so far as possible, and to stay clear of the discussions – important though they are – on what exactly constitutes an early modern experiment, and of those concerning the connection between experience and experiment. I have used the term in the everyday sense of the word, much as Porret did (see Chapter 10).
4. While agreeing with much of the substance of Ogilvie's essay on 'information overload' and sixteenth-century naturalists, I do not find that term itself helpful, mainly because it suggests that these naturalists were *confronted* with masses of information by developments outside themselves (e.g. voyages of discovery), while I believe (and it often seems that Ogilvie does too) that they *themselves produced* vast amounts of detailed descriptive information given their involvement in the collective enterprise discussed here. See B. Ogilvie 'The Many Books of Nature: Renaissance Naturalists and Information Overload', *Journal of the History of Ideas*, 64 (2003), pp. 29–40.

5. In this respect I put more emphasis on construction than on circulation-reception-transfer, but the issues are much the same as those dealt with in Secord, 'Knowledge in Transit'. See also Smith, 'Art, Science and Visual Culture', who presents a similar argument for the role of artists with respect to the sciences.
6. Rudwick, 'Charles Darwin in London'.
7. Ibid., esp. pp. 190–2.
8. I am wary of attributing natural philosophies to practitioners of either the learned or the not so learned kind unless they themselves made explicit remarks about them.
9. See Huppert, *The Style of Paris*. For stimulating ideas on the connection between first-person observation, autoptic proof and the visual culture of early modern science, see Smith, 'Art, Science and Visual Culture', esp. pp. 89–91.
10. I am basing this on Clusius's printed works, his letters to Matteo Caccini (published in Ginori Conti, *Letter inedite*), and the general assessments of Clusius's work by Hunger, *Charles de l'Escluse*, vol. 1; Lewis, 'Clusius in Montpellier', esp. pp. 91, 97; Ogilvie, *The Science of Describing*, *passim*, but esp. pp. 182–92; and Cook, *Matters of Exchange*, pp. 90–132. On the especially important topic of conveying reliability via image and text, see S. Kusukawa, 'Uses of Pictures in Printed Books: The Case of Clusius's *Exoticorum libri decem*', in Egmond et al. (eds), *Carolus Clusius*, pp. 221–46.
11. The parallels with the styles of reporting called characteristic of the scientific revolution are striking; see esp. P. Dear, '*Totius in Verba*: Rhetoric and Authority in the Early Royal Society', *Isis*, 76 (1985), pp. 144–61. Cf. note 19 to Chapter 12, above, pp. 257–8.
12. Hopper, 'Clusius' World', esp. pp. 13–29. On Clusius and aesthetics, see also Ogilvie, *The Science of Describing*, p. 186.
13. Mason, *Before Disenchantment*, esp. pp. 124–48.
14. In his *Revolutionizing the Sciences: European Knowledge and Its Ambitions, 1500–1700* (Basingstoke: Palgrave, 2001), Peter Dear has argued for a two-phase model of early modern science: a *Scientific Renaissance* of the fifteenth and sixteenth centuries, focused on the restoration of the natural knowledge of the ancients; and a *Scientific Revolution* of the seventeenth century, when scientists shifted from recovery to innovation. Though useful as a very schematic characterization, this does pull apart what coexisted in many naturalists of the sixteenth century.
15. Cf. Palmer ('Medical Botany', p. 156), who uses the term backward-looking for sixteenth-century Italian botany. It is meant here in a literal sense, not as a negative qualification.

WORKS CITED

Acidini Luchinat, C. (ed.), *Giardini Medicei: Giardini di palazzo e di villa nella Firenze del Quattrocento* (Milan: Federico Motta, 2000).

Alessandrini, A., and A. Ceregato (eds), *Natura Picta: Ulisse Aldrovandi* (Bologna: Editrice Compositori, 2007).

Alpini, P., *De Plantis Aegypti Liber* (Venice: De Francisis, 1592).

Amico, L., *A la recherche du paradis terrestre: Bernard Palissy et ses continuateurs* (Paris: Flammarion, 1996).

Andrews, M., 'Theobalds Palace: The Gardens and Park', *Garden History*, 21 (1993), pp. 129–49.

Azzi Visentini, M., *L'orto botanico di Padova e il giardino del Rinascimento* (Milan: Il Polifilo, 1984).

—, 'The Gardens of Villas in the Veneto from the 15th to the 18th Centuries', in J. D. Hunt (ed.), *The Italian Garden: Art, Design, and Culture* (Cambridge: Cambridge University Press, 1996), pp. 93–126.

Azzolini, M., 'Leonardo da Vinci's Anatomical Studies in Milan: A Re-Examination of Sites and Sources', in J. A. Givens, K. Reeds and A. Touwaide (eds), *Visualizing Medieval Medicine and Natural History, 1200–1550* (Aldershot: Ashgate, 2006), pp. 147–76.

Baas, P., 'De VOC in Flora's lusthoven', in L. Blussé and I. Ooms (eds), *Kennis en Compagnie: De Verenigde Oost-Indische Compagnie en de moderne wetenschap* (Amsterdam: Balans, 2002), pp. 124–37.

Backer, A., 'Tuinkunst tijdens de Nederlandse Opstand: De prinses van Chimay: Marie de Brimeu en de "humanisering" van de bloem 1550–1605', *Cascade Bulletin voor Tuinhistorie*, 15 (2006), pp. 14–39.

Baldriga, I., 'The Influence of Clusius in Italy: Federico Cesi and the Accademia dei Lincei', in Egmond et al. (eds), *Carolus Clusius*, pp. 249–65.

Balis, J. (ed.), *Hortus belgicus* (Brussels: Bibliothèque Albert I, 1962).

Bangs, J., and G. Winius, 'Daniël van der Meulen (= 1600), First Midas of Holland's Golden Age', in *Atti del II° Congresso Internazionale di studi storici: Rapporti Genova – Mediterraneo – Atlantico – nell'età moderna* (Genoa: Istituto di Scienze Storiche, 1985), pp. 345–71.

Barbarics, Z., and R. Pieper, 'Handwritten Newsletters as a Means of Communication in Early Modern Europe', in Bethencourt and Egmond (eds), *Correspondence and Exchange*, pp. 53–79.

Barisi, I., M. Fagiolo and M. Madonna, *Villa d'Este* (Rome: De Luca Editori, 2003).

Barona, J. L., 'Clusius' Exchange of Botanical Information with Spanish Scholars', in Egmond et al. (eds), *Carolus Clusius*, pp. 99–116.

Barona, J. L., and X. Gómez i Font, *La correspondencia de Carolus Clusius con los científicos españoles* (Valencia: Universitat de Valencia, 1998).

Barrera Osorio, A., *Experiencing Nature: The Spanish American Empire and the Early Scientific Revolution* (Austin, TX: University of Texas Press, 2006).

Bartels, K., 'Exotica et Transmarina: Arzneidrogen-handel und Apotheke in Mittelalter und früher Neuzeit', in C. Friedrich and S. Bernschneider-Reif (eds), *Rosarium Litterarum: Beiträge zur Pharmazie- und Wissenschaftsgeschichte* (Eschborn: Govi-Verlag, 2003), pp. 33–48.

Bassani Pacht, P., et al., *Marie de Médici, un gouvernement par les arts* (Paris: Somogy editions, 2003).

Bath, M., *Emblems for a Queen: The Needlework of Mary Queen of Scots* (London: Archetype Publications, 2008).

Belloni Speciale, G., 'La ricerca botanica dei Lincei a Napoli: Corrispondenti e luoghi', in Lomonaco and Torrini (eds), *Galileo a Napoli*, pp. 59–79.

Berendts, A., 'Carolus Clusius (1526–1609) and Bernardus Paludanus (1550–1633): Their Contacts and Correspondence', *Lias*, 5 (1978), pp. 49–64.

Berkel, K. van, 'Citaten uit het Boek der Natuur: Zeventiende-eeuwse Nederlandse naturalienkabinetten en de ontwikkeling van de natuurwetenschap', in Kistemaker and Bergvelt (eds), *De wereld binnen handbereik*, pp. 169–91.

—, 'Een onwillige Mecenas? De rol van de VOC bij het natuurwetenschappelijk onderzoek in de 17e eeuw', in J. Bethlehem and A. Meijer (eds), *VOC en Cultuur: Wetenschappelijk onderzoek en culturele relaties tussen Europa en Azië ten tijde van de VOC* (Amsterdam: Schiphouwer & Brinkman, 1993), pp. 39–58.

Berkvens-Stevelinck, C., H. Bots and J. Häseler (eds), *Les grands intermédiaires culturels de la République des Lettres: Etudes de réseaux de correspondance du XVIe au XVIII siècles* (Paris: Honoré Champion, 2005).

Beschi, L., *Onorio Belli Accademico Olimpico: Scritti d' Antiquaria e Botanica 1586–1602* (Rome: Viella, 2000).

Bethencourt, F., and F. Egmond (eds), *Correspondence and Exchange in Europe, 1400–1700* (Cambridge: Cambridge University Press, 2007).

Biagioli, M., 'Scientific Revolution, Social Bricolage, and Etiquette', in R. Porter and M. Teich (eds), *The Scientific Revolution in National Context* (Cambridge: Cambridge University Press, 1992), pp. 11–54.

—, *Galileo Courtier: The Practice of Science in the Age of Absolutism* (Chicago, IL: University of Chicago Press, 1993).

—, 'Etiquette, Interdependence, and Sociability in Seventeenth-Century Science', *Critical Inquiry*, 22 (1996), pp. 193–238.

Bleichmar, D., 'Books, Bodies, and Fields: Sixteenth-Century Transatlantic Encounters with New World Materia Medica', in Schiebinger and Swan (eds), *Colonial Botany*, pp. 83–99.

Blunt, W., and W. Stearn, *The Art of Botanical Illustration* (Woodbridge: Antique Collectors Club, 1994).

Bobory, D., *The Sword and the Crucible: Boldizsár Batthyány and Natural Philosophy in Sixteenth-Century Hungary* (Cambridge: Cambridge Scholars Publishing, 2009).

Boersma, O., *Vluchtig voorbeeld: De Nederlandse, Franse en Italiaanse vluchtelingenkerken in Londen, 1568–1585* (Kampen: Theologische Universiteit, 1994).

Boogaart, E. van den, *Civil and Corrupt Asia: Image and Text in the 'Itinerario' and the 'Icones' of Jan Huygen van Linschoten* (Chicago, IL: University of Chicago Press, 2003).

Boogert, B. van den, and J. Kerkhoff, *Maria van Hongarije: Koningin tussen keizers en kunstenaars* (Zwolle: Waanders, 1993).

Bosman-Jelgersma, H., 'Dirck Outgaertsz Cluyt', *Farmaceutisch Tijdschrift voor België*, 53 (1976), pp. 525–48.

—, 'Augerius Clutius (1578–1636), apotheker, botanicus en geneeskundige', *Farmaceutisch Tijdschrift voor België*, 56 (1982), pp. 167–74.

—, 'De lotgevallen van een apothekersleerling in het 17de-eeuwse Leiden', *Leids Jaarboekje*, 79 (1987), pp. 62–81.

—, 'Dirck Cluyt: De eerste Leidse hortulanus', *Leids Jaarboekje*, 83 (1991), pp. 74–88.

— (ed.), *Pieter van Foreest; de Hollandse Hippocrates* (Heiloo: Vereniging Oud Heiloo, 1996).

Bots, H., and F. Waquet, *La République des Lettres* (Paris: Belin / De Boeck, 1997).

Bouza Alvarez, F., 'Letters and Portraits: Economy of Time and Chivalrous Service in Courtly Culture', in Bethencourt and Egmond (eds), *Correspondence and Exchange*, pp. 145–62.

Browne, J., *Charles Darwin: Voyaging* (London: Jonathan Cape, 1995).

—, *Charles Darwin: The Power of Place* (London: Jonathan Cape, 2002).

Bruijn, J., 'Scheepvaart en overheid omstreeks 1600', in Lesger and Noordegraaf (eds), *Ondernemers en Bestuurders*, pp. 77–83.

Bussadori, P., 'Gli orti botanici privati padovani', in Bussi (ed.), *Di Sana Pianta*, pp. 47–54.

Bussi, R. (ed.), *Di Sana Pianta: Erbari e taccuini di sanità: Le radici storiche della nuova farmacologia* (Padua: Panini, 1988).

Campitelli, A., *Gli Horti dei Papi: I giardini Vaticani dal Medioevo al Novecento* (Milan: Jaca Book / Musei Vaticani, 2009).

Cañizares-Esguerra, J., 'Iberian Science in the Renaissance: Ignored How Much Longer', *Perspectives on Science*, 12 (2004), pp. 86–124.

—, 'Iberian Colonial Science', *Isis*, 96 (2005), pp. 64–70.

Cap, M. de, 'Un apothicaire belge au XVIe siècle: Pierre Coudenberg', *Journal de Pharmacie et de Chimie*, 3rd series, 39 (1861), pp. 448–57.

Cappelletti, E., 'Plants Cultivated at the time of Aguillara', in Minelli (ed.), *The Botanical Garden of Padua*, pp. 163–71.

Carasso-Kok, M. (ed.), *Geschiedenis van Amsterdam. I: Een stad uit het niets: tot 1578* (Amsterdam: SUN, 2004).

Chandler, J., et al. (eds), *Questions of Evidence: Proof, Practice and Persuasion Across the Disciplines* (Chicago, IL: Chicago University Press 1994).

Cima, M., and E. Talamo, *Gli horti di Roma antica* (Rome: Musei Capitolini / Mondadori-Electa, 2008).

Clusius, C., *Rariorum plantarum historia* (Antwerp: Officina Plantiniana, 1601).

—, *Exoticorum libri decem: quibus animalium, plantarum, aromatum, aliorumque peregrinorum fructuum historiae describuntur* (Leiden: Officina Plantiniana, 1605).

—, *Curae posteriores, seu plurimarum non antè cognitarum, aut descriptarum stirpium, peregrinorumque aliquot animalium novae descriptiones* (Leiden & Antwerp: Officina Plantiniana, 1611).

Cobb, R., *A Sense of Place* (London: Duckworth, 1975).

Coffin, D., *Gardens and Gardening in Papal Rome* (Princeton, NJ: Princeton University Press, 1991).

Contant, P., *Le Jardin, et cabinet poétique* (1609), ed. M. Marache-Gouraud and P. Martin (Rennes: Presses Universitaires, 2004).

Cook, H., 'The Cutting Edge of a Revolution? Medicine and Natural History near the Shores of the North Sea', in Field and James (eds), *Renaissance and Revolution*, pp. 45–61.

—, 'Global Economies and Local Knowledge in the East Indies', in Schiebinger and Swan (eds), *Colonial Botany*, pp. 100–18.

—, *Matters of Exchange: Commerce, Medicine, and Science in the Dutch Golden Age* (New Haven, CT: Yale University Press, 2007).

Coomans, H., 'Schelpenverzamelingen', in Kistemaker and Bergvelt (eds), *De wereld binnen handbereik*, pp. 192–203.

Cooper, A., *Inventing the Indigenous: Local Knowledge and Natural History in Early Modern Europe* (Cambridge: Cambridge University Press, 2007).

Cooter, R., and S. Pumfrey, 'Separate Spheres and Public Places: Reflections on the History of Science Popularization and Science in Popular Culture', *History of Science*, 32 (1994), pp. 237–67.

Dal Piaz, V., and M. Rippa Bonati, 'The Design and Form of the Padua *Horto medicinale*', in Minelli (ed.), *The Botanical Garden of Padua*, pp. 33–54.

Daston, L., and K. Park, *Wonders and the Order of Nature, 1150–1750* (New York: Zone, 1998).

Dawbarn, F., 'Patronage and Power: The College of Physicians and the Jacobean Court', *British Journal for the History of Science*, 31 (1998), pp. 1–19.

Dear, P., '*Totius in Verba*: Rhetoric and Authority in the Early Royal Society', *Isis*, 76 (1985), pp. 144–61.

—, 'Narratives, Anecdotes, and Experiments: Turning Experience into Science in the Seventeenth Century', in P. Dear (ed.), *The Literary Structure of Scientific Argument* (Philadelphia, PA: University of Pennsylvania Press, 1991), pp. 135–63.

—, *Revolutionizing the Sciences: European Knowledge and Its Ambitions, 1500–1700* (Basingstoke: Palgrave, 2001).

Dodoens (Dodonaeus), R., *Florum, et coronariarum odoratarumque nonnullarum herbarum historia* (Antwerp: Plantin 1568).

Dübber, I., *Zur Geschichte des Medizinal-und Apothekenwesens in Hessen-Kassel und Hessen-Marburg von den Anfängen bis zum Dreissigjährigen Krieg* (Marburg: Marburg University, 1969).

Duits, H., and T. van Strien (eds), *Een intellectuele activist: Studies over leven en werk van Philips van Marnix van Sint Aldegonde* (Hilversum: Verloren, 2001).

Eamon, W., 'Court, Academy, and Printing House: Patronage and Scientific Careers in Late-Renaissance Italy', in Moran (ed.), *Patronage and Institutions*, pp. 25–50.

—, 'Unmannered Science: Natural Philosophy and Medical Practice in the Piazza', in Guzzetti (ed.), *Science and Power*, pp. 63–8.

Eamon, W., and F. Paheau, 'The Accademia Segreta of Girolamo Ruscelli. A Sixteenth-Century Italian Scientific Society', *Isis*, 75 (1984), pp. 327–42.

Eichberger, D., *Leben mit Kunst: Wirken durch Kunst: Sammelwesen und Hofkunst unter Margarete von Österreich, Regentin der Niederlande* (Turnhout: Brepols, 2002).

Egmond, F., 'Clusius, Cluyt, Saint Omer: The Origins of the Sixteenth-Century Botanical and Zoological Watercolours in Libri Picturati A. 16–30', *Nuncius*, 20 (2005), pp. 11–67.

—, *Het Visboek: De wereld volgens Adriaen Coenen* (Zutphen: Walburg Pers, 2005).

—, 'Clusius and Friends: Cultures of Exchange in the Circles of European Naturalists', in Egmond et al. (eds), *Carolus Clusius*, pp. 9–48.

—, 'Apothecaries as Experts and Brokers in the Sixteenth-Century Network of the Naturalist Carolus Clusius', *History of Universities*, 23 (2008), pp. 59–91.

—, 'The Making of the Libri Picturati A.16–30', in De Koning et al. (eds), *Drawn After Nature*, pp. 13–21.

Egmond, F., and P. Mason, *The Mammoth and the Mouse: Microhistory and Morphology* (Baltimore, MD: Johns Hopkins University Press, 1997).

—, '"These are People who Eat Raw Fish": Contours of the Ethnographic Imagination in the Sixteenth Century', *Viator*, 31 (2000), pp. 311–60.

— (eds), *The Whale Book: Whales and other Marine Animals as Described by Adriaen Coenen in 1585* (London: Reaktion Books, 2003).

Egmond, F., P. Hoftijzer and R. Visser (eds), *Carolus Clusius: Towards a Cultural History of a Renaissance Naturalist* (Amsterdam: Edita-KNAW, 2007).

Eldering, P., 'Caspar Pelletier: Ein Botaniker in den Anfängen des 17. Jahrhunderts', *Janus*, 64 (1977), pp. 263–78.

Enthoven, V., 'Een symbiose tussen koopman en regent: De tweetrapsraket van de opkomst van de Republiek en Zeeland', in Lesger and Noordegraaf (eds), *Ondernemers en Bestuurders*, pp. 203–36.

Felipe II. El Rey intimo: Jardín e Naturaleza en el siglo XVI (Madrid: Sociedad Estatal Para la conmemoración de los centenarios de Felipe II y Carlos V, 1998).

Fernández de Oviedo y Valdes, G., *Historia general y natural de las Indias* (Seville: J. Cromberger, 1535).

Festschrift anlässlich der 400 jährigen Wiederkehr des wissenschaftlichen Tätigkeit von Carolus Clusius (Charles de l'Escluse) im pannonischen Raum (Eisenstadt: Amt der Burgenländischen Landesregierung, 1973).

Field, J., and F. James (eds), *Renaissance and Revolution: Humanists, Scholars, Craftsmen and Natural Philosophers in Early Modern Europe* (Cambridge: Cambridge University Press, 1993).

Filardi, D., *Orto de' Pitti: The Architects, Gardeners and Botanical Design of the Boboli Gardens* (Florence: Centro Di, 2007).

Findlen, P., 'The Economy of Scientific Exchange in Early Modern Italy', in Moran (ed.), *Patronage and Institutions*, pp. 5–24.

—, *Possessing Nature: Museums, Collecting and Scientific Culture in Early Modern Italy* (Berkeley, CA: University of California Press, 1994).

—, 'Francis Bacon and the Reform of Natural History in the Seventeenth Century', in D. Kelley (ed.) *History and the Disciplines: The Reclassification of Knowledge in Early Modern Europe* (Rochester, NY: University of Rochester Press, 1997), pp. 239–60.

—, 'The Formation of a Scientific Community: Natural History in Sixteenth-Century Italy', in A. Grafton and N. Siraisi (eds), *Natural Particulars: Nature and the Disciplines in Renaissance Europe* (Cambridge, MA: MIT Press, 1999), pp. 369–400.

—, 'Inventing Nature: Commerce, Art, and Science in the Early Modern Cabinet of Curiosities', in P. Smith and P. Findlen (eds), *Merchants and Marvels: Commerce, Science and Art in Early Modern Europe* (New York and London: Routledge, 2002), pp. 297–323.

Fontaine Verweij, H. de la, 'De eerste "private press" in de Nederlanden: Marcus Laurinus en de "Officina Goltziana"', in H. de la Fontaine Verweij, *Uit de wereld van het boek*, 2 vols (Amsterdam: Nico Israel, 1976), vol. 1, pp. 69–83.

Fontes Da Costa, P., 'Secrecy, Ostentation and the Illustration of Exotic Animals in Sixteenth-Century Portugal', *Annals of Science*, 66 (2009), pp. 59–82.

Frasca-Spada, M., and N. Jardine (eds), *Books and the Sciences in History* (Cambridge: Cambridge University Press, 2000).

Fuhring, P. (ed.), *De wereld is een tuin: Hans Vredeman de Vries en de tuinkunst van de Renaissance* (Ghent: Ludion, 2002).

Gabrieli, G., *Contributi alla storia della Accademia dei Lincei*, 2 vols (Rome: Accademia Nazionale dei Lincei, 1989).

—, *Il Carteggio Linceo* (Rome: Accademia Nazionale dei Lincei, 1996).

Garbari, F., 'The Prefetti of the Botanical Gardens from their Origins', in Garbari et al., *Giardino dei Semplici*, pp. 107–47.

Garbari, F., L. Tongiorgi Tomasi and A. Tosi, *Giardino dei Semplici / Garden of Simples* (Pisa: Edizioni Plus, 2002).

Geirnaert, N., and L. Vandamme, 'Cultuur en mentaliteit', in M. Martens (ed.), *Brugge en de Renaissance: Van Memling tot Pourbus* (Bruges: Ludion, 1998), pp. 33–42.

Gelder, E. van, 'Tussen hof en keizerskroon: Carolus Clusius en de ontwikkeling van de botanie aan Midden-Europese hoven (1573–1593)' (PhD dissertation, Leiden University, 2010).

Gelder, R. van, 'De wereld binnen handbereik: Nederlandse kunst- en rariteitenverzamelingen, 1585–1735', in Kistemaker and Bergvelt (eds), *De wereld binnen handbereik*, pp. 15–38.

—, 'Liefhebbers en geleerde luiden: Nederlandse kabinetten en hun bezoekers', in Kistemaker and Bergvelt (eds), *De wereld binnen handbereik*, pp. 259–92.

—, 'Paradijsvogels in Enkhuizen: De relatie tussen Van Linschoten en Bernardus Paludanus', in Van Gelder et al. (eds), *Souffrir pour parvenir*, pp. 30–50.

Gelder, R. van, J. Parmentier and V. Roeper (eds), *Souffrir pour parvenier: De wereld van Jan Huygen van Linschoten* (Haarlem: Arcadia, 1998).

Gelderblom, O., 'From Antwerp to Amsterdam: The Contribution of Merchants from the Southern Netherlands to the Rise of the Amsterdam Market', *Review: A Journal of the Fernand Braudel Center for the Study of Economies, Historical Systems, and Civilizations*, 26 (2003), pp. 247–82.

Ginori Conti, P., *Lettere inedite di Charles de l'Escluse (Carolus Clusius) a Matteo Caccini, Floricultore Fiorentino* (Florence: Olschki, 1939).

Gogelein, A., et al., *Leidse Universiteit 400: Stichting en eerste bloei 1575–ca. 1650* (Amsterdam: Rijksmuseum, 1975).

Goldgar, A., *Tulipomania: Money, Honor, and Knowledge in the Dutch Golden Age* (Chicago, IL: University of Chicago Press, 2007).

Gouw, J. van der, 'Marie de Brimeu: Een Nederlandse prinses uit de eerste helft van de tachtigjarige oorlog', *De Nederlandsche Leeuw*, 64 (1947), pp. 5–49.

Grendler, M., 'A Greek Collection in Padua: The Library of Gian Vincenzo Pinelli', *Renaissance Quarterly*, 33 (1980), pp. 386–416.

Grote, A. (ed.), *Macrocosmos in Microcosmo – Die Welt in der Stube: Zur Geschichte des Sammelns, 1450 bis 1800* (Opladen: Leske & Budrich, 1994).

Grove, R., 'Indigenous Knowledge and the Significance of South-West India for Portuguese and Dutch Constructions of Tropical Nature, *Modern Asian Studies*, 30 (1996), pp. 121–43.

Guerra, F., 'Drugs from the Indies and the Political Economy of the Sixteenth Century', *Analecta Medico-Historia*, 1 (1966), pp. 29–54.

Guicciardini, L., *Descrittione di tutti i paesi bassi* (Antwerp: Plantin, 1581).

Gunther, R., *Early British Botanists and their Gardens* (Oxford: Oxford University Press, 1922).

Guzzetti, L. (ed.), *Science and Power: The Historical Foundations of Research Policies in Europe* (Luxembourg: European Communities, 2000).

Hakluyt, R., *The Principall Navigations, Voiages, and Discoveries of the English Nation*, 2nd expanded edn (London: G. Bishop, 1598–1600).

Halikowski Smith, S., '"The Physician's Hand": Trends in the Evolution of the Apothecary and his Art across Europe, 1500–1700', *Nuncius*, 24 (2008), pp. 33–60.

Hall, A. R., 'The Scholar and the Craftsman in the Scientific Revolution', in M. Clagett (ed.), *Critical Problems in the History of Science* (Madison, WI: University of Wisconsin Press, 1962), pp. 3–23.

Hannay, M., *Philip's Phoenix: Mary Sidney, Countess of Pembroke* (New York: Oxford University Press, 1990).

—, '"How I These Studies Prize": The Countess of Pembroke and Elizabethan Science', in Hunter and Hutton (eds), *Women, Science, and Medicine*, pp. 108–21.

Harkness, D., *The Jewel House: Elizabethan London and the Scientific Revolution* (New Haven, CT: Yale University Press, 2007).

Härting, U., 'Over doel, gebruik en vorm van Nederlandse tuinen in de zestiende en zeventiende eeuw', in Fuhring (ed.), *De wereld is een tuin*, pp. 89–105.

— (ed.), *Gärten und Höfe der Rubenszeit: Im Spiegel der Malerfamilie Brueghel und der Künstler um Peter Paul Rubens* (Munich: Hirmer Verlag, 2000).

Hawkins, R., 'The Friendship of Joseph Scaliger and François Vertunien', *Romanic Review*, 8 (1917), pp. 117–44, 307–27.

Heijer, H. den, 'Zuid-Nederlanders en het ontstaan van de Noord-Nederlandse handel buiten Europa', in Parmentier (ed.), *Noord-Zuid in Oost-Indisch Perspectief*, pp. 59–80.

Heniger, J., 'De Eerste Nederlandse wetenschappelijke reis naar Oost-Indië, 1599–1601', *Jaarboekje voor geschiedenis en oudheidkunde van Leiden en Omstreken*, 65 (1973), pp. 27–49.

—, *Hendrik Adriaan van Reede tot Drakenstein (1636–1691) and Hortus Malabaricus: A Contribution to the History of Dutch Colonial Botany* (Rotterdam: Balkema, 1986).

Henrey, B., *British Botanical and Horticultural Literature before 1800*, 3 vols (Oxford: Oxford University Press, 1975).

Hessels, J. (ed.), *Ecclesiae Londino-Batavae archivum*, 4 vols (Cambridge: Typis Academiae, 1887–98).

Heyning, K., *Return to Zealand: Masterpieces from the 16th and 17th Centuries* (Middelburg: Zeeuws Museum, 2008).

Hochlenert, D., *Das "Tagebuch" des Felix Platter: Die Autobiographie eines Arztes und Humanisten* (Tübingen: University of Tübingen, 1996).

Hoeniger, F., and J. Hoeniger, *The Development of Natural History in Tudor England* (Charlottesville, VA: University Press of Virginia, 1969).

Holton, D. (ed.), *Literature and Society in Renaissance Crete* (Cambridge: Cambridge University Press, 1991).

Hondius, P., *Dapes inemptae, oft de Mouffe-Schans, dat is de soeticheyt des buyten-levens, vergeselschapt met de boecken* (n.p., *c.* 1619).

Hopper, F., 'Clusius' World: The Meeting of Science and Art', in Tjon Sie Fat and De Jong (eds), *The Authentic Garden*, pp. 13–36.

Houtzager, H., 'Opleiding in Italië en verblijf in Frankrijk', in Bosman-Jelgersma (ed.), *Pieter van Foreest*, pp. 81–5.

Huber, O., 'Die "Geographie der Pflanzen"', in *Alexander von Humboldt: Netzwerke des Wissens* (Ostfildern-Ruit: Hantje Catze Verlag, 1999), pp. 100–3.

Huisman, T., *The Finger of God: Anatomical Practice in 17th-Century Leiden* (Leiden: Primavera Pers, 2009).

Hulthem, C. van, *Discours sur l'état ancien et moderne de l'agriculture et de la botanie dans les Pays Bas*, 2nd edn (Ghent: Goesin-Verhaeghe, 1837).

Hunger, F., 'Acht brieven van Middelburgers aan Carolus Clusius', *Archief Zeeuws Genootschap der Wetenschappen* (1925), pp. 110–33.

—, *Charles de l'Escluse (Carolus Clusius) Nederlandsch Kruidkundige, 1526–1609*, 2 vols (The Hague: M. Nijhoff, 1927–42).

—, 'Bernardus Paludanus (Berent ten Broecke) 1550–1633', *Nederlands Tijdschrift voor Geneeskunde*, 72 (1928), pp. 2–11.

—, 'Bernardus Paludanus (Berent ten Broecke) 1550–1633: Zijn verzamelingen en zijn werk', in C. Burger and F. Hunger (eds), *Itinerario: Voyage ofte Schipvaert van Jan Huygen van Linschoten*, 5 vols (The Hague: Nijhoff, 1910–39), vol. 3, pp. 249–68.

Hunt, J. D., and E. de Jong (eds), *The Anglo-Dutch Garden in the Age of William and Mary* (London: Taylor & Francis, 1988).

Hunter, L., 'Women and Domestic Medicine: Lady Experimenters, 1570–1620', in Hunter and Hutton (eds), *Women, Science, and Medicine*, pp. 89–107.

Hunter, L., and S. Hutton (eds), *Women, Science, and Medicine 1500–1700: Mothers and Sisters of the Royal Society* (Phoenix Mill: Sutton, 1997).

Huppert, G., 'Antiquity Observed: A French Naturalist in the Aegean Sea in 1547', *International Journal of the Classical Tradition*, 2 (1995), pp. 275–83.

—, *The Style of Paris: Renaissance Origins of the French Enlightenment* (Bloomington, IN: Indiana University Press, 1999).

Iliffe, R., '"In the Warehouse": Privacy, Property and Priority in the Early Royal Socety', *History of Science*, 30 (1992), pp. 29–68.

Imperato, F., *Dell' Historia Naturale* (Naples: C. Vitale, 1599).

Impey, O., and A. Macgregor (eds), *The Origins of Museums: The Cabinets of Curiosities in Sixteenth- and Seventeenth-Century Europe* (Oxford: Clarendon Press, 1985).

Istvánffi, G., *Études et commentaires sur le code de l'Escluse* (Budapest: private publication, 1900).

Jaffé, D., 'Peiresc – Wissenschaftlicher Betrieb in einem Raritäten-Kabinett', in Grote (ed.), *Macrocosmos in Microcosmo*, pp. 301–22.

Jardine, N., J. Secord and E. Spary (eds), *Cultures of Natural History* (Cambridge: Cambridge University Press, 1996).

Jelsma, A., and O. Boersma (eds), *Acta van het consistorie van de Nederlandse gemeente te London 1569–1585* (The Hague: ING, 1993).

Jones, W., *William Turner: Tudor Naturalist, Physician, Divine* (London: Routledge, 1988).

Jong, E. de, 'Nature and Art: The Leiden Hortus as "musaeum"', in Tjon Sie Fat and De Jong (eds), *The Authentic Garden*, pp. 37–60.

Jongbloet-Van Houtte, G. (ed.), *Brieven en andere bescheiden betreffende Daniël van der Meulen, 1584–1600* (The Hague: Nijhoff, 1986).

Jonge, K. de, 'Les jardins de J. du Broeucq et de J. Hollebecque à Binche, Mariemont et Boussu', in *Felipe II. El Rey intimo*, pp. 191–220.

Jorink, E., *Het Boeck der Natuere: Nederlandse geleerden en de wonderen van Gods Schepping 1575–1715* (Leiden: Primavera, 2006). [English translation forthcoming]

Keevil, J., *Medicine and the Navy 1200–1900*, 4 vols (Edinburgh: Livingstone, 1957).

Keller, K., 'Kommunikationsraum altes Reich: Zur Funktionalität des Korrespondenznetze von Fürstinnen im 16. Jahrhundert', *Zeitschrift für Historische Forschung*, 31 (2004), pp. 205–30.

Kerkhoff, J., *Maria van Hongarije en haar hof 1505–1558* (Hilversum: Verloren, 2008).

Kirk, R., and E. Kirk (eds), *Returns of Aliens, Dwelling in the City and Suburbs of London, from the Reign of Henry VIII to that of James I*, 4 vols (Aberdeen: University Press, 1900–8).

Kistemaker, R., and E. Bergvelt (eds), *De wereld binnen handbereik: Nederlandse kunst- en rariteitenverzamelingen 1585–1735* (Zwolle: Waanders, 1992).

Klooster, W., *The Dutch in the Americas 1600–1800* (Providence, RI: John Carter Brown Library, 1997).

Koerner, L., *Linnaeus: Nature and Nation* (Cambridge MA: Harvard University Press, 1999).

Koning, J. de, G. van Uffelen, A. Zemanek and B. Zemanek (eds), *Drawn After Nature: The Complete Botanical Watercolours of the 16th-Century Libri Picturati* (Zeist: KNNV-Publishing, 2008).

Kusukawa, S., 'Illustrating Nature', in Frasca-Spada and Jardine (eds), *Books and the Sciences*, pp. 90–113.

—, 'Uses of Pictures in Printed Books: The Case of Clusius' *Exoticorum libri decem*', in Egmond et al. (eds), *Carolus Clusius*, pp. 221–46.

Lack, H., 'Lilac and Horse-Chestnut: Discovery and Rediscovery', *Curtis's Botanical Magazine*, 17 (2000), pp. 109–37.

Landtsheer, J. de, 'Justus Lipsius and Carolus Clusius, a Flowering Friendship', *Bulletin of the Belgian Institute*, 68 (1998), pp. 273–95.

Lazzaro, C., *The Italian Renaissance Garden: From the Conventions of Planting, Design and Ornament to the Grand Gardens of Seventeenth-Century Central Italy* (New Haven, CT: Yale University Press, 1990).

Legré, L., *La botanique en Provence au XVIe siècle* (Marseille: H. Aubertin & Cie, 1901).

Leith-Ross, P., 'A Seventeenth-Century Paris Garden', *Garden History*, 21 (1993), pp. 150–7.

Leith-Ross, P., and H. McBurney, *The Fiorilegium of Alexander Marshal* (London: Royal Collection Enterprises, 2000).

Le Loup, W., 'Biografie van Hubertus Goltzius', in W. Le Loup (ed.), *Hubertus Goltzius en Brugge 1583–1983* (Bruges: Stad Brugge, 1983), pp. 9–12.

Lesger, C., and L. Noordegraaf (eds), *Ondernemers en Bestuurders: Economie en politiek in de Noordelijke Nederlanden in de Late Middeleeuwen en vroegmoderne tijd* (Amsterdam: NEHA, 1999).

Lewis, G., 'Clusius in Montpellier, 1551–1554: A Humanist Education Completed?', in Egmond et al. (eds), *Carolus Clusius*, pp. 65–98.

Lietzmann, H., *Irdische Paradiese: Beispiele höfische Gartenkunst der 1. Hälfte des 16. Jahrhunderts* (Munich: Deutscher Kunstverlag, 2007).

Lobel, M. de, *Plantarum seu Stirpium historia* (Antwerp: Plantin, 1576).

—, *Kruydtboeck oft Beschryvinghe van allerleye ghewassen, kruyderen, hesteren ende gheboomten* (Antwerp: Plantin, 1581).

Lomonaco, F., and M. Torrini (eds), *Galileo a Napoli* (Naples: Guida, 1987).

Longeon, C. (ed.), *Gesner, Conrad, vingt lettres á Jean Bauhin fils (1563–1565)* (Saint Etienne: Université de Saint Etienne, 1976).

Louis, A., *Mathieu de l'Obel (1538–1616): Épisode de l'histoire de la botanie* (Ghent: Story-Scientia, 1980).

Lugli, A., *Naturalia et mirabilia: Les cabinets de curiosités en Europa* (Paris: Adam Biro, 1998).

— (ed.), *Albrecht Dürer: Viaggio nei Paesi Bassi*, 2nd edn (Reggio Emilia: Diabasis, 2005).

Lundquist, K., 'Lilies to Norway and Cloudberry Jam to the Netherlands – On the Relationship, Correspondence and Exchange of Naturalia between Carolus Clusius and Henrik Høyer, 1597–1604', in Egmond et al. (eds), *Carolus Clusius*, pp. 145–69.

Lunsingh Scheurleer, T., 'Un amphithéatre d'anatomie moralisée', in T. Lunsingh Scheurleer and G. Posthumus Meyjes (eds), *Leiden University in the Seventeenth Century: An Exchange of Learning* (Leiden: Brill, 1975), pp. 216–77.

MacDougall, E., *Fountains, Statues, and Flowers: Studies in Italian Gardens of the Sixteenth and Seventeenth Centuries* (Washington, DC: Dumbarton Oaks, 1994).

MacGregor, A., *Curiosity and Enlightenment: Collectors and Collections from the Sixteenth to the Nineteenth Century* (New Haven, CT: Yale University Press, 2007).

Maegd, C. de, '"En ung sien jardin de plaisance au faubourgs de ceste ville": Het hof van plaisantie van Karel van Croÿ in Sint-Joost-ten-Node rond 1600', *Tijdschrift van Dexia bank*, 55:218 (2001), pp. 45–68.

—, 'Tuinbezit, tuinen en tuinlui: Een licht op de praktijk in de tijd van Vredeman de Vries', in Fuhring (ed.), *De wereld is een tuin*, pp. 69–87.

Martels, Z. van, 'On his Majesty's Service: Augerius Busbequius, Courtier and Diplomat of Maximilian II', in F. Edelmayer and A. Kohler (eds), *Kaiser Maximilian II: Kultur und Politik im 16. Jahrhundert* (Vienna: Oldenbourg, 1992), pp. 169–81.

Maselis, M., A. Balis and R. Marijnissen, *De Albums van Anselmus de Boodt (1550–1632): Geschilderde natuurobservatie aan het Hof van Rudolf II te Praag* (Tielt: Lannoo, 1989).

Mason, P., *Infelicities: Representations of the Exotic* (Baltimore, MD: Johns Hopkins University Press, 1998).

—, '*Americana* in the *Exoticorum libri decem* of Charles de l'Écluse', in Egmond et al. (eds), *Carolus Clusius*, pp. 195–219.

—, *Before Disenchantment: Images of Exotic Animals and Plants in the Early Modern World* (London: Reaktion Books, 2009).

Masson, G., 'Italian Flower Connoisseurs', *Apollo: A Journal of the Arts*, 88 (1968), pp. 164–71.

—, 'Italian Flower Collector's Gardens in Seventeenth-Century Italy', in D. Coffin (ed.), *The Italian Garden* (Washington, DC: Dumbarton Oaks, 1972), pp. 61–80.

Matthews, L., 'Royal Apothecaries of the Tudor Period', *Medical History*, 8 (1964), pp. 170–80.

—, 'Herbals and Formularies', in Poynter (ed.), *The Evolution of Pharmacy*, pp. 187–213.

—, *The Royal Apothecaries* (London: Wellcome, 1967).

Meertens, P., *Letterkundig leven in Zeeland in de zestiende en de eerste helft der zeventiende eeuw* (Amsterdam: Noord-Hollandsche Uitgeversmaatschappij, 1943).

Miller, P., 'Description Terminable and Interminable: Looking at the Past, Nature and Peoples in Peiresc's Archive', in G. Pomata and N. Siraisi (eds), *Historia: Empiricism and Erudition in Early Modern Europe* (Cambridge, MA: MIT Press, 2005), pp. 355–97.

Minelli, A. (ed.), *The Botanical Garden of Padua 1545–1995* (Venice: Marsilio, 1995).

Mirek, Z., and A. Zemanek (eds), *Studies in Renaissance Botany* (Kraków: W. Szafer Institute, 1998).

Moens, W. (ed.), *The Marriage, Baptismal, and Burial Registers 1571 to 1874 and Monumental Inscriptions of the Dutch Reformed Church, Austin Friars* (Lymington: King, 1884).

Moes, E., 'David Mostart en zijn album amicorum', *Amsterdamsch Jaarboekje* (1904), pp. 29–41.

—, 'De sonderling-heden oft rariteyten ende wtgelesen sinnelickheden van Christiaen Porret', *Leids Jaarboekje*, 2 (1905), pp. 93–100.

Molhuysen, P., *Bronnen tot de geschiedenis der Leidsche Universiteit deel I (1574–1610)* (The Hague: Nijhoff, 1913).

Montaigne, M. de, *Journal de Voyage en Italie: Oeuvres Complètes* (Paris: Gallimard, 1962).

Moran, B., 'Wilhelm IV of Hesse-Kassel: Informal Communication and the Aristocratic Context of Discovery', in T. Nickles (ed.), *Scientific Discovery: Case Studies* (Dordrecht: Reidel, 1980), pp. 67–96.

—, 'Prince-Practitioning and the Direction of Medical Roles at the German Court: Maurice of Hesse-Kassel and his Physicians', in Nutton (ed.), *Medicine at the Courts of Europe*, pp. 95–116.

—, 'Patronage and Institutions: Courts, Universities, and Academies in Germany; an Overview: 1550–1750', in Moran (ed.), *Patronage and Institutions*, pp. 169–83.

— (ed.), *Patronage and Institutions: Science, Technology, and Medicine at the European Court 1500–1700* (Rochester, NY: Boydell Press, 1991).

Morán, J., and F. Checa, *El coleccionismo en España: De la cámara de maravillas a la galería de pinturas* (Madrid: Cátedra, 1985).

Moree, P. (ed.), *Dodo's en galjoenen: De reis van het schip Gelderland naar Oost-Indië, 1601–1603* (Zutphen: Walburg Pers, 2001).

Mosley, A., 'Astronomical Books and Courtly Communication', in Frasca-Spada and Jardine (eds), *Books and the Sciences*, pp. 114–31.

Munck, G. de, and H. Wille (eds), *Peeter van Coudenberghe: Apotheker-botanicus (1517–1599) en tijdgenoten: De natuur als verzamelobject in de 16de eeuw* (Antwerp: KAVA, 1996).

Nave, F. de, and D. Imhof (eds), *De Botanica in de Zuidelijke Nederlanden (einde 15e eeuw–ca. 1650)* (Antwerp: Snoeck-Ducaju, 1993).

Neviani, A., *Ferrante Imperato: Speziale e naturalista Napolitano con documenti inediti* (Rome: Accademia di Storia dell'Arte Sanitaria, 1936).

Nichols, L., 'Jan Govertz van der Aar: On the Identification of Goltzius's Patron', *Nederlands Kunsthistorisch Jaarboek*, 38 (1987), pp. 241–55.

Noordegraaf, L., and T. Wijsenbeek-Olthuis, 'De wereld ontsloten', in Kistemaker and Bergvelt (eds), *De wereld binnen handbereik*, pp. 39–51.

Nutton, V. (ed.), *Medicine at the Courts of Europe, 1500–1837* (London: Routledge, 1990).

Ogilvie, B., 'Travel and Natural History in the Sixteenth Century', in B. Ogilvie, A. te Heesen and M. Gierl, *Sammeln in der Fruhen Neuzeit* (Berlin: Max-Planck-Institut für Wissenschaftsgeschichte, 1996), pp. 3–28.

—, 'The Many Books of Nature: Renaissance Naturalists and Information Overload', *Journal of the History of Ideas*, 64 (2003), pp. 29–40.

—, *The Science of Describing: Natural History in Renaissance Europe* (Chicago, IL: Chicago University Press, 2006).

Olmi, G., *Scienza e natura del secondo cinquecento* (Trento: Libera Università degli Studi, 1976).

—, 'Science – Honour – Metaphor; Italian Cabinets of Curiosity of the Sixteenth and Seventeenth Centuries', in Impey and MacGregor, *The Origins of Museums*, pp. 3–16.

—, 'La colonia Lincea di Napoli', in Lomonaco and Torrini (eds), *Galileo a Napoli*, pp. 23–59.

—, '"Molti amici in varii luoghi": Studio della natura e rapporti epistolari nel secolo XVI', *Nuncius*, 6 (1991), pp. 3–31.

—, *L'Inventario del mondo: Catalogazione della natura e luoghi del sapere nella prima età moderna* (Bologna: Mulino, 1992).

—, 'Science and the Court: Some Comments on "Patronage" in Italy', in Guzzetti (ed.), *Science and Power*, pp. 25–45.

—, 'I cappucini e la scienza nell'età moderna', in G. Pozzi and P. Prodi (eds), *I Cappucini in Emilia-Romagna: Storia di una presenza* (Bologna: EDB, 2002), pp. 288–328.

—, 'Lettere di fra Gregorio da Reggio, cappuccino e botanico del tardo rinascimento', in M. Beretta, P. Galluzzi and C. Triarico (eds), *Musa Musaei, Studies on Scientific Instruments and Collections in Honour of Mara Miniati* (Florence: Olschki, 2003), pp. 117–39.

—, 'Per la storia dei rapporti scientifici fra Italia e Germania: Le lettere di Francesco Calzolari a Joachim Camerarius II', in G. Brizzi and G. Olmi (eds), *Dai cantieri della storia: Liber amicorum per Paolo Prodi* (Bologna: CLUEB, 2007), pp. 343–61.

Ommen, K. van (ed.), *The Exotic World of Carolus Clusius 1526–1609: Catalogue of an Exhibition on the Quatercentenary of Clusius' Death, 4 April 1609* (Leiden: Leiden University Library, 2009).

Opsomer, J., 'Un botaniste trop peu connu, Willem Quackelbeen (1527–1561)', *Bulletin de la Société royale de botanique de Belgique*, 93 (1961), pp. 113–30.

—, 'Notes complémentaires sur les plantes envoyées de Turquie en 1557 par le botaniste Quackelbeen', *Bulletin de la Société royale de botanique de Belgique*, 103 (1970), pp. 5–10.

Pächt, O., 'Early Italian Nature Studies and the Early Calendar Landscape', *Journal of the Warburg and Courtauld Institutes*, 13 (1950), pp. 13–47.

Paganelli, A., 'L'Horto Medicinale dell'Università di Padova nella storia della medicina', in Bussi (ed.), *Di Sana Pianta*, pp. 41–5.

Palmer, R., 'Medical Botany in Northern Italy in the Renaissance', *Journal of the Royal Society of Medicine*, 78 (1985), pp. 149–57.

—, 'La botanica medica nell'Italia del Nord durante il Rinascimento', in Bussi (ed.), *Di Sana Pianta*, pp. 55–60.

—, 'Medicine at the Papal Court in the Sixteenth Century', in Nutton (ed.), *Medicine at the Courts of Europe*, pp. 49–78.

Pardo Tomás, J., 'Obras españolas sobre historia natural y materia medica americana en la Italia del siglo XVI', *Asclepio: Revista de Historia de la medicina y de la Ciencia*, 43 (1991), pp. 561–94.

—, 'Tra "Oppinioni" e "Dispareri": La flora americana nell'erbario di Pier'Antonio Michiel (1510–1576)', in G. Olmi and G. Papagno (eds), *La natura e il corpo: Studi in memoria di Attilio Zanca* (Florence: Olschki, 2005), pp. 73–100.

—, 'Two Glimpses of America from a Distance: Carolus Clusius and Nicolás Monardes', in Egmond et al. (eds), *Carolus Clusius*, pp. 173–93.

Pardo Tomás, J., and J. Lopez Pinero, *La influencia de Francisco Hernández (1515–1587) en la constitución de la botánica y de la materia médica modernas* (Valencia: Universitat de València, CSIC, 1996).

Parmentier, J. (ed.), *Noord-Zuid in Oost-Indisch Perspectief* (Zwolle: Walburg Pers, 2005).

Pasquale, G. Di, and F. Paolucci (eds), *Il giardino antico da Babilonia a Roma: Scienza, arte e natura* (Livorno: Sillabe, 2007).

Pelletier, C., *Plantarum tum patriarum tum exoticarum in Walachria, Zeelandiae insula, nascentium synonymia* (Middelburg: R. Schilders, 1610).

Pelling, M., and C. Webster, 'Medical Practitioners', in Webster (ed.), *Health, Medicine and Mortality*, pp. 165–235.

Pérez de Tudela, A., and A. Jordan Schwendt, 'Renaissance Menageries: Exotic Animals and Pets at the Habsburg Courts in Iberia and Central Europe', in K. Enenkel and P. Smith (eds), *Early Modern Zoology: The Construction of Animals in Science, Literature and the Visual Arts*, Intersections: Yearbook for Early Modern Studies, 7 (Leiden and Boston, MA: Brill, 2007), pp. 419–47.

Pettegree, A., *Foreign Protestant Communities in Sixteenth-Century London* (Oxford: Clarendon Press, 1986).

Pieper, R., *Die Vermittlung einer neuen Welt: Amerika im Nachrichtennetz des habsburgischen Imperiums 1493–1598* (Mainz: Philip von Zabern, 2000).

Pimentel, J., 'The Iberian Vision: Science and Empire in the Framework of a Universal Monarchy, 1500–1800', *Osiris*, 15 (2001), pp. 17–30.

Pollitt, R., 'The Wounding and Treatment of William of Orange in 1582', *Journal of the History of Medicine and Allied Sciences*, 37 (1982), pp. 135–58.

Pomian, K., *Collectioneurs, amateurs et curieux: Paris, Venise, xvie–xviiie siècle* (Paris: Gallimard, 1987).

Pona, G., *Monte Baldo descritto da Giouanni Pona veronese* (Venice: R. Meietti, 1617).

Postma, A., 'Een Zuidnederlandse hovenier in Spanje: Over Filips II, Jehan Holbecq en een nieuwe tuinkunst', *Tuinkunst: Nederlands Jaarboek voor de Geschiedenis van Tuin- en Landschapsarchitectuur*, 1 (1995), pp. 9–22.

Poynter, F. (ed.), *The Evolution of Pharmacy in Britain* (London: Pitman Medical, 1965).

Puerto, J., *La Leyende Verde: Naturaleza, sanidad e ciencia en la corte de Felipe II, 1527–1598* (Castilla y León: Junta de Castilla y León, 2003).

Pumfrey, S., and F. Dawbarn, 'Science and Patronage in England, 1570–1625: A Preliminary Study', *History of Science*, 42 (2004), pp. 137–88.

Pumfrey, S., P. Rossi and M. Slawinski (eds), *Science, Culture and Popular Belief in Renaissance Europe* (Manchester: Manchester University Press, 1991).

Ramón-Laca, L., 'Las plantas americanas en la obra de Charles de l'Écluse: Primeras citas en las cartas de Juan de Castañeda', *Anales del Jardín Botánico de Madrid*, 57:1 (1999), pp. 97–107.

—, 'Charles de l'Écluse and *Libri Picturati* A. 16–30', *Archives of Natural History*, 28 (2001) pp. 195–243.

—, 'The Additions by the Count of Arenberg', in De Koning et al. (eds), *Drawn After Nature*, pp. 28–32.

Ramón-Laca, L., and R. Morales Valverde (eds), *Charles de L'Écluse de Arras, Descripción de algunas plantas raras encontradas en España y Portugal* (Castilla y León: Junta de Castilla y León, 2005).

Rankin, A., 'Becoming an Expert Practitioner: Court Experimentalism and the Medical Skills of Anna of Saxony (1532–1585)', *Isis*, 98 (2007), pp. 23–53.

Raugei, A. (ed.), *Pinelli, Gian Vincenzo – Dupuy, Claude: Une correspondance entre deux humanistes*, 2 vols (Florence: Olschki, 2001).

Raven, C., *English Naturalists from Neckam to Ray: A Study of the Making of the Modern World* (Cambridge: Cambridge University Press, 1947).

Reeds, K., *Botany in Medieval and Renaissance Universities* (New York: Garland, 1991).

Reznicek, E., 'De achtste tronie van de schelpenverzamelaar', in A. Logan (ed.), *Essays in Northern European Art Presented to Egbert Haverkamp-Begemann* (Doornspijk: Davaco, 1983), pp. 209–12.

Roberts, L., S. Schaffer and P. Dear (eds), *The Mindful Hand: Inquiry and Invention from the Late Renaissance to Early Industrialisation* (Amsterdam: Edita-KNAW, 2007)

Roberts, R., 'The Personnel and Practice of Medicine in Tudor and Stuart England. Part I: The Provinces', *Medical History*, 6 (1962), pp. 362–82.

—, 'The London Apothecaries and Medical Practice in Tudor and Stuart England' (PhD dissertation, University of London, 1964).

—, 'The Personnel and Practice of Medicine in Tudor and Stuart England. Part II: London', *Medical History*, 8 (1964), pp. 217–34.

—, 'The Early History of the Import of Drugs into England', in Poynter (ed.), *The Evolution of Pharmacy*, pp. 165–86.

Roegiers, J. (ed.), *Arenberg in de Lage Landen: Een hoogadellijk huis in Vlaanderen en Nederland* (Louvain: Universitaire Pers, 2002).

Roegiers, J., and H. Wille, 'Natuur en techniek', in Roegiers (ed.), *Arenberg in de Lage Landen*, pp. 371–7.

Rudwick, M., 'Charles Darwin in London: The Integration of Public and Private Science', *Isis*, 73 (1982), pp. 186–206.

Rydén, M., 'William Turner the Botanist in a European Context', in G. Sorelius and M. Srigley (eds), *Cultural Exchange between European Nations during the Renaissance* (Uppsala: Uppsala University, 1994), pp. 173–83.

Salzmann, C., 'Francesco Calzolari, Student of the Flora of Monte Baldo, and his Shipments of Plants Addressed to Conrad Gessner in Zurich', *Gesnerus*, 16 (1959), pp. 81–103.

Sandrini, G. (ed.), *Francesco Calzolari, il viaggio di Monte Baldo, con la Testimonianza sul Museo Calzolari di Ulisse Aldrovandi* (Verona: Alba Pratalia, 2007).

Sarasohn, L., 'Nicolas-Claude Fabri de Peiresc and the Patronage of the New Science in the Seventeenth Century', *Isis*, 84 (1993), pp. 70–90.

Schaap, E., *Dutch Floral Tiles in the Golden Age and their Botanical Prints / Bloemen op tegels in de Gouden Eeuw: Van prent tot tegel* (Haarlem: Becht, 1994).

Schiebinger, L., *The Mind Has No Sex? Women in the Origins of Modern Science* (Cambridge, MA: Harvard University Press, 1989).

—, 'Women of Natural Knowledge', in K. Park and L. Daston (eds), *The Cambridge History of Science, vol. III. Early Modern Science* (Cambridge: Cambridge University Press, 2006), pp. 192–205.

Schiebinger, L., and C. Swan, 'Introduction', in Schiebinger and Swan (eds), *Colonial Botany*, pp. 1–16.

— (eds), *Colonial Botany: Science, Commerce and Politics in the Early Modern World* (Philadelphia, PA: University of Pennsylvania Press, 2004).

Schlosser, J. von, *Die Kunst- und Wunderkammern der Spätrenaissance: Ein Beitrag zur Geschichte des Sammelwesens*, 2nd edn (Braunschweig: Klinkhardt and Biermann, 1978).

Schnap, J., and Y. Hersant (eds), *Nicandre de Corcyre, le voyage d'Occident* (Toulouse: Anacharsis Editions, 2002).

Schnapp, A., *The Discovery of the Past: The Origins of Archaeology* (London: British Museum Press, 1996).

Schnapper, A., *Le géant, la licorne et la tulipe: Collections et collectionneurs dans la France du XVIIe siècle* (Paris: Flammarion, 1988).

Secord, J., 'Knowledge in Transit', *Isis*, 95 (2004), pp. 654–72.

Séronie-Vivien, M., P. Coudroy de Lille and M. Favreau, 'Le plus ancien botaniste bordelais connu: Joachim Levenier dit Venerius (ca 1565–1619), *Bulletin de la Société Linnéenne de Bordeaux*, 140:33 (2005), pp. 163–9.

Sikkens-De Zwaan, M., 'Magdalena Poulle (1632–99): A Dutch Lady in a Circle of Botanical Collectors', *Garden History*, 30 (2002), pp. 206–20.

Sloan, K. (ed.), *European Visions: American Voices* (London: British Museum, 2009).

Smith, P., *The Body of the Artisan: Art and Experience in the Scientific Revolution* (Chicago, IL: Chicago University Press, 2004).

—, 'Art, Science and Visual Culture in Early Modern Europe', *Isis*, 97 (2006), pp. 83–100

—, 'Making and Knowing in a Sixteenth-Century Goldsmith's Workshop', in Roberts et al. (eds), *The Mindful Hand*, pp. 20–37.

Somer, J., *Zee en Landt reyse naer de Levante, als Italien, Candyen, Cypres, Egypten, Rhodes, Archipelago, Turckyen en wederom door Duytslant*, 2nd edn (Amsterdam: Joost Hartgens, 1649).

Spary, E., 'Botanical Networks Revisited', in R. Dauser et al. (eds), *Wissen im Netz: Botanik und Pflanzentransfer des 18. Jahrhunderts* (Berlin: Akademie Verlag, 2008), pp. 1–18.

Stendardo, E., *Ferrante Imperato: Collezionismo e studio della natura a Napoli tra Cinque e Seicento* (Naples: Accademia Pontaniana, 2001).

Strong, R., *The Renaissance Garden in England* (London: Thames & Hudson, 1979).

—, 'The Renaissance Garden in England Reconsidered: A Survey of Two Decades of Research on the Period 1485–1642', *Garden History*, 27 (1999), pp. 2–9.

Sweert, E., *Florilegium ... tractans de variis floribus et aliis indicis plantis ad vivum delineatum in duabus partis et quatuor linguis concinnatum* (Frankfurt: A. Kempner, 1612–14).

Symcox, G. (ed.), *Italian Reports on America 1493–1522, Letters, Dispatches, and Papal Bulls*, Repertorium Columbianum, 10 (Turnhout: Brepols, 2001).

Tagliaferri, M., S. Tommasini and S. Tugnoli Pattaro, 'Ulisse Aldrovandi als Sammler: Das Sammeln als Gelehrsamkeit oder als Methode wissenschaftlichen Forschens?', in Grote (ed.), *Macrocosmos in Microcosmo*, pp. 265–81.

Tebeaux, E., 'Women and Technical Writing, 1475–1700: Technology, Literacy and Development of a Genre', in Hunter and Hutton (eds), *Women, Science, and Medicine*, pp. 29–62.

Theunisz, J., 'B. Paludanus en de brieven uit zijn Zwolse tijd', *Verslagen en Mededeelingen Vereeniging tot beoefening van Overijselsch Regt en Geschiedenis*, 52 (1936), pp. 108–24.

Tjon Sie Fat, L., 'Clusius' Garden: A Reconstruction', in Tjon Sie Fat and De Jong (eds), *The Authentic Garden*, pp. 3–12.

Tjon Sie Fat, L., and E. de Jong (eds), *The Authentic Garden: A Symposium on Gardens* (Leiden: Clusius Foundation, 1991).

Tongiorgi Tomasi, L., 'L'isola dei semplici', *Kos*, 1:5 (1984), pp. 61–78.

—, 'L'immagine naturalistica; tecnica e invenzione', in G. Olmi, L. Tongiorgi Tomasi and A. Zanca (eds), *Natura-Cultura: L'interpretazione del mondo fisico nei testi e nelle immagini* (Florence: Olschki, 2000), pp. 133–51.

Tongiorgi Tomasi, L., and F. Garbari, *Il giardiniero del Granduca: Storia e immagini del Codice Casabona* (Pisa: Edizioni ETS, 1995).

Tongiorgi Tomasi, L., and G. A. Hirschauer, *The Flowering of Florence* (Aldershot: Lund Humpries, 2002).

Toni, G. B. de, *Il carteggio degli Italiani col botanico Carlo Clusio nella biblioteca Leidense* (Modena: Società Tipografica Modenese, 1911).

—, *Pietro Antonio Michiel: I cinque libri di piante: Codice Marciano, Trascrizione e Commento* (Venice: Reale Istituto Veneto di Scienze, Lettere ed Arti, 1940).

Trevor-Roper, H., 'The Sieur de la Rivière: Paracelsian Physician of Henry IV', in A. Debus (ed.), *Science, Medicine and Society in the Renaissance: Essays to Honor Walter Pagel*, 2 vols (New York: Heinemann, 1972), vol. 2, pp. 227–50.

Ubrizsy Savoia, A., 'Environmental Approach in the Botany of the 16th Century', in Mirek and Zemanek (eds), *Studies in Renaissance Botany*, pp. 73–86.

—, 'Some Aspects of Clusius' Hungarian and Italian Relations', in Egmond et al. (eds), *Carolus Clusius*, pp. 267–92.

Ubrizsy Savoia, A., A. Zemanek and B. Zemanek, 'The Beginnings of Ecological Thought', in De Koning et al. (eds), *Drawn After Nature*, pp. 90–7.

Uffelen, G. van, 'The *Libri Picturati* and the Early History of the *hortus botanicus* Leiden', in De Koning et al. (eds), *Drawn After Nature*, pp. 54–8.

Vallet, P., *Le Jardin du Roy tres chrestien Henry IV* (Paris: by the author, 1608).

Vandenbroeck, P., 'Amerindian Art and Ornamental Objects in Royal Collections, Brussels, Mechelen, Duurstede, 1520–30', in *America: Bride of the Sun: 500 years Latin America and the Low Countries* (Antwerp: Imschoot, 1991), pp. 99–119.

Vandewalle, W., and K. Kasaya (eds), *Dodonaeus in Japan: Translation and the Scientific Mind in the Togukawa Period* (Louvain and Kyoto: Leuven University Press and International Research Center for Japanese Studies, 2001).

Vandewiele, L., 'Wat groeide in de tuin van Pieter van Coudenberghe', in De Nave and Imhof (eds), *De Botanica in de Zuidelijke Nederlanden*, pp. 23–31.

Vasbinder-Ouwendijk, E., 'Carolus Clusius en Jacques Plateau' (MA thesis, University of Utrecht, 1979).

Veen, J. van der, 'Dit klain vertrek bevat een weereld vol gewoel: Negentig Amsterdammers en hun kabinetten', in Kistemaker and Bergvelt (eds), *De wereld binnen handbereik*, pp. 223–58.

Veendorp, H., and L. Baas Becking, *Hortus Academicus Lugduno Batavus 1587–1937* (Haarlem: Enschedaiana, 1938).

Verburgt, J., 'Uit het particuliere leven van Daniël Van der Meulen, heer van Ranst', *Leids Jaarboekje*, 29 (1936–7), pp. 73–88.

Versprille, A., 'De geschiedenis van het huis van Daniël van der Meulen (Rapenburg 19)', *Leids Jaarboekje*, 35 (1943), pp. 158–69.

—, 'Hester della Faille', *Leids Jaarboekje*, 67 (1975), pp. 77–97.

Verweij, M., 'Une lettre retrouvée d'un botaniste bordelais mal connu (Bruxelles KBR, ms. III 893/160)', *Bulletin de la Société Linnéenne de Bordeaux*, 140:33 (2005), pp. 159–62.

Voet, L., *The Golden Compasses: A History and Evaluation of the Printing and Publishing Activities of the Officina Plantiniana at Antwerp*, 2 vols (Amsterdam: Van Gendt, 1969–72).

Vries, W. de, 'Marnix als intellectuele tuinliefhebber', in Duits and Van Strien (eds), *Een intellectuele activist*, pp. 87–95.

Wallis, P., 'Apothecaries and the Consumption and Retailing of Medicines in Early Modern London', in L. H. Curth (ed.), *From Physick to Pharmacology: Five Hundred Years of British Drug Retailing* (Ashgate: Aldershot, 2006), pp. 13–27.

Warner, M., 'Jean and Vespasien Robin, "Royal Botanists," and North American Plants, 1601–1635', *National Horticultural Magazine*, 35 (1956), pp. 214–20.

Webster, C., 'The Recognition of Plant Sensitivity by English Botanists in the Seventeenth Century', *Isis*, 57 (1966), pp. 5–23.

—, 'Alchemical and Paracelsian Medicine', in Webster (ed.), *Health, Medicine and Mortality*, pp. 301–33.

— (ed.), *Health, Medicine and Mortality in the Sixteenth Century* (Cambridge: Cambridge University Press, 1979).

Welti, M., 'English-baslerische Beziehungen zur Zeit der Renaissance in der Medizin, den Naturwissenschaften und der Naturphilosophie', *Gesnerus*, 20 (1963), pp. 105–30.

Wijk, L. van, 'Het album amicorum van Bernardus Paludanus', *Het Boek*, 29 (1948), pp. 265–86.

Wijnands, O., 'Commercium Botanicum: The Diffusion of Plants in the 16th Century', in Tjon Sie Fat and De Jong (eds), *The Authentic Garden*, pp. 74–84.

Wille, H., 'The Discovery of the Scientific Heritage of Karel van Sint Omaars (1533–1569): Libri Picturati A 16–31 in the Jagiellon Library in Krakow', *Scientiarum historia*, 22 (1996), pp. 67–80.

—, 'The Albums of Karel van Sint Omaars (1533–1569) (Libri Picturati A 16–31, in the Jagiellon Library in Krakow)', *Archives of Natural History*, 24 (1997), pp. 423–37.

Witkam, H., *De Financiën van de Leidse Universiteit in de zestiende eeuw*, 5 parts (Leiden: University of Leiden, 1979–82).

Wittop-Koning, D., *De handel in geneesmiddelen te Amsterdam tot omstreeks 1637* (Purmerend: J. Muusses, 1942).

—, 'De voorgeschiedenis van het collegium medicum (1638) te Amsterdam', *Jaarboek Amstelodamum*, 41 (1947), pp. 51–63.

Woodhouse, E., 'Spirit of the Elizabethan Garden', *Garden History*, 27 (1999), pp. 10–31.

Zalum Cardon, M., *Passione e cultura dei fiori tra Firenze e Roma nel XVI e XVII secolo* (Florence: Olschki, 2008).

Zandvliet, K., *Mapping for Money: Maps, Plans and Topographic Paintings and their Role in Dutch Overseas Expansion during the 16th and 17th Centuries* (Amsterdam: Batavian Lion International, 1998).

Zemanek, A., 'Renaissance Botany and Modern Science', in Mirek and Zemanek (eds), *Studies in Renaissance Botany*, pp. 9–47.

Zucchi, L., 'Brunfels e Fuchs: L'illustrazione botanica quale ritratto della singola pianta o immagine della specie', *Nuncius*, 18 (2004), pp. 411–66.

INDEX